Shock Wave and High Pressure Phenomena

L.L. Altgilbers, M.D.J. Brown, I. Grishnaev, B.M. Novac, I.R. Smith, I. Tkach, and Y. Tkach: Magnetocumulative Generators

T. Antoun, D.R. Curran, G.I. Kanel, S.V. Razorenov, and A.V. Utkin: Spall Fracture

J. Asay and M. Shahinpoor (Eds.): High-Pressure Shock Compression of Solids

S.S. Batsanov: Effects of Explosion on Materials: Modification and Synthesis Under High-Pressure Shock Compression

R. Cherét: Detonation of Condensed Explosives

L. Davison, D. Grady, and M. Shahinpoor (Eds.): High-Pressure Shock Compression of Solids II

L. Davison and M. Shahinpoor (Eds.): High-Pressure Shock Compression of Solids III

L. Davison, Y. Horie, and M. Shahinpoor (Eds.): High-Pressure Shock Compression of Solids IV

L. Davison, Y. Horie, and T. Sekine (Eds.): High-Pressure Shock Compression of Solids V

A.N. Dremin: Toward Detonation Theory

Y. Horie, L. Davison, and N.N. Thadhani (Eds.): High-Pressure Shock Compression of Solids VI

R. Graham: Solids Under High-Pressure Shock Compression

J.N. Johnson and R. Cherét (Eds.): Classic Papers in Shock Compression Science

V.F. Nesterenko: Dynamics of Heterogeneous Materials

M. Sućeska: Test Methods of Explosives

J.A. Zukas and W.P. Walters (Eds.): Explosive Effects and Applications

G.I. Kanel, S.V. Razorenov, and V.E. Fortov: Shock-Wave Phenomena and the Properties of Condensed Matter

V.E. Fortov, L.V. Altshuler, R.F. Trunin, and A.I. Funtikov: High-Pressure Shock Compression of Solids VII

L.C. Chhabildas, L. Davison, and Y. Horie (Eds.): High-Pressure Shock Compression of Solids VIII

R.P. Drake: High-Energy-Density Physics

D. Grady: Fragmentation of Rings and Shells

M.V. Zhernokletov and B.L. Glushak (Eds.): Material Properties under Intensive Dynamic Loading

G. Ben-Dor: Shock Wave Reflection Phenomena

S.M. Peiris and G.J. Piermarini (Eds.): Static Compression of Energetic Materials

Suhithi M. Peiris · Gasper J. Piermarini

Static Compression of Energetic Materials

With 156 Figures

Springer

Suhithi M. Peiris
Defense Threat Reduction Agency
8725 Kingman Road
Fort Belvoir, VA 22060-6201
USA
E-mail: Suhithi.Peiris@dtra.mil

Gasper J. Piermarini
National Institute of Standards
and Technology
100 Bureau Drive
Gaithersburg, MD 20899-1070
USA
E-mail: GJPiermarini@aol.com

Series Editors-in-Chief:

Lee Davison
39 Cañoncito Vista Road
Tijeras, NM 87059, USA
E-mail: leedavison@aol.com

Yasuyuki Horie
AFRL/MNME Munitions Directorate
2306 Perimeter Road
Eglin AFB, FL 32542, USA
E-mail: yasuyuki.horie@eglin.af.mil

ISSN: 8063-7200
ISBN: 978-3-540-68146-5 e-ISBN: 978-3-540-68151-9

Library of Congress Control Number: 2008930164

Cover design: WMXDesign GmbH, Heidelberg, Germany

Printed on acid-free paper

5 4 3 2 1 0

springer.com

Preface

This volume, belonging to the series "High-Pressure Shock Compression of Solids", focuses on the utility of static-compression studies of energetic materials to understand shock processes such as detonation. Static compression, initially based on P. W. Bridgman's work during the first half of the 20th century, launched into its own science in 1959 when the diamond anvil cell (DAC) was developed and subsequently utilized as a primary tool to achieve very high pressures. One motivation for this book is to document the history of the development of the DAC and its associated analytical techniques to the study of energetic materials at static high pressures. Another incentive is to present recent work utilizing the DAC in several areas relevant to understanding the behavior of energetic materials at high static pressures.

In the late 1960s, three scientists at the National Bureau of Standards (NBS) in Washington, DC, S. Block, C. E. Weir, and G. J. Piermarini (the co-editor of this book), were actively involved in the development of a single crystal x-ray diffraction technique utilizing the DAC. Several papers describing the progress they were making in that area had been published, demonstrating promising results. However, one major obstacle had not been overcome, and that was the ability to measure accurate sample pressure in the DAC in a simple, routine, and rapid way. Research scientists at the Explosives Laboratory, Picatinny Arsenal, Dover NJ, were aware of the NBS work and were following its progress with great interest. In 1969, the Picatinny scientists approached the NBS group and asked if they would try the new high-pressure single-crystal x-ray diffraction technique on several inorganic azides that were of interest to them. They desired information on pressure-induced polymorphism and anisotropic compressibility of these azides, data that were unavailable at that time.

The NBS scientists initial response, to say the least, was not very enthusiastic, because they pointed out that such data required a knowledge of sample pressure and also a hydrostatic pressure environment due to the desired anisotropic properties of the crystals. Both pressure and hydrostaticity were quite unreliable at that time because of the lack of exact measurement capabilities with respect to those properties. The Picatinny scientists were more optimistic than the NBS group and suggested that they undertake the study to see where it would lead. After some deliberation, the NBS group agreed to take a look at these azides, but only when they could fit the work into an already busy schedule. They cautioned the Picatinny scientists that it might take months before they could get meaningful results, if at all. The Picatinny scientists presented a convincing argument to study these materials. They noted that little work on explosive materials at very high pressures and also at elevated temperatures had been published because of the nature of the experimental conditions available at that time, i.e., relatively large volume presses had to be used to generate the pressures needed to obtain useful data such as compressibility, thermal stability, pressure-induced polymorphism, etc. Few contemporary experimental scientists were willing to conduct such experiments owing to the danger of the sample detonating inside the anvils of the press, jeopardizing the safety of the researchers. So it was that the NBS scientists were introduced to studying energetic materials at high pressures in a DAC. Little did they know or appreciate the far-reaching consequences of their decision to study pressure effects on these azides.

In 1970, the DAC was an unsophisticated piece of equipment and the technology associated with it for pressure measurement was rudimentary (known freezing pressures of the pressure-transmitting media were used to define the pressure), but nevertheless the NBS scientists were able to obtain useful data with it, and, in the case of the azides, some of the data were reported for the first time. Since then, DAC technology and the means for measuring sample pressure in the instrument (e.g., the ruby fluorescence method and now other more recently developed techniques) have dramatically progressed. The DAC has been transformed from a rather crude qualitative instrument to the sophisticated quantitative research tool that it is today, capable of routinely producing and measuring sustained static pressures in the multi-megabar range and readily adaptable to numerous scientific measurement techniques because of its optical window, miniature size, and portability.

Today, the DAC is the premier instrument of choice for conducting experiments of all kinds and in all disciplines which utilize the static high pressure and temperature variables. Indeed, for static high pressure/temperature studies on energetic materials, the DAC has become a widely used indispensable tool especially for studying these very sensitive energetic materials with multiple phase transitions at small pressure/temperature changes. Common energetic materials such as ionic ammonium nitrate, ammonium perchlorate, sodium perchlorate, and the molecular solids, NM, TNT, HMX, RDX, and TATB when initiated, undergo rapid decomposition resulting in a high-pressure high-temperature shock wave or detonation. This is a complex process to accurately describe, being a combination of almost-instantaneous chemical, mechanical, and physical changes, requiring estimations of high-pressures, high-temperature thermodynamic properties, chemical kinetics, and reaction mechanisms. Fortunately, as described in this book, DACs allow the design of well-defined static experimental conditions with accurate analytical measurements, leading to a much better understanding of the behavior of energetic materials with respect to initiation, reaction, and detonation processes. The various chapters in this book will attest to this fact, because much of the experimental data represented in these chapters were obtained with the DAC.

Because of its enormous popularity in the scientific community, and because many users of the DAC are unaware of its origin due to its many modifications, the history behind the invention of the DAC at NBS is presented in Chapter 1. This chapter highlights the development of the original lever-arm type DAC together with a description of four other generic types of DACs developed later. Also described are several analytical techniques – optical polarizing light microscopy, infrared absorption spectroscopy, and powder- and single-crystal x-ray diffraction – that were applied successfully for the first time to the study of energetic materials at static high pressures.

In addition, Chapter 1 describes a process of invention, which illustrates a concept that unique and original research often is not planned by the organization, but originates from the scientists themselves, who must be highly qualified and motivated to do research. Planned research assumes that certain predictable results will be achieved, and therefore, funding can be justified by the institution. Unfortunately, this line of reasoning frequently produces mediocre scientific results. The quintessential message proffered in Chapter 1 is that the development of the DAC was not planned in advance; but, after its conception by the scientists themselves, the work was allowed to simply progress, until ultimately its importance was recognized and financially supported.

Chapter 2 explains the utility of static high pressures to synthesize energetic materials. The ultimate intent of the research described in this chapter is to create high-nitrogen or polymeric nitrogen materials. Thus, it describes the creation of single-bonded crystalline nitrogen, synthesized from N_2 at $P > 100$ GPa and $T > 2{,}000$ K. The material has a cubic-gauche (cg-N) structure, which was determined from x-ray diffraction and Raman measurements.

Chapter 3 elucidates experimental measurements of the pressure–volume equations of state of energetic materials and their phase transitions at high pressure and temperature, while Chapter 4 explains the same for binders and polymers used in the formulation of explosives. The adaptation of the DAC for x-ray diffraction experiments, Raman and Brillouin spectroscopy, and other analytical techniques is detailed. The influence of nano- to micron-scale features of binders and polymers, on the equation of state of formulated explosives is explained.

Chapter 5 highlights dynamic techniques with the DAC, utilizing time-resolved spectroscopy to study complex processes of rapid mechanical, physical, and chemical changes that occur during the detonation of an energetic material. Chapter 6 discusses static experiments that allow a detailed understanding of these complex processes at the molecular level to illustrate the interplay between static and shock experiments.

Chapter 7 describes molecular dynamics simulation methods and applications to energetic materials under high compression. Applications using both electronic-structure-based methods and conventional moleculardynamics methods employing empirically derived force fields are given, as well as a discussion on advantages and limitations/deficiencies of the various methods and models. Chapter 8 addresses how defects affect the initiation and detonation of energetic materials. Ab initio modeling of several energetic materials is used to elucidate the effect of electronic and spatial structure of defects and their complex involvement in initiation of the explosion process.

The editors wish to acknowledge the demanding efforts of all the contributing authors to this book, and also to recognize their personal contributions to static compression and its utility in studying energetic materials. This is a dynamic and diverse field with many questions still to be investigated, leaving exciting possibilities for future work.

April, 2008
Virginia, USA

Suhithi M. Peiris
Gasper J. Piermarini

Contents

Preface v

1 **Diamond Anvil Cell Techniques** 1
Gasper J. Piermarini

2 **Synthesis of High-Nitrogen Energetic Material** 75
Mikhail I. Eremets, Ivan A. Trojan, Alexander G. Gavriliuk, and Sergey A. Medvedev

3 **Equations of State and High-Pressure Phases of Explosives** 99
Suhithi M. Peiris and Jared C. Gump

4 **Equations of State of Binders and Related Polymers** 127
Dana M. Dattelbaum and Lewis L. Stevens

5 **Reaction Kinetics** 203
Suhithi M. Peiris

6 **Understanding Shock-Induced Changes in Molecular Crystals** 219
Zbigniew A. Dreger

7 **Equilibrium Molecular Dynamics Simulations** 255
Betsy M. Rice and Thomas D. Sewell

8 **Modeling Defect-Induced Phenomena** 291
Maija M. Kuklja and Sergey N. Rashkeev

Index 327

Contributors

Dana M. Dattelbaum Dynamic and Energetic Materials Division, Los Alamos National Laboratory, Los Alamos, NM 87544, USA
danadat@lanl.gov

Zbigniew Dreger Institute for Shock Physics, Washington State University, PO Box 642816, Pullman, WA 99164, USA; dreger@wsu.edu

Mikhail I. Eremets Max Planck Institute for Chemistry, Postfach 3060, 55020 Mainz, Germany; eremets@mail.ru

Alexander G. Gavriliuk A. V. Shubnikov Institute of Crystallography, Russian Academy of Sciences, Leninskii Pr. 59, Moscow 117333, Russia, Institute for High Pressure Physics, Russian Academy of Sciences, Troitsk, Moscow 142190, Russia

Jared C. Gump Indian Head Division - Naval Surface Warfare Center, Indian Head, MD 20640, USA

Maija M. Kukla Office of Integrative Activities, National Science Foundation, 4201 Wilson Boulevard, Arlington, VA 22230, USA; mkukla@nsf.gov

Sergey A. Medvedev Max Planck Institute for Chemistry, Postfach 3060, 55020 Mainz, Germany

Suhithi M. Peiris Indian Head Division - Naval Surface Warfare Center, Indian Head, MD 20640, USA, now at: Defense Threat Reduction Agency, 8725 Kingman Road, Fort Belvoir, VA 22060, USA; Suhith.Peiris@dtra.mil

Gasper J. Piermarini National Institute of Standards and Technology, 100 Bureau Drive, Gaithersburg, MD 20899, USA; GJPiermarini@aol.com

Sergey N. Rashkeev Center for Advanced Modeling and Simulation, Idaho National Laboratory, P.O. Box 1625, Idaho Falls, ID 83415, USA

Betsy M. Rice U. S. Army Research Laboratory, AMSRL-WM-BD (Bldg. 4600), Aberdeen Proving Ground, MD 21005, USA; betsyr@arl.army.mil

Thomas D. Sewell Theoretical Division, Los Alamos National Laboratory, Los Alamos, NM 87545, USA, Now at: Department of Chemistry, University of Missouri-Columbia, Columbia, MO 65211, USA

Lewis L. Stevens Dynamic and Energetic Materials Division, Los Alamos National Laboratory, Los Alamos, NM 87544, USA

Ivan A. Trojan Max Planck Institute for Chemistry, Postfach 3060, 55020 Mainz, Germany

Chapter 1
Diamond Anvil Cell Techniques

Gasper J. Piermarini

1.1 Introduction

It has often been said that scientific advances are made either in a dramatic and revolutionary way, or, as in the case of the diamond anvil cell (DAC), in a slow and evolutionary manner over a period of several years. For more than 2 decades, commencing in 1958, the DAC developed stepwise from a rather crude qualitative instrument to the sophisticated quantitative research tool it is today, capable of routinely producing sustained static pressures in the multi-megabar range and readily adaptable to numerous scientific measurement techniques because of its optical accessibility, miniature size, and portability.

During the last several decades the amount and quality of research with DACs has increased enormously with new instrument modifications, new applications, and experimental results being reported in literature frequently. In fact, the DAC now has become the premier instrument of choice for conducting experiments of all kinds and in all disciplines that utilize static high pressure and temperature variables. Indeed, for static high pressure/temperature studies on energetic materials, the DAC has become a widely used indispensable tool. Never in my thoughts did I ever expect this simple, yet elegant, device to become so widely used and accepted throughout the world for conducting scientific research in many diverse disciplines. Because of its enormous popularity in the scientific community, and also because I have perceived that many scientists who use the instrument today are unaware of its origin, probably because so many different modifications of the instrument have been made and are currently in use, I think it is appropriate to narrate the history behind the invention of the DAC in the introductory chapter of this book.

The DAC, as we know it today, was invented at the National Bureau of Standards (NBS), renamed the National Institute of Standards and Technology (NIST) in 1986. Its development illustrates a process of invention almost unique in the scientific research community. I believe it will be of noteworthy value to present in detail the motivations behind this invention. I will also present first simple and later

S.M. Peiris, G.J. Piermarini (eds.), *Static Compression of Energetic Materials,*
Shock Wave and High Pressure Phenomena,

more refined experimental techniques and results permitting the reader to appreciate and assess the merits of research on energetic materials with the DAC. In the process, I will discuss four fundamental topics: (1) an in-depth history of how and why the DAC was invented; (2) descriptions of five basic generic models of the DAC and their characteristic merits and disadvantages in various research applications; (3) the development of the ruby fluorescence method at NBS to measure sample pressures in the DAC; (4) the development of pressure-transmitting media for specific applications; and (5) the development of high- and low-temperature capability. I will follow with descriptions of several innovative DAC applications, the result of research at NIST and the Naval Surface Warfare Center (NSWC, White Oak), which are especially useful in the study of energetic materials. They are: (1) optical polarizing microscopy for detecting birefringence changes in single crystals resulting from pressure-induced polymorphism and measuring phase transition pressures; (2) Fourier transform infrared absorption (FTIR) spectroscopy for detecting phase transformations and obtaining kinetic data on thermal decomposition reactions at high pressures; and (3) x-ray diffraction techniques, including single crystal, powder, and energy dispersive methods for confirming the existence of pressure-induced polymorphism and identifying the new phases crystallographically. Experimental results obtained with the aid of these techniques will be presented on the thermal decomposition kinetics, chemical reactivity, and phase behavior of RDX and nitromethane (NM).

1.2 Invention

The invention of the original DAC (lever-arm type with 180° optical transmission window) took place in the late 1950s at the NBS in Washington, DC [1]. At that time I was a young budding scientist attached to the Constitution and Microstructure Section (Howard F. McMurdie, Section Chief), where the DAC was invented and later developed into a sophisticated scientific research instrument. My own research activity at that time was unrelated to high-pressure work. I was fulfilling my 2-year military obligation (1956–1958) in the US Army and was assigned to Aaron S. Posner's crystallography laboratory to develop a rapid x-ray fluorescence technique for quantitative analysis of dental amalgams used by the military services. Because my office and measurement apparatus were near the high-pressure laboratory, I was able to develop close relationships with the individuals involved in the DAC's development, so I did, indeed, witness on a daily basis the process of invention. It was not until 1961, about 2 years after the first paper describing the pioneering work on the DAC was published, that I began my own collaboration with C. E. Weir on the development of a modified (hydraulically loaded) version of the lever-arm type DAC for x-ray powder diffraction measurements [2].

As I reflect on it now, it was the coincidental meeting of four scientists, C. E. Weir, A. Van Valkenburg, E. R. Lippincott, and E. N. Bunting (Fig. 1.1), each with his own technical and scientific expertise, which provided the impetus

Fig. 1.1 The four scientists associated with the invention of the Diamond Anvil Pressure Cell at the National Bureau Standards, Washington, DC, in 1957. From the left: Charles Edward Weir, 1911–1987; Alvin Van Valkenburg, 1913–1991; Ellis Ridgeway Lippincott, 1920–1974; Elmer Newman Bunting, 1892–1966 [NIST Archives]

necessary for the invention of the DAC. Interestingly enough, the activity leading up to the invention had not been officially defined or even recognized in its early stages. Rather, it was the innate intuition of these four scientists which permitted them to recognize a fertile area of high-pressure research, and moreover, each one knew how he could contribute to its realization. It was not planned that way, for, in the beginning, there was little or no collaboration or even interaction among these individuals.

In the early to mid-1950s, Weir, a chemist by training, was assigned to the Leather Section, studying the physical properties of leather and how they affected wear. For example, he used the pressure variable to measure porosity in leather by Hg intrusion using a piston/cylinder hydraulic press [3–5]. He initiated these studies because porosity in leather was known qualitatively to affect wear. Weir conducted these experiments in an attempt to quantify that affect. Apparently Weir was so efficient and productive in performing his official tasks, that he had additional time during which he could pursue his own personal research interests, discretionary research which most of us did at NBS and still do today at NIST. In the 1950s, it was common for NBS scientists to nurture unofficial discretionary research of personal interest provided it did not interfere with performing official duties, and, furthermore, was related to them in a somewhat general way. Weir adapted his piston/cylinder press used for the leather porosity measurements to measure the compressibility of many solids, including borosilicate glasses, teflon, natural and synthetic rubbers as well as to study other phenomena such as pressure-induced polymorphism (Fig. 1.2). He published the results of his piston-displacement measurements in the archival scientific literature, some of which were industrially important and appeared in print for the first time [6–9].

As a result of his early high pressure work, Weir received the Department of Commerce Silver Medal for his outstanding accomplishments. While performing his studies, Weir developed an intense interest and appreciation for the effect of pressure on the properties and structure of materials. In fact, he expressed to the

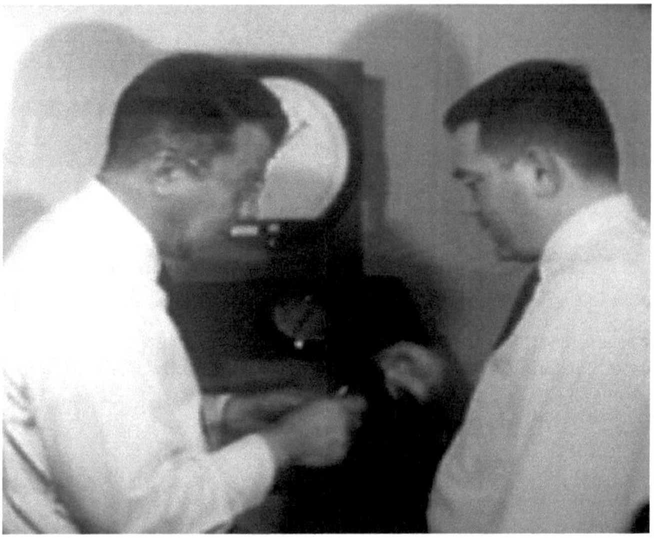

Fig. 1.2 Weir (left) explains to G. Piermarini the details of the encapsulation method for samples used in his early award-winning work with the hydraulic press located in the background (*c.*1968) [W. P. Richardson, NIST Photo Lab]

management his desire to pursue this kind of research, rather than continue his work on leather. In recognition of his accomplishments, Weir was able to continue his high-pressure work in McMurdie's Section. In the course of discussing his results with other NBS scientists, Weir met Van Valkenburg (known as Van by his colleagues), and many discussions related to high-pressure studies took place between them.

In 1958, Allen Astin, the Director of NBS, announced that several research areas in high pressure would be emphasized in current and new NBS programs. Three activities relevant to DAC were in that announcement: (1) to develop new methods and technologies to achieve high pressures, (2) to develop measurement techniques for the study of the structure and properties of materials at high pressure, and (3) to establish a pressure scale to measure them [10]. Thus, in 1959 work on the development of the DAC became an official NBS program.

In the early to mid-1950s, Van, also working in McMurdie's Section, was engaged officially in crystal synthesis by hydrothermal techniques, particularly in the re-crystallization of mica by the Morey bomb method. At that time mica was classified as a strategic material because it was universally used as an insulator in vacuum tubes. Solid state devices were not in commercial use in the mid-1950s. Van was a mineralogist by training and had an intense appreciation for the effect of pressure on the physics and chemistry of materials. Like Weir, Van had a discretionary research activity, which he pursued in his spare time. He hoped to develop a high temperature–high pressure method for the synthesis of diamond, which, incidentally, he was never able to achieve. Enthusiasm for this endeavor stemmed from the fact that a few years earlier the General Electric Company had announced

Fig. 1.3 Van Valkenburg (ca 1957) adjusting the parameters on his high-temperature furnace used for the synthesis, crystallization and reconstitution of mica [W. P. Richardson, NIST Photo Lab]

the first successful production of synthetic diamond by a high temperature–high pressure method – a technological achievement prominent on the world scientific scene at that time. Van used a piston/cylinder press in combination with a high-temperature NBS-type self-aligning tetrahedral anvil device for his diamond synthesis studies [11] (Fig. 1.3).

As I recall, Weir expressed a strong desire to join Van so they could collaborate in this newly emerging area of high-pressure research. With the introduction of solid state devices, mica was no longer classified a strategic material and Van's research project was terminated. Also, the fact that P. W. Bridgman had been awarded the Nobel Prize for his work in high-pressure physics a few years earlier may have influenced their motivation. Weir convinced McMurdie that his proposed move to join Van would be beneficial for both of them and for the section. So, in 1957 Weir and Van were working together on high-pressure phenomena in the same laboratory room. As it turned out, Bunting, also in McMurdie's section, was working in that same large laboratory room on the second floor of the Industrial Building on the NBS site on Van Ness Street in Northwest Washington, DC (now demolished and replaced by the University of the District of Columbia) (Fig. 1.4). Bunting, Weir, and Van collaborated with Professor Lippincott, a guest scientist and consultant from the University of Maryland, on the measurement of infrared absorption spectra of

Fig. 1.4 Bunting's office was in the high-temperature furnace laboratory where there was also a Beckman infrared (IR) spectrophotometer that he used to study polymorphs of silica and germania in collaboration with Lippincott. Bunting is shown sitting at his desk facing the spectrophotometer located behind Van's high-temperature furnace on the left (*c.*1959) [G. J. Piermarini, NBS]

inorganic materials. Lippincott's expertise was in absorption spectroscopy, particularly in the infrared (IR) region, and they published results of their collaborative work on temperature-induced polymorphism in silica and germania [12]. As I recall, Earl Schoonover, Chief of the Mineral Products Division at the time, promoted collaborations with outside organizations such as local universities, e.g., the University of Maryland and other government agencies to assist the NBS scientific staff in developing new research areas to investigate.

These circumstances placed all four scientists in the same room where they performed their own independent research work and also collaborated with Lippincott on infrared spectroscopy studies. These conditions also permitted them to interact with each other daily, exchanging ideas of common scientific interest. In the relatively free laboratory environment of the post-Sputnik era, funds for research were not lacking. In fact, I recall one high-level manager, Earl Schoonover, Chief of the Mineral Products Division, speaking to a bench scientist, "Don't worry about where the funds are coming from, just do good science." In this kind of environment, which is rare today, it was not long before one of these four individuals came up with the notion that a major scientific breakthrough could be achieved in the area of high-pressure research by developing a miniature high-pressure device with 180° optical transmission, permitting the measurement of infrared absorption spectra of materials under high pressure. This inspiration came about because of Lippincott's spectroscopy background. During their many conversations, he frequently brought the subject up and commented that such a device would be an effective tool for probing the behavior of atomic forces in matter. He argued that unlike temperature, which cannot exceed the limits imposed by the thermal stability of the solid under study and the absolute zero of temperature, the only limitation imposed by the use of pressure is the failure or limit of the pressure-producing vessel itself. Lippincott was excited over the possibility of being able to measure and interpret the effects of pressure on infrared spectra, something which had not been done before. If they succeeded in doing this, he said, it would be a major breakthrough in the areas of IR spectroscopy and high-pressure research. That was enough to convince all four of them. They decided to undertake the development of such a device. However, unlike prior discretionary activity, usually carried out individually, this one would be a combined effort of these four scientists and it would take over a year of their discretionary time to accomplish the goal. It is important to note that their main impetus was not the development of a pressure scale, although knowledge of the pressure was important to them, but the measurement of the infrared spectra of solids at high pressures as a means of studying interatomic forces and bonding. This spectroscopic technique placed an immediate and immense constraint on their high-pressure apparatus, for whatever they designed and fabricated had to be small enough to fit into the sample chamber of an IR unit and also had to be transparent to infrared radiation.

Their first attempt to do this failed miserably. It was based on a design first published in 1957 [13], in which the principle of pressing a powder in a cylinder by inserting two opposed pistons in the cylinder and forcing them together to compress the sample between them was used. A large single-crystal diamond was employed with a cylindrical hole to contain the sample for x-ray diffraction measurements. For infrared absorption measurements, Van was able to obtain a 7 $^1/_2$ carat (1 metric carat = 200 mg) single-crystal Type II (low IR absorption properties) diamond from the Smithsonian Institution free of charge. The justification for the gratis transfer was that the rough stone would be used specifically for research purposes at the NBS and would very likely be destroyed in the process of experimentation. A cylindrical hole (0.4 mm) was bored through the diamond perpendicular to two parallel polished faces. Tight-fitting drill rods were inserted to compress the powdered sample within the cylinder. After a few excursions to elevated pressures by applying a load to the ends of the drill rods using a simple clamp device hand-made by Weir, the diamond developed radial fissures around the hole which increased with subsequent load applications as shown in Fig. 1.5. The diamond eventually split into pieces with subsequent applications of load because axial alignment of the drill rods was difficult to maintain, leading to further crack propagation and ultimate splitting. This development quickly put an end to the experiment and also any future experiments of this kind because no one or no organization was going to donate a diamond of that size again even for the benefit of scientific research.

Thus, they had to come up with a fundamentally new approach. After much discussion and deliberation, they finally agreed on a very simple device suggested by Weir which made use of the principle of Bridgman-opposed anvils [14, 15]. Percy W. Bridgman, a Nobel Laureate and the father of high-pressure research, was a pioneer in the pressing of materials between two loaded flat-faced anvils (Fig. 1.6). As noted earlier, Weir had been studying the effect of pressure on leather for several years and had acquired a wealth of knowledge in the experimental techniques employed by researchers in the field. He concluded that opposed anvils, transparent to infrared radiation, would be the simplest approach, especially because they would have to make the device themselves during their spare time. All agreed to proceed in that way.

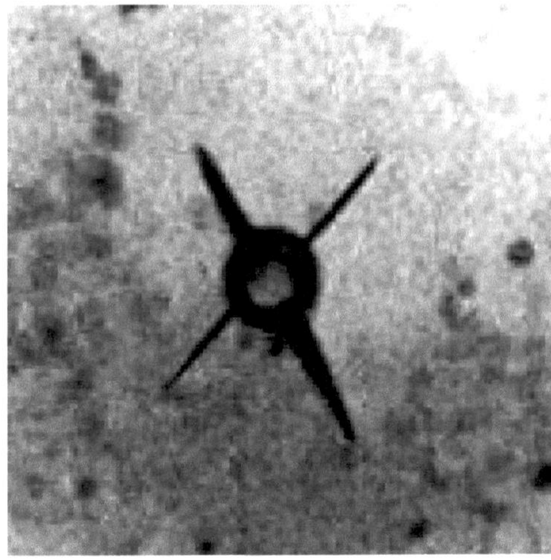

Fig. 1.5 This photo depicts the radial cracks that develop around the bored hole through the diamond. These cracks increase in size as the load on the pistons increases ultimately leading to complete fracture of the diamond into several pieces [13]

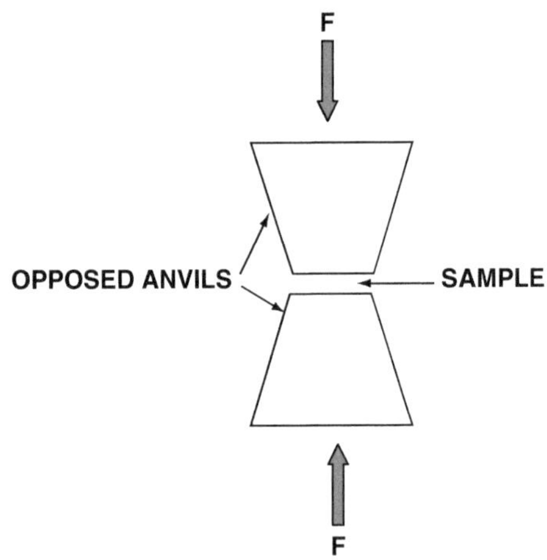

Fig. 1.6 This schematic depicts the fundamental concept of Bridgman-opposed anvils. A load (F) forces two opposed anvils (usually a very hard material like WC) together with the sample squeezed in the small area between them. The load is generated usually by utilizing a hydraulic press. For a given load, the smaller the anvil area, the greater the pressure generated on the sample

The NBS scientists also knew that diamond, because of its extreme hardness, high compressive strength, and IR transmission properties, was the obvious material to use for the anvils. For optimum IR transmission, particularly in the fingerprint region for the study of interatomic forces, they would have to obtain rare Type II diamonds or at least those possessing mainly Type II character. Most diamonds exhibit gradations between Types I and II properties [16].

Rare Type II diamonds could be used for infrared microspectroscopy studies because they could withstand the loads necessary to generate high pressures. Also, by utilizing a miniature opposed-anvil arrangement, only microgram quantities of sample are required, so a very small portable apparatus for generating load was all that was needed. But at this stage, they had no idea how or where to obtain Type II diamonds. The work was still discretionary and no supporting funds were available.

McMurdie certainly would not agree to their purchase, especially if the diamonds were going to be squeezed together and very likely crushed in the process, ending up like their earlier experience with the $7^1/_2$ carat stone. Whatever they were going to do had to be done on their own initiative and without official support. Van, whose mineralogy background was critical at this stage, came up with a wonderful solution. Through his efforts, he succeeded in obtaining 1,100 brilliant-cut gem diamonds from the General Services Administration (GSA), which maintained a large accumulation of contraband diamonds confiscated by the US Customs Service from smugglers attempting to enter the country with them illegally. Unlike today, where confiscated diamonds are periodically auctioned off to the general public by the GSA, in 1957 disposing the stones by public sale was not permitted and the contraband gems accumulated in the GSA coffers and remained there indefinitely. With the assistance of McMurdie, who provided an NBS document declaring that these diamonds would be used specifically in high-pressure experiments at the NBS and would very likely be destroyed in the process of experimentation, the GSA agreed to transfer to NBS about 1,100 brilliant-cut gems, all roughly one third of a carat in weight. I do believe that without the availability of these cost-free gems, the effort to build this high-pressure instrument would have been seriously curtailed, if not entirely terminated. In any case, the venture would not have proceeded easily beyond this point without the availability of these gratis gems.

Bunting and Van proceeded to measure the IR transmission and other properties of the 1,100 gems and found that only 55 of them were primarily of Type II character and suitable as anvils for IR transmission experiments [16] (Fig. 1.7, Fig. 1.8). They now had the proper diamonds, but they still had to shape them into anvils and then fabricate the mechanism for applying load to them. Both procedures had to be simple because Weir was going to use only an old South Bend$^{\diamond}$ lathe, a drill press, a hack saw, files, an old high-speed diamond wheel obtained from surplus, and any other equipment readily available to him in his laboratory to do all this. These simple tools dictated how the instrument was going to be designed and fabricated. Thus, what was initially thought to be a great hindrance turned out, in the end, to be a great boon, because these simple fabrication tools and the constraints imposed by the infrared apparatus, determined the ultimate simplicity in the design of the device.

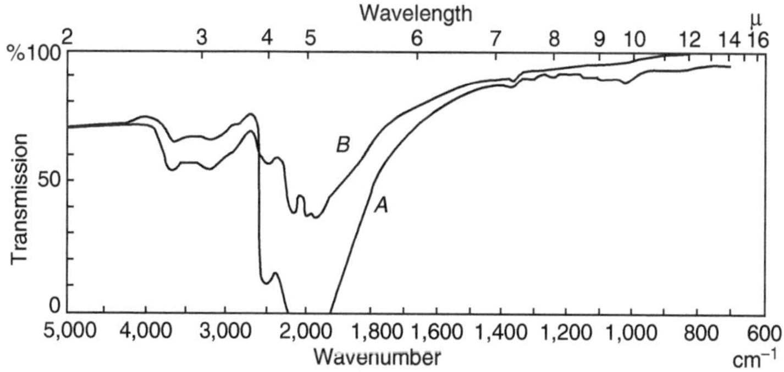

Fig. 1.7 A. Infrared spectrum of typical type-II diamonds (2–4 mm thick). B. Infrared spectrum of a thin type-II diamond (1 mm thick) [20, 22]

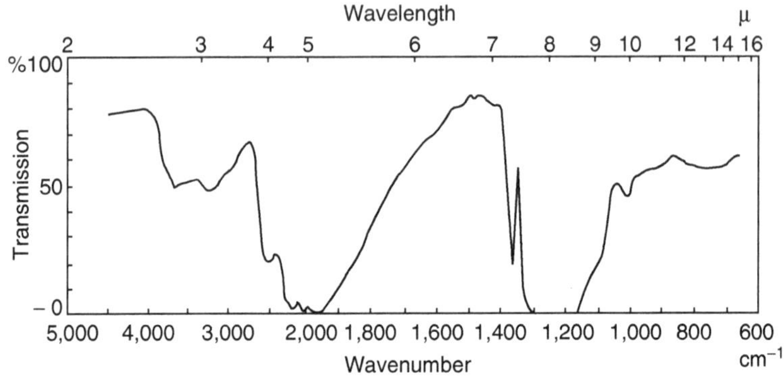

Fig. 1.8 Infrared spectrum of a typical type I diamond (2–4 mm thick) [20, 22]

The actual construction of the first DAC was literally left in the hands of Weir, who had a special proclivity for design and fabrication and excelled at it. Weir preferred to do things himself and rarely involved the Shops Division, even when carrying out his official work. He often was heard to say, "I don't have time to wait for this job to get done because it may take a week or so, and, furthermore, I will have to make a careful drawing of the object I need made to give to the Shops. I think I'll just do it myself right now and be over with it." And he usually did, as was the case with the DAC. Because readymade anvils had not been obtained, the Type II brilliant-cut gem stones, which had been selected by Van and Bunting, had to be modified into anvils. The simplest thing to do was to grind down the tip or culet of the brilliant-cut stone until a small flat area about 0.5 mm^2 was formed. Weir did this himself on the selected type II gems by grinding down the culets on a high-speed grinding wheel impregnated with diamond grit with a home-made jig to keep the ground flat parallel to the table of the gem. Several stones, often as many as ten, were mounted in the jig simultaneously, held firmly in place with low-melting solder. This procedure permitted several anvils to be made concurrently.

Although each scientist had input into the overall design of the clamping device, the final approval was left up to Weir because he was going to have to fabricate it using the equipment he had available to him. Once they all agreed on the lever-arm design, which has remained essentially the same even to today, Weir proceeded to fabricate the device by himself. He used tools only available to him in the laboratory, i.e., an old South Bend® lathe with sloppy, noisy bearings as I recall, an electric drill press, a hack saw, and whatever else was handy to get the job done. Finally, after many weeks of trials and failures, the final instrument was created (Fig. 1.11). It was used successfully to demonstrate for the first time that pressure shifts and other pressure effects on the infrared absorption spectra of materials could be observed and measured [1]. The device, a crude, homemade instrument, became the prototype for all succeeding designs of the opposed diamond anvil high-pressure cell with 180° optical transmission, initiating a new era for high-pressure research. The design of the original instrument is shown in schematic cross section in Fig. 1.9 and Fig. 1.10.

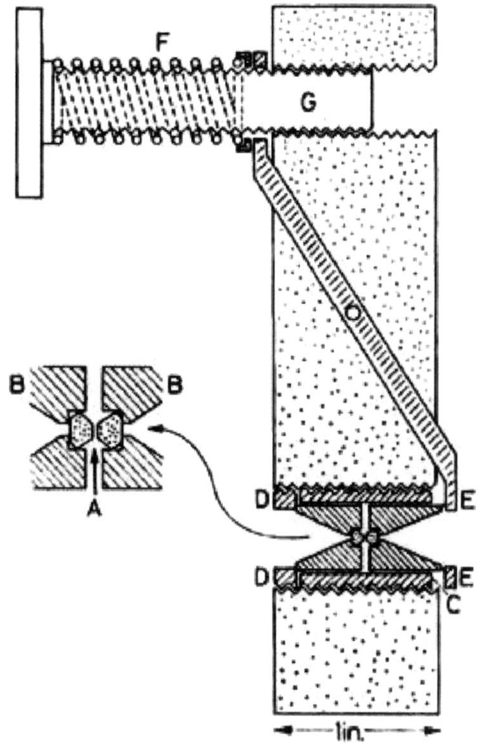

Fig. 1.9 This is a schematic diagram of the original lever-arm type DAC fabricated by C. E. Weir in 1957/58. The simplicity in the design is clearly demonstrated. In operation, rotation of the screw G compresses the spring F, which transmits a force through the lever arms to E. The opposed pistons B, containing the diamond anvils A, are fixed in position in the massive block C by the entablature ring D. Thus the diamonds are squeezed together under an applied load. Note that the thickness of the block C is only 1 inch, permitting the device to be positioned in the focusing mirrors of the IR instrument as shown in Fig. 1.10 below [20, 22]

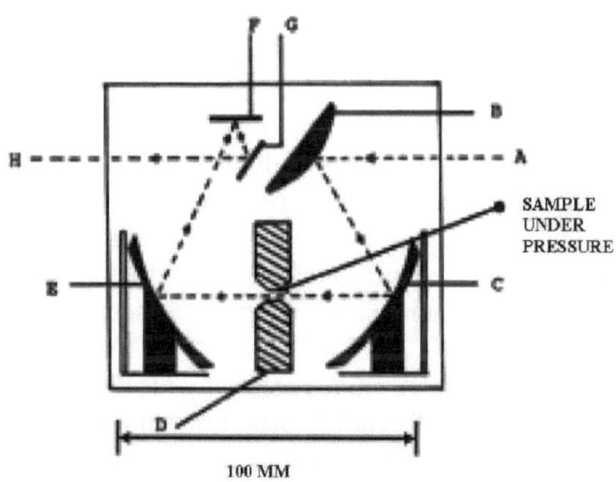

Fig. 1.10 The IR beam path is designated by the letters A through H. The pressure cell is designated by the letter D. B is a convex mirror. C and E are concave focusing mirrors. F and G are planar reflecting mirrors [20]

In what appears to be a case of simultaneous discovery, because to my knowledge neither group was aware of the other's work until just before publication, a similar device was developed at the University of Chicago (UC) for x-ray powder diffraction measurements at high pressures [17, 18]. Like the NBS DAC, this instrument

Fig. 1.11 This photograph shows the original instrument, homemade by C. E. Weir. On close inspection its crude fabrication is readily apparent. Because of its importance in the field of high-pressure research, this instrument, small enough to fit in the palm of one's hand and capable of producing very high pressures, is on display today in the NIST Museum at Gaithersburg, MD [NIST Archives]

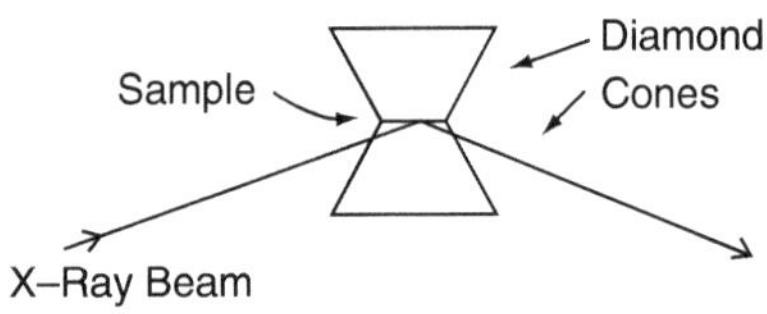

Fig. 1.12 The University of Chicago instrument was designed to replace the sample mount of a standard 2θ scan x-ray diffractometer. The entablatures for the two diamond cones were Carbaloy inserts in steel, thus precluding the possibility of 180° transmission

made use of two opposed diamonds in a Bridgman flat-face anvil arrangement, but the fundamental difference between the two instruments is in the direction of the incident and reflected x-ray beams. Unlike the NBS design, which has 180° transmission so that the x-ray beam enters one anvil and exits through the other anvil, in the UC device the x-ray beam enters and exits the same anvil as shown in the Fig. 1.12.

With official NBS funding more refined pressure cells (Fig. 1.13) were built in the NBS fabrication shops specifically for IR absorption studies which were performed on many materials of interest and published in the scientific literature over the next several years [19–23].

This instrument with 180° optical transmission was used for infrared absorption measurements on powders squeezed between the two opposed diamond anvils. Some of the very first materials studied were $NaNO_3$, KNO_3, $AgNO_3$, ferrocene, ice, and $CaCO_3$.. These studies demonstrated for the first time that pressure-dependent shifts in the IR absorption bands could be measured, in addition to the observation of large changes in spectra as a consequence of pressure-induced phase transformations [19–22].

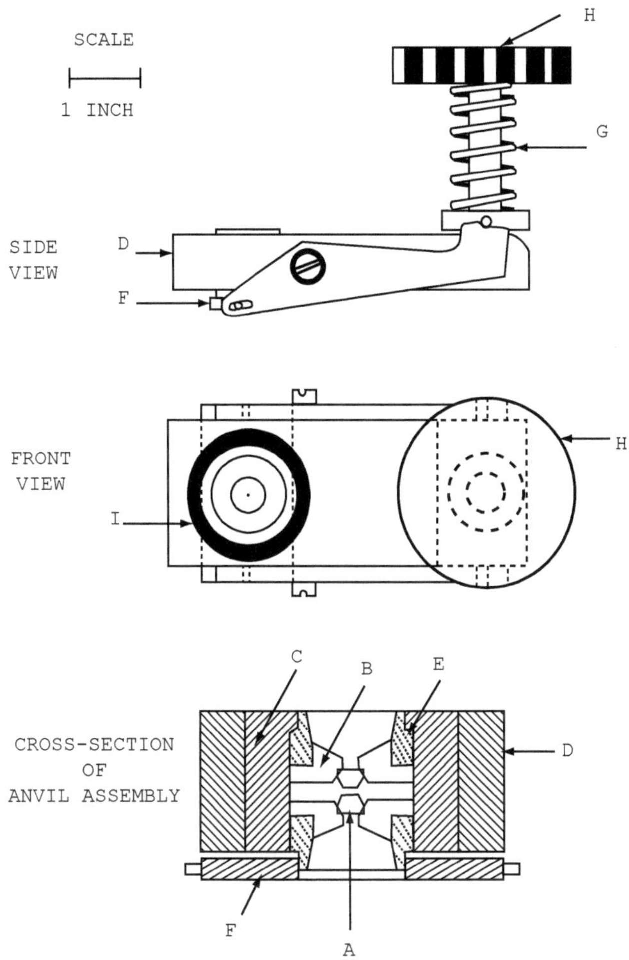

Fig. 1.13 A schematic view of the refined pressure cell is shown on the right. Critical components are identified as follows: A – the diamond anvils, B – the supporting steel entablatures, C – steel guiding cylinder, D – aluminum cell body, E – piston, F – steel pressure plate, G – tappet spring, H – screw knob, I – circular cavity for insertion of a resistance coil furnace. Except for the Al cell body and the diamond anvils, all other components were fabricated from a stainless steel. Two push/pull screws, 120° apart, located in each piston (E), provided a mechanism for tilting the keyed supporting steel entablatures (B) permitting parallel alignment of the diamond anvil faces. In a later similar design, the keyed entablature feature was removed and only one piston was fitted with these screws because alignment could be achieved by a combination of a tilt of one anvil and a rotation of the other [Courtesy of High Pressure Diamond Optics, Inc. Tucson, AZ 85719]

Van was born in Schenectady, NY in 1913, earned a B.S. degree in geology from Union College in 1936, and, 2 years later, an M.S. degree in mineralogy and petrology from the University of Colorado. During World War II, he supervised degaussing ships at the Charlestown Navy Yard in Boston, MA. He joined NBS in 1945 and left in 1964 to become a Program Director in Geochemistry at the National Science Foundation. Later he became Chief of University Relations at the Bureau of Mines until his retirement in 1980. All the while, Van continued to work with the DAC at the Geophysical Laboratory until 1980. Van continued to be associated with the DAC in the years after he departed NBS through High Pressure Diamond Optics (HPDO), a company which manufactures and sells DACs to researchers in the high-pressure community. HPDO originally was a partnership consisting of Weir, Lippincott, and Van Valkenburg, with Van being the active partner and Weir and Lippincott silent partners. Later, however, Van became the sole owner of the company. Over the years, Van provided many DAC instruments to the high-pressure community through the vehicle of HPDO. This too, had an important impact in the field of high pressure for it made the DAC readily available to many scientists who would otherwise have had to design and fabricate the instrument in their own laboratories. Many would have chosen not to do so. In retirement, Van moved to Tucson, AZ, where he continued to manufacture and sell DACs and anvils with his son Eric. Van passed away in 1991. His son Eric continues to operate HPDO [23].

Weir retired from NBS in 1970 after the development and refinement of the high-pressure single-crystal x-ray diffraction technique, and moved to San Luis Obispo, CA. He was born in Washington, DC in 1911, the son of a government employee. He attended Dunbar High School, graduating at the head of his class. In 1929 he was one of four young men appointed to the US Naval Academy, Annapolis, by Oscar De Priest, representative in Congress from Chicago. However, even though he passed the mental examination, he was not accepted to the Academy because of poor vision [24]. Weir went on to study chemistry at the University of Chicago and earned a B.S. degree in 1932, an M.S. degree in physical chemistry from Howard University in 1934 and taught there until 1937, when he entered the California Institute of Technology (CalTech) to study for the Ph.D. in physics. Owing to illness he terminated his studies in 1940 and joined NBS in 1943, where he remained until his retirement after an outstanding and prolific scientific career. Weir passed away in 1987 [25].

Bunting retired from NBS in 1962 after a long and productive career (37 years) at NBS. He was born in Chicago in 1892. He was one of the few scientists during those early years that had a doctorate degree (from UC in physics, mathematics, and chemistry, granted in 1918). He passed away in 1966 [26]. Lippincott was born in Philadelphia in 1920 and attended Earlham College in Richman, IN. He earned a doctorate from Johns Hopkins University in 1947. In 1974, Lippincott passed away at an early age (54 years) from Hodgkin's disease [27].

1.3 Introduction of the Gasket Technique

In the years immediately following the pioneering work on the development of the DAC, Van made an extremely valuable contribution to DAC technology in 1962 [28, 29]. He had the clever idea of encapsulating pure liquid samples in the DAC by using a thin metal gasket (Fig. 1.14) containing a very small hole placed between the flat surfaces of the opposed anvils. The hole was filled with liquid and

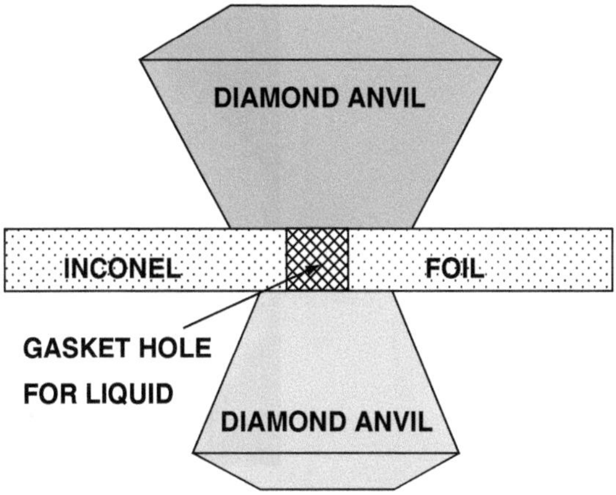

Fig. 1.14 This schematic diagram shows a thin metal foil (gasket) separating the two diamond anvils. The foil is typically an Inconel alloy about 2–3 mm thick, and contains a hole approximately 0.2–0.3 mm in diameter. Note that the face of the lower anvil has less area than the face of the upper anvil, permitting it to indent the foil more effectively, thereby reducing the volume of the hole and thus increasing the pressure on the confined liquid

the anvils were squeezed together while observing the liquid with the aid of a polarizing light microscope. Prior to 1962, only powders were pressed between the diamond anvils and were studied mainly by IR absorption spectroscopy and x-ray powder-diffraction methods.

In his initial experiment, Van filled the cavity with water, quickly assembled the pressure cell and then applied a force to the anvils while observing the confined water in the gasket hole with the aid of a polarizing microscope. The metal foil thinned down reducing the volume of the hole and thus the confined water also, thereby increasing the pressure. At about 0.98 GPa, he found that the water crystallized to ice VI. He observed for the first time the crystallization of water at room temperature (RT) to solid ice VI. What he saw is shown in Fig. 1.15.

Van was able to observe, for the first time, crystallization and polymorphism in water as the pressure was increased at room temperature [28] (Fig. 1.16). Furthermore, by employing this technique on a single crystal of calcite and using glycerine as the pressure-transmitting medium, Van was able to observe two pressure-induced polymorphic transformations in calcite when polarized light was used to illuminate the crystal [29]. The original work on calcite was done soon after the gasket technique was developed, but the results of that study were not published until 1970, long after Van had left NBS in 1964. The development of the gasket technique was very important for it permitted hydrostatic pressure environments to be achieved thereby laying the foundation for quantitative measurements at high pressures because of the reduction and even elimination of pressure gradients in some liquids. One of the early applications derived from this gasket technique was high-pressure single-crystal x-ray diffraction. In fact, it was Van's gasket technique and his observations of the transformations in calcite in glycerine that led Weir, Stanley Block,

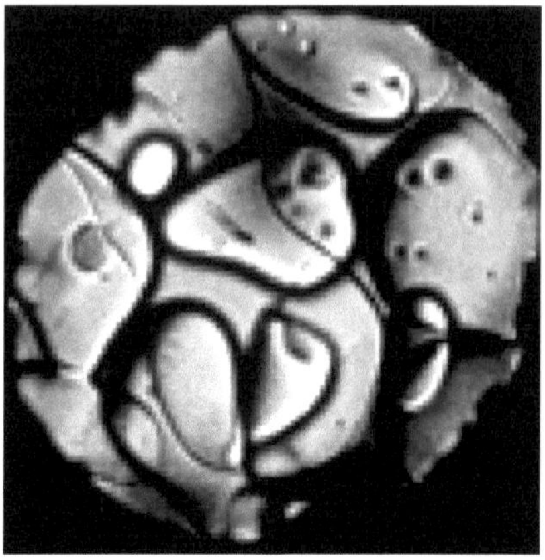

Fig. 1.15 These are amorphous shapes or globules of crystalline ice VI in equilibrium with water at approximately 0.98 GPa at RT as viewed through the DAC with the aid of a polarizing light microscope. This photograph is the original one that Van took when he saw for the first time ice VI in equilibrium with water at room temperature. To my knowledge, no one had ever seen this before

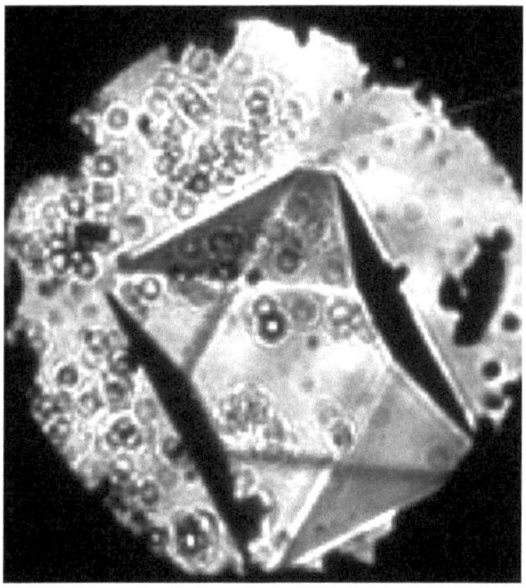

Fig. 1.16 A single crystal of ice VI was grown from one of the globular crystals shown in Fig. 1.15. The procedure Van followed was to melt all but one of these faceless crystals by carefully warming the cell slightly above RT and then allowing the temperature to return to ambient. This procedure was done at approximately 0.98 GPa, the freezing pressure of ice VI at RT

and me in 1964 to begin a long collaboration, this time with NBS support, resulting in the development of the high-pressure single-crystal x-ray diffraction technique utilizing the precession camera [30].

By 1971, the DAC had undergone several stages of refinement both at NBS and other laboratories and also was adapted to other measurement techniques by several laboratories in the US and abroad. However, the DAC was still not fully appreciated by many scientists including those in the NBS management because there was no rapid, convenient, and accurate method available to measure the sample pressure. At that time pressures were generally determined by a tedious and time-consuming x-ray diffraction method measuring the volume compression of NaCl based on an equation of state. The technique often required as much as 15 h to make one pressure measurement. Thus, the DAC's acceptance as a tool for high-pressure research was limited and was used primarily in laboratories engaged in research of geological interest where very high pressures were desired, but high accuracy at these very high pressures was not considered a serious issue at the time.

1.4 Application to X-Ray Diffraction Techniques

The NBS opposed-anvil DAC became the prototype for all subsequent diamond cells with 180° optical transmission. While the first application of the DAC was in IR spectroscopy, we proceeded to investigate the possibility of applying the instrument to x-ray powder diffraction because we were part of the Crystallography Group. At the conclusion of my military service obligation, I joined the Constitution and Microstructure Section, headed by H. F. McMurdie. A year after joining that Section, Weir and I initiated a collaboration on applying the DAC to x-ray powder diffraction. In 1962 we published a description of a hydraulically loaded miniature x-ray powder diffraction high-pressure camera along with results demonstrating the usefulness of the technique [2] (Fig. 1.18).

When it became obvious that single crystals could be grown in a DAC from liquids, the possibility of performing single-crystal x-ray diffraction experiments at high pressures was obvious to us. Weir and I had published several successful x-ray diffraction studies on powders at high pressures using the hydraulic pressure cell (Fig. 1.17). It was only natural for us to extend the technique to single crystals. In 1963, Stanley Block, head of the Crystallography Group, joined us in this endeavor. As before with the original pressure cell, Weir modified a Buerger-type

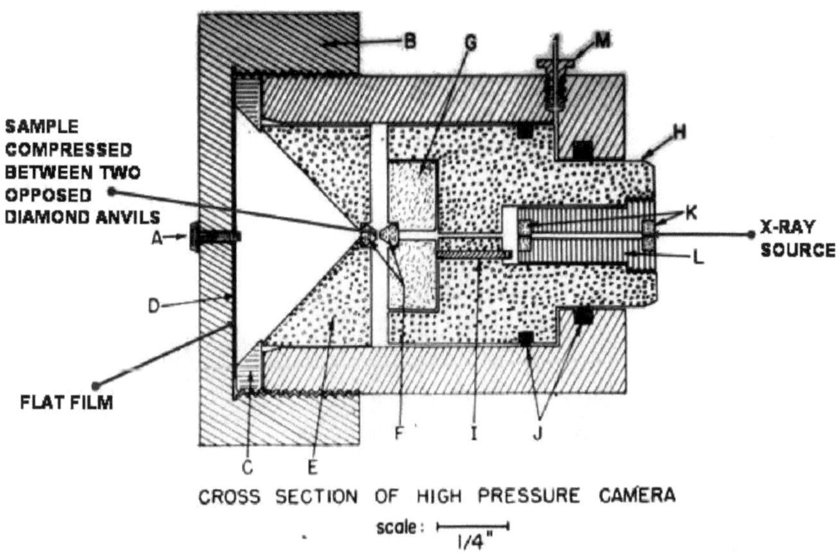

Fig. 1.17 A schematic diagram of the hydraulically loaded x-ray diffraction high-pressure camera. It demonstrated for the first time that useful powder patterns could be recorded on film from samples under pressure. The number of observed *d* spacings was limited, but often could be indexed according to a crystal system, thus yielding volume compression data. Also pressure-induced phase transformations could be detected. Components are identified in [2]

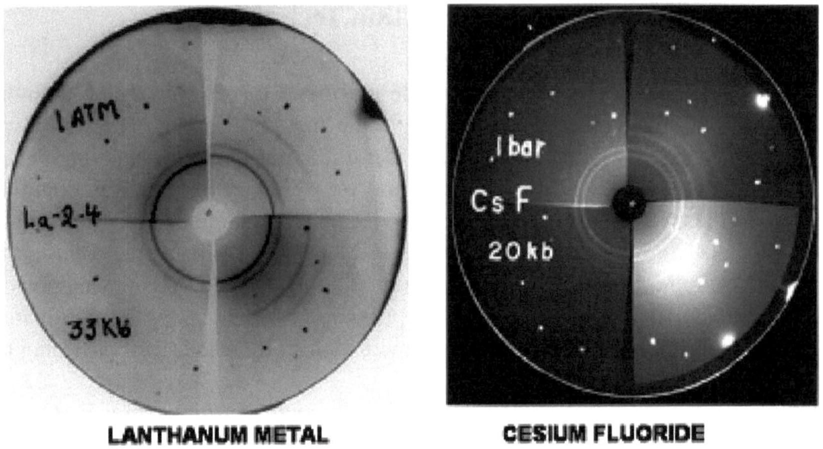

Fig. 1.18 Two examples of phase transitions are shown in this figure. Opposite quadrants (La and CsF) of x-ray powder diffraction films demonstrate pressure-induced phase transformations, e.g., the hexagonal structure of α-La transforms to a face-centered cubic structure (fcc); CsF transition is from a NaCl-type-to-CsCl-type structure [31, 32]. These films were taken with the instrument shown in Fig. 1.19

Fig. 1.19 This is a photograph of the crudely modified Buerger-type precession camera that Weir built to accommodate the lever-arm type DAC to determine if meaningful x-ray diffraction patterns could be obtained from a single crystal under pressure. Be metal inserts were used to support the anvils because of the metal's relatively high transmission properties for Ag K_α radiation [33]

precession camera to accommodate a lever-arm type pressure cell fitted with beryllium (Be) metal plates to support the diamond anvils. Beryllium was used because of its relative transparency to Ag K_α ($\lambda = 0.5609$ Å) radiation and diffracted beams from the crystal could penetrate it without much loss in intensity (Fig. 1.19).

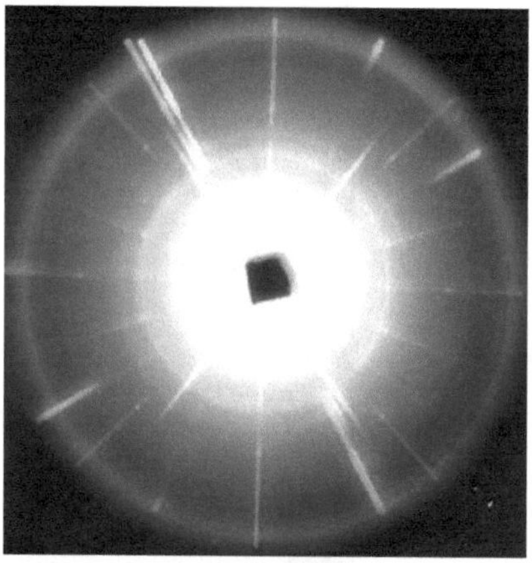

Fig. 1.20 This is one of the very first precession patterns ($\mu = 16°$, Pd filtered Ag radiation) obtained of the hk0 level of ice VI at RT and 0.98 GPa. The heavy diagonal streaks are from the single-crystal diamonds and the rings are from the metal gasket. The very small spots in a regular geometric pattern arise from diffraction from the single crystal of ice VI. This photograph is one of the very first ever taken of a single crystal at RT and these high pressures. A detailed analysis of this film related to the structure of ice VI can be found in [34]

Fig. 1.21 This is a typical 24 h exposure using for a zero-level photograph obtained from a bromine single crystal. Bromine was used for this analysis because of its relatively intense diffraction spots as shown in the photograph [35]

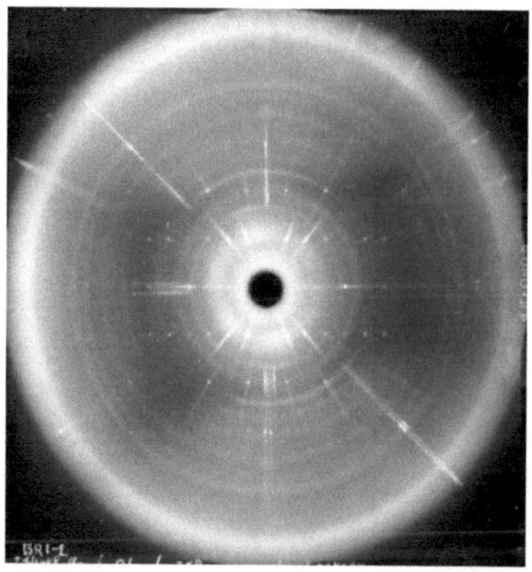

Because of the encouraging x-ray data obtained for ice VI (Fig. 1.20), a complex set of absorption and polarization corrections for the Be inserts and the crystal under precession motion were evaluated and tested on a single crystal of Br_2 whose structure was known (Fig. 1.21). The results of the analytical procedure showed a dramatic improvement in intensities and that a structure factor (R-Factor) as low as 9.1% was possible when these corrections were applied [35]. Utilizing these absorption corrections, the first structure determination ever done on an unknown single crystal by this DAC technique was accomplished on a high-pressure form of benzene, benzene II [36]. At a pressure of about 2.5 GPa and RT, benzene II was found to crystallize in the monoclinic system (space group $P2_1/c$; Fig. 1.22). The

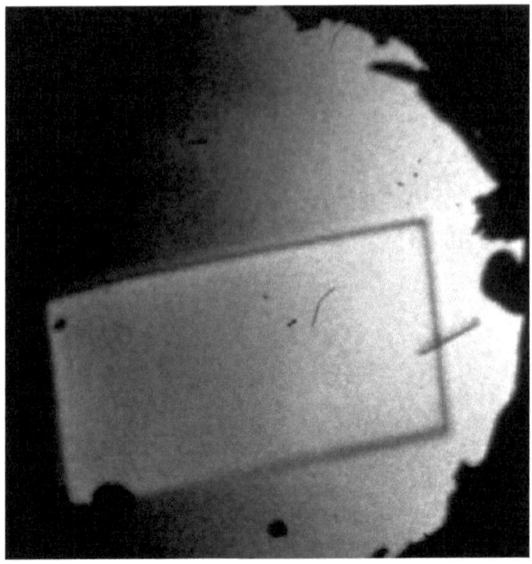

Fig. 1.22 A single crystal of benzene II in equilibrium with liquid at about 310°C and 3 GPa, showing well-defined crystal morphology. This crystal was slowly grown to fill the gasket cavity by gradually reducing the temperature to RT. A crystal such as this one was used to study the structure of benzene II [36]

Table 1.1 A comparison of crystal data for benzene I and benzene II. The numbers in parentheses are standard deviations on the reported values in units of the last place, obtained from least-squares refinement from 20 experimental values of 2θ. The increase in density (ρ_c) for benzene II illustrates the effect of 2.5 GPa [36]

BENZENE I (21°C, 0.07 GPa)	BENZENE II (21°C, 2.5 GPa)
$a = 7.17$	$a = 5.417(5)$ Å
$b = 9.28$	$b = 5.376(19)$
$c = 6.65$	$c = 7.532(7)$
	$\beta = 110.00(8)°$
Space group *Pbca*	Space group $p2_1/c$
$Z = 4$	$Z = 2$
$\rho_c = 1.18\,\mathrm{g\,cm^{-3}}$	$\rho_c = 1.258\,\mathrm{g\,cm^{-3}}$

structure was solved by generating all possible molecular packing configurations and calculating structure factors, reliability factors, and packing energies for each configuration. The procedure produced a unique solution for the molecular packing of benzene II, yielding a conventional reliability index (R factor) of 7.6% based on 19 unique reflections. Later, the method was used in a similar approach to determine the crystal structure of an unknown high-pressure form of carbon tetrachloride (CCl_4-III). The space group was found to be $P2_1/c$ with four molecules per unit cell [37]. The crystal data on benzene are listed in Table 1.1

In the intervening years between 1964 and 1968 refinements to the DAC and the precession camera were made to facilitate the operation of the high-pressure single-crystal x-ray diffraction technique. A new DAC was designed and fabricated almost entirely of Be metal except for high-strength components needed to support the required loads, e.g., the load-generating screw, the tappet spring, and the spring pressure plate [38]. Also a Buerger-type precession camera, designed to accommodate the Be DAC, was built in the NBS fabrication shop (Fig. 1.23).

With this improved instrumentation crystal data on high-pressure phases of benzene, carbon disulfide, bromine, carbon tetrachloride, and potassium nitrate were obtained [39]. In these experiments, single crystals were grown at elevated pressures either from the liquid, e.g., C_6H_6-I, CS_2, CCl_4-I, CCl_4-II, CCl_4-III, or from the solid, e.g., KNO_3–III and KNO_3-IV(?). Results are shown in Table 1.2

With the publication of our successes utilizing the DAC in single-crystal x-ray diffraction at high pressures, military research laboratories such as the Explosives

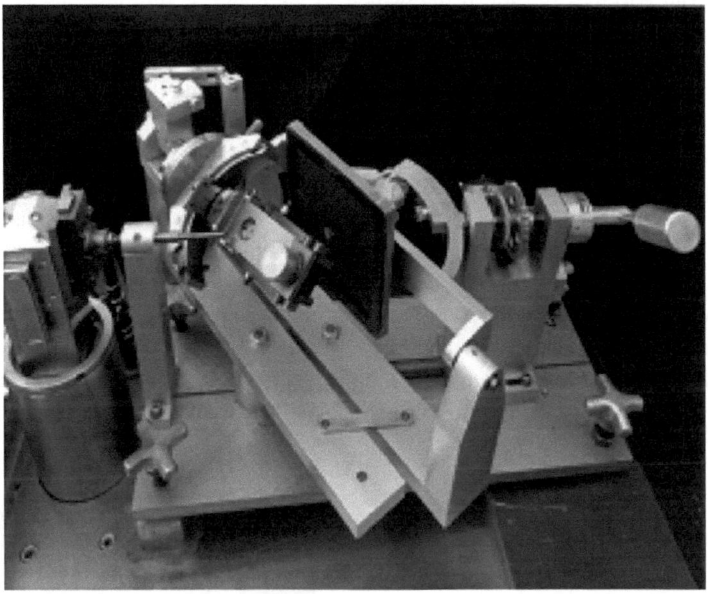

Fig. 1.23 The redesigned Be DAC and precession camera are shown here mounted on an x-ray unit ready for operation

Table 1.2 From single-crystal x-ray diffraction data obtained from high-pressure phases at room temperature, unit cell and space group data are listed. All unit cell dimensions are given in angstroms (Å) with estimated uncertainties of ±2 in the last decimal place given and uncertainties of ±0.5° in angles

C_6H_6 I–orthorhombic, $a = 7.17, b = 9.28, c = 6.65, bca$,
CS_2–orthorhombic, $a = 6.16, b = 5.38, c = 8.53, Cmca$,
Br_2–orthorhombic, $a = 8.54, b = 6.75, c = 8.63, Cmca$,
CCl_4, II–rhombohedral, $a = 14.27, \alpha = 90°$,
CCl_4, II–monoclinic, $a = 22.10, b = 11.05, c = 25.0$, $\beta = 114°$, Ca or $C2/c$,
CCl_4, III–orthorhombic, $a = 11.16, b = 14.32, c = 5.74, C222_1$,
KNO_3, III–rhombohedral, $a = 4.31, \alpha = 78°54$,
KNO_3, IV(?)–orthorhombic, $a = 5.58, b = 7.52, c = P2_1nb$ or $Pmnb$.

Laboratory, Picatinny Arsenal, Dover, NJ, and The Army Materials Technology Laboratory, Watertown, MA asked us if we could investigate the feasibility of carrying out such studies on energetic materials because compression data as well as pressure-induced phase transformations and structures of high-pressure phases were literally nonexistent on explosive materials at that time and were highly desired. The reason for the paucity of this data was that safe experimental techniques to study the behavior of explosives at high pressures were unavailable until we introduced this new DAC technique. Until the request from military laboratories came, we had no intention of studying energetic materials in the DAC. It was only because of the interest of these military laboratories that we got involved in this activity. Scientists in those laboratories recognized before we did the usefulness of the DAC in studying explosives under pressure. They noted that because of the relatively small amount of material required for study, the danger of explosion is minimal and often even nonexistent. So, in 1969 we initiated single-crystal studies for Picatinny Arsenal to determine if we could measure the compressibility of several single-crystal azides of military interest, e.g., $Pb(N_3)_2$, $Ba(N_3)_2$, KN_3, TlN_3, and NaN_3. For these studies, the measurements had to be made in a chemically inert hydrostatic pressure environment to preserve the single-crystal character of the sample. Nonhydrostatic stresses in the pressure environment would produce cracks and other kinds of defects in the crystal, destroying its single-crystal character. In addition to the chemically inert, hydrostatic-pressure-transmitting medium requirements, the pressure on the sample had to be known accurately to obtain meaningful compression measurements. We developed a novel methodology for solving these problems simultaneously. We used accurately known RT freezing pressures of *n*-hexane (10395 bar) and ethanol (22210 bar) known to be chemically inert to the azide crystals studied. Figure 1.24 shows in detail the procedure followed to solve these issues [40].

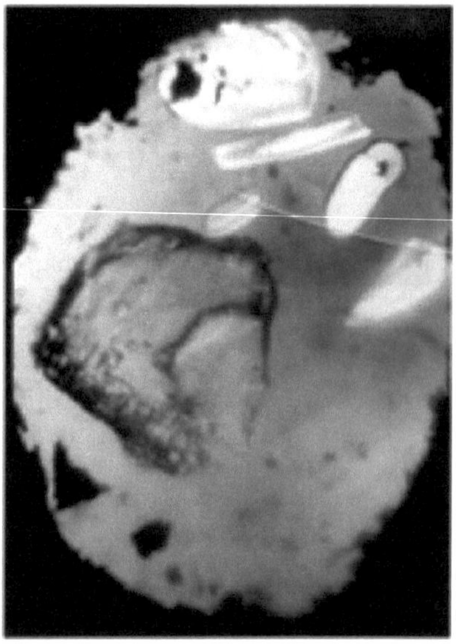

Fig. 1.24 This is a single crystal of Pb $(N_3)_2$ at approximately 2.2 GPa and RT. The matrix surrounding the crystal is chemically inert liquid ethanol in equilibrium with multiple small crystals of ethanol floating in its liquid (upper right) [40]

This condition describes the freezing pressure of ethanol at RT, which is known from previously published results of measurements made by P. W. Bridgman. The single-crystal x-ray diffraction measurements were made under these conditions utilizing the Buerger-type precession camera. The RT freezing pressures of other chemically inert liquids were similarly used to obtain data at other pressures. In this way, compression, and therefore, compressibilites of several inorganic azides were determined for the first time because previously no one was eager to compress explosives in conventional presses where relatively large volumes were required.

As illustrated in Tables 1.3 and 1.4, the compressibility of α- and β-lead azide, barium azide, potassium azide, sodium azide, and thallium azide were successfully determined by single-crystal x-ray diffraction techniques for the first time in a new application of the diamond anvil pressure cell. Because single-crystal data were obtained in this study, both anisotropic and volume compressibilities were able to be determined. A pressure-induced phase transformation was discovered for the first time in thallium azide at RT at a pressure between the freezing points of chloroform (5390 bar) and *n*-decane (2990 bar). Pressure–temperature observations of lead azide using a polarizing microscope with the DAC were carried out to 300°C and about 3 GPa. No phase transition was observed. In the overall study, it was determined that radiation damage to azide crystals under high pressures was reduced significantly compared with radiation damage observed at RT. However, a great

Table 1.3 Anisotropic and volume compression of six inorganic azides determined by single-crystal x-ray diffraction measurements in a new application of the DAC [40]

Substance	Crystal system	Presure (bar)	$a(10^{-8}\,\mathrm{cm})$	$b(10^{-8}\,\mathrm{cm})$	$c(10^{-8}\,\mathrm{cm})$	β(deg)	$V(10^{-24}\,\mathrm{cm}^3)$
$Pb(N_3)_2$	Orthorhombic	1	11.31	16.25	6.63		1218.
		10 395	11.08(5)[a]	16.16(3)	6.630(5)		1187.(5)
		22 210	10.83(1)	16.14(1)	6.601(1)		1154.(2)
$Pb(N_3)_2$	Monoclinic	1	18.46(8)	8.909(8)	5.093(6)	106.2(2)	804.4(4)
		10 395	18.01(9)	8.774(8)	5.065(5)	105.9(2)	770.(4)
$Ba(N_3)_2$	Monoclinic	1	5.435(4)	4.401(1)	9.611(4)	99.67(8)	226.2(2)
		10 395	5.395(4)	4.345(3)	9.553(6)	99.8(2)	220.2(3)
		22 210	5.375(9)	4.316(3)	9.47(2)	101.2(5)	215.4(6)
KN_3	Tetragonal	1	6.0727		7.144		263.4
		10 395	6.034		6.828		248.6
		22 210	5.992		6.638		238.3
TIN_3	Tetragonal	1	6.196(8)		7.376(7)		283.2(6)
		2 990	6.178(8)		7.316(7)		279.9(2)
NaN_3	"Monoclinc"[b]	1	6.630(2)	3.640(2)	5.299(2)	111.5(5)	118.9(2)
		10 395	6.098	3.593	5.288	106.0	111.3

[a] Numbers in parentheses represent standard deviations in the last significant figure shown resulting from a least-squares fitting process. Where standard deviations are not shown, either the parameters were obtained from the literature or there was insufficient data to do a meaningful least-squares refinement.

[b] NaN_3 assumed monoclinic at 9 bar and the parameters listed derived on that basis.

Table 1.4 Compressibilities of six inorganic azides (10^{-4}bar^{-1}) [40]

Substance	$-a_0^{-1}$ $(\Delta a/\Delta P)$	$-b_0^{-1}$ $(\Delta b/\Delta P)$	$-c_0^{-1}$ $(\Delta c/\Delta P)$	$-V_0^{-1}$ $(\Delta V/\Delta P)$
$Pb(N_3)_2$ (ortho)	1.96[a]	0.53		2.44
	1.91	0.30	0.20	2.36
$Pb(N_3)_2$ (mono)	2.34	1.51	0.57	4.12
$Ba(N_3)_2$	0.71	1.22	0.58	2.53
	0.50	0.97	0.65	2.15
KN_3	0.60		4.25	5.41
	0.59		3.17	4.30
TlN_3	0.97[b]		2.17[b]	4.63[b]
NaN_3	7.69	1.32		6.15

[a]Where two values are given, the first refers to the interval 1 bar to 10,395 bar. The second, to the interval 1 bar to 22,210 bar. One value only refers to the interval 1bar to 10,395 bar.
[b]These values refer to the interval 1bar to 2,990 bar.

disadvantage to the method is that pressure measurement is cumbersome and time-consuming because the cell has to be reloaded each time a different pressure point is needed for a different liquid.

1.5 The Ruby Fluorescence Pressure Measurement Technique

By 1971, the DAC had undergone several stages of refinement and had been adapted to other measurement techniques by NBS and other laboratories in the United States. It was used primarily for x-ray powder diffraction studies, optical polarizing microscopy, and IR absorption spectroscopy. At that time single-crystal x-ray measurements were not of interest because of the difficulty in measuring the sample pressure as discussed in our compression of azides measurements. As a consequence of this limitation, the DAC was not fully appreciated by many research scientists because there was no convenient and rapid method available for measuring the sample pressure. Pressures, determined by calculating force per unit area, were inaccurate and associated uncertainties imprecise because of the unknown distribution of force over the gasket and sample areas. This situation persisted for several years. Until 1971, pressures were either calculated (force per unit area) or measured from compression data (utilizing an equation of state, e.g., NaCl) obtained by x-ray powder diffraction measurements. But the former procedure, as already stated, was very inaccurate, and the latter was tedious and time-consuming, often requiring as many as 15 h of exposure time to produce a measurable diffraction pattern on film for one pressure determination. Thus, acceptance of the DAC by the scientific community as a tool in high-pressure research was somewhat limited. It was used in laboratories engaged in research of geological interest where very high pressures simulating the earth's interior were desired, while a knowledge of the pressure was not yet of primary importance.

In 1971, however, the situation changed dramatically. Stanley Block and I (Weir had retired from NBS and had moved to San Luis Obispo, CA in 1970) were under constant prodding by the management to develop a better method for measuring pressures in the DAC. The problem was discussed in the NBS cafeteria one day among John (Jack) B. Wachtman Jr., Chief of the Inorganic Materials Division, Stanley Block, Chief of the Crystallography Section, and Professor Dean Barnett, a Guest Scientist at NBS on sabbatical leave from Brigham Young University (BYU). (Dean had worked with large presses in his research at BYU and wanted to learn how to use the DAC.) While having lunch together, their conversation turned to the problem of measuring pressure in the DAC. During the conversation, Wachtman asked a pivotal question, "Have you considered fluorescence spectroscopy?" Among the many possible techniques we had already tested over the years, all were found to be unsuitable. However, we had never considered fluorescence spectroscopy, so they replied they would look into it. Encouraged by this meeting, Stanley Block, Dean Barnett, Richard Forman, and I undertook to measure the pressure-dependence of several fluorescing materials thought to be good candidates for a pressure sensor [41, 42]. Some of the materials investigated happened to be lying on the shelves in Forman's spectroscopy laboratory, while others were provided by H. S. Parker and W. S. Brower of the Solid State Chemistry Section. Those materials included Al_2O_3(ruby), $YAlO_3$, YAG, MgO, and a few others. In our investigation, some were found to have observable pressure dependences, but ruby exhibited by far the most promising results as revealed in Table 1.5. The main fluorescence lines in ruby (the R_1 and R_2 doublet) were intense and sharp and both lines shifted appreciably toward the red end of the spectrum with increasing pressure indicating that ruby would be a sensitive pressure sensor compared to the other materials studied. With ruby in a gasketed DAC, pressures could be detected in situ using only a very small ruby crystal (only 1% of the available volume is needed because of ruby's high R-lines fluorescent intensity). Furthermore, because ruby is chemically quite inert to most substances, including liquids, it is almost an ideal material for an internal pressure sensor and, after accurate calibration, would be useful as a secondary pressure standard. Moreover, the chemical inertness of the ruby makes it likely that experiments on many substances at temperatures above 200°C will not destroy the usefulness of this measurement technique. The very small ruby crystal can be present in the sample chamber to detect pressures without interfering with any other specimen under investigation in the chamber, including the pressure-transmitting liquid itself. From the significance of these observations, it became apparent that the pressure dependence of ruby needed to be calibrated immediately.

In 1972, a preliminary calibration of the ruby R-lines pressure dependencies (Fig. 1.25) was accomplished by using the method employed previously for the measurement of the azide compressibilities, i.e., the freezing pressures of liquids, as described in Table 1.6 [41].

This tentative result, showing linearity in the R_1 and R_2 pressure shifts, was very promising. In 1974 a more precise calibration was done at NBS utilizing an x-ray powder diffraction method to measure the compression of NaCl and relating this to

Table 1.5 Comparison of various fluorescing crystalline materials for potential use in a DAC

Material[a]	Line Designation and Description	Wavelength λ (Å)	Relative Intensity	d λ/dP[b] (Å/kbar)	dλ/dP[b] (Å/kbar)	Linewidth[c] (Å)	Background Ratio[d]
Ruby, Al_2O_3 (0.5% Cr)	R-lines Doublet	6928 6942	very strong	0.36	0.068	7.5	0.01
$YAlO_3$ (0.2% Cr)	R-lines Doublet	7228 7251	strong	0.70	0.076	10	0.28
YAG[e] (0.38% Cr_2O_3)	R-lines Doublet	6887 6878	medium	0.31	0.105	8	0.53
$YAlO_3$ (2.5% Cr)	Pair line Singlet	7320	strong	0.84	0.093	21	0.39
MgO (Cr unknown)	R-line Singlet	6992	weak	0.35	0.090	10	0.92
$YAlO_3$ (Nd unknown)	R_2-Z_2 line[f] Multiplet	8753	strong	−0.13	0.010	20	0.22

[a] Amount of dopant is designated as a weight percent
[b] These values were measured and are recorded here for comparison purposes only and are not reported as accurate measurements of the derivatives with respect to pressure and temperature. We consider the values to be accurate to approximately 10%
[c] Linewidths indicated are those measured with the apparatus described herein. They represent full width at half-maximum values (FWHM)
[d] Ratio of background intensity near the peak to intensity of the peak above the background. In MgO and the Nd-doped $YAlO_3$ samples' the background may be due to poor crystal quality and/or other impurities
[e] Yttrium Aluminum Garnet
[f] As designated in M. J. Weber and T. E. Varitimos, J. Appl. Phys. **42,** 4996 (1971)

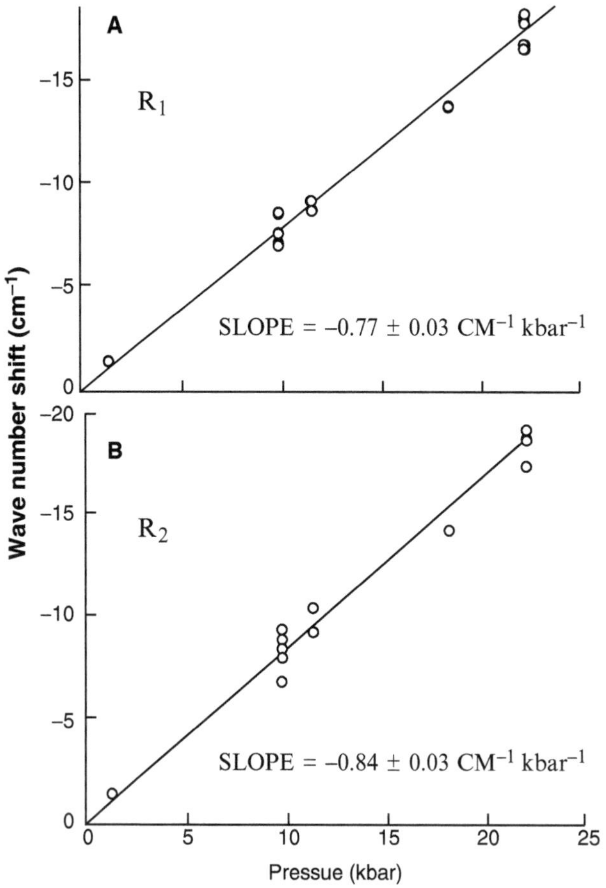

Fig. 1.25 Pressure-dependencies of the ruby R_1 and R_2 fluorescence lines determined from the pressure points listed in Table 1.6. These pressures are approximate because the temperature at which the measurement was made was not measured or controlled

Table 1.6 Freezing pressure points

Substance	Transition[a]	Transition pressure (kbar)
CCl_4	L-I	1.3
CCl_4	III – IV[b]	40
H_2O	L-VI	9.6
H_2O	VI-VII	22.3
n-C_7H_{16}	L-I	11.4
n-C_2H_5Br	L-I	18.3

[a] L refers to the liquid state and the Roman numerals designate solid phases.
[b] The III–IV transition point of CCL_4 was excluded from the calculation because of the nonhydrostatic character of the medium surrounding the ruby crystal [41].

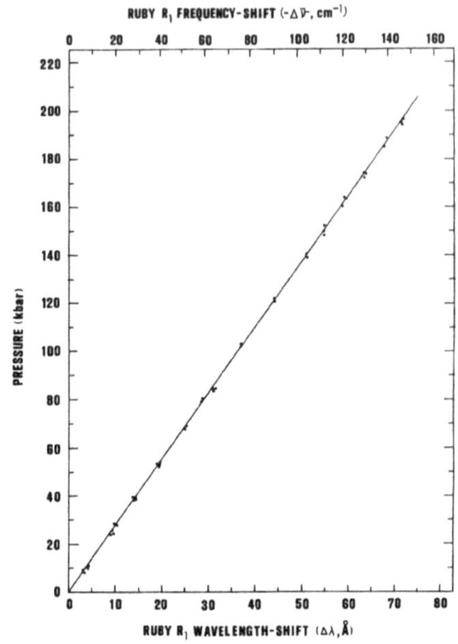

Fig. 1.26 The pressure dependence at 25°C of the ruby R1 fluorescence line at 6,942 Å as a function of wavelength (Å) and also wave number or frequency (cm^{-1}). Values for pressures are based on the Decker equation of state for NaCl. The dependence is linear to 195 kbar following the equation $P_{NaCl} = 2.746(\Delta\lambda)$, where P is in kbar and $\Delta\lambda$ in Å. The uncertainty in the slope, $dp/d\lambda$ in terms of a 95% confidence interval is 2.746 ± 0.014 kbar $Å^{-1}$. Including the reported uncertainty associated with the Decker EOS for NaCl, the value of the slope is 2.740 ± 0.016 kbar $Å^{-1}$ within a 95% confidence interval [43]

the measured shift in the wavelength of the ruby R_1 line under the same conditions (Fig. 1.26). The corresponding pressure was calculated from the compression data utilizing an established equation of state for NaCl (the Decker EOS) [43].

1.6 Hydrostatic Pressure-Transmitting Media

One ramification of the ruby fluorescence method of pressure measurement is the very useful property of line broadening of the R_1, R_2 doublet peak, resulting from the initiation of inhomogeneous stresses in the pressure-transmitting medium surrounding the ruby as shown in Fig. 1.27. Because knowledge of the nature of the pressurized environment surrounding a sample under study is extremely important, the phenomenon was studied further to corroborate the assumption that the broadening was due to inhomogeneous stresses [44]. Other liquids, known to have hydrostatic pressure limits from earlier-published press experiments, were studied by the ruby fluorescence method, e.g., isopropyl alcohol, 1:1 pentane:isopentane, and ethanol, the results of which are shown in Fig. 1.28. Included are observations on 4:1 methanol:ethanol. The results were quite interesting. First, they demonstrated that under increasing hydrostatic pressures, the R-lines sharpen significantly until a pressure is reached where an abrupt discontinuity occurs and line broadening sets in severely. The discontinuities occurred at pressures in agreement with published hydrostatic limits, corroborating our initial assumption that line broadening was the result of non-uniform stresses. Second, our measurements on 4:1 methanol:ethanol,

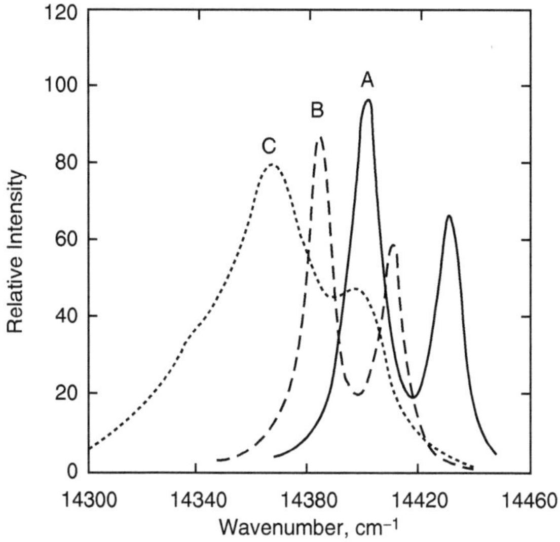

Fig. 1.27 Luminescence spectra from a crystal of ruby (0.05% Cr by wt) in a gasketed DAC. A is at ambient pressure; B, at 2.23 GPa (mixture of ices VI and VII); C, in a nonhydrostatic environment at about 4 GPa (mixture of CCl_4 III and IV) illustrating a line broadening affect [41]

Fig. 1.28 Line-broadening effects due to the onset of nonhydrostatic stresses on the ruby crystal relative to the 1 atmosphere line-width. The line-width decreases continuously until the discontinuity points, P_1, P_2, P_3, and P_4, which indicate the pressures at which the ruby senses nonhydrostatic stresses in a given pressure-transmitting medium. For example, the discontinuity in 4:1 methanol: ethanol occurs at P_4, about 104 kbar, for methanol, P_3 at about 70 kbar, etc. [44]

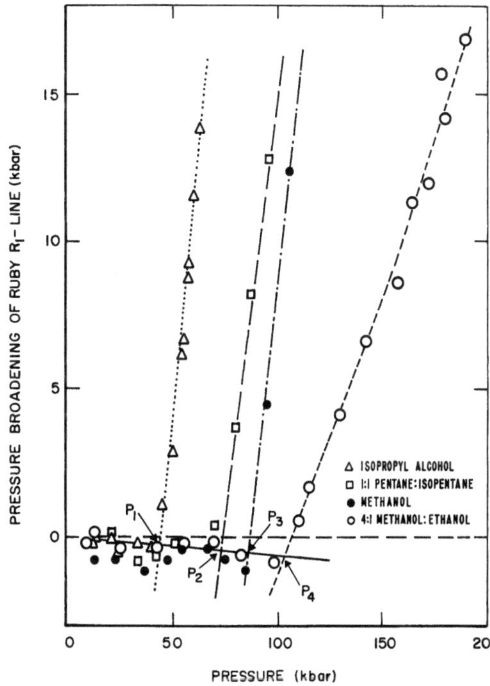

carried out for the first time on that liquid mixture, produced a discontinuity (and thus a hydrostatic limit) at a pressure far beyond any known liquid or liquid mixture at that time.

To advance our knowledge on this phenomenon, and to further support our argument that line broadening of the ruby R-lines is the result of the onset of nonhydrostaticity, another experiment was devised. Fine particles of crystalline ruby chips were dispersed throughout two solids, AgCl and NaCl (both commonly used as pressure-transmitting media in press experiments), water, and 4:1 methanol:ethanol (to corroborate our own ruby fluorescence experiments), were pressed in a gasketed DAC. A unique system of diaphragms and fine slits built into our microscope permitted us to measure fluorescence spectra from isolated small areas along the gasket diameter. Position sensitive sampling of the R_1 ruby lines revealed the pressure distribution in the compressed media [44].

Ruby R-line broadening data such as that shown in Fig. 1.28 is extremely important because it permits one to readily assess the critical property of hydrostaticity in liquids that are used as pressure-transmitting media. Confirmation of the hydrostatic limit of liquids and powders is shown in Fig. 1.29. To illustrate, the 4:1 ethanol:methanol mixture begins to show nonhomogeneity in the stress at a pressure of about 10 GPa, in agreement with the result shown in Fig. 1.28.

Our interpretation of the discontinuity in the line broadening graphs in Fig. 1.28, i.e., the onset of nonhydrostaticity in the medium, initially was not fully appreciated by some skeptical scientists. To satisfy the skeptics, we provided additional experimental evidence to support our argument. We developed a simple Stokes Falling Ball method to measure the pressure-dependence of viscosity of liquids [45]. The DAC was mounted on an optical goniometer modified to permit viewing the sam-

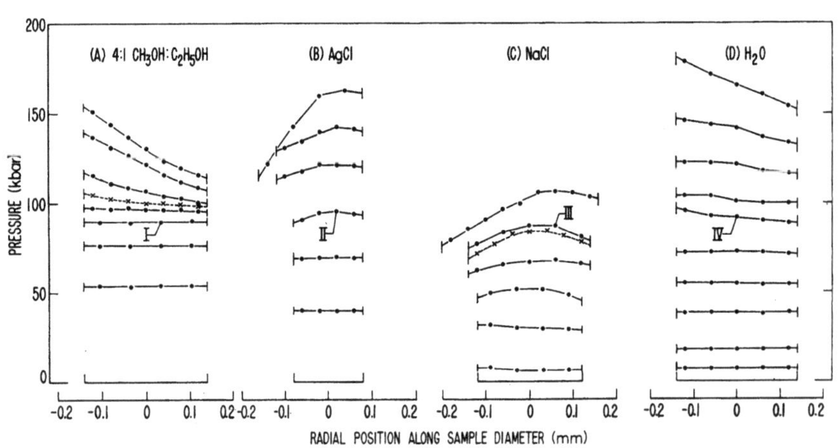

Fig. 1.29 Position sensitive line broadening measurements in some commonly used pressure-transmitting media. Note that the 4:1 methanol:ethanol mixture begins to show nonuniform stresses above 10 GPa [44]

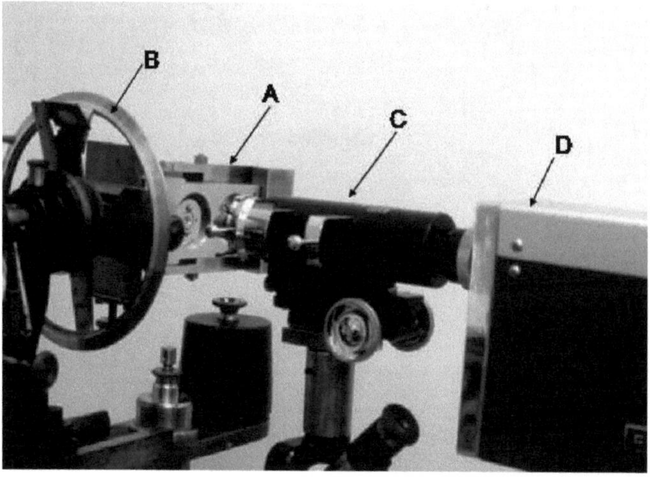

Fig. 1.30 A view of the viscosity apparatus, showing the mounted DAC (**A**), the optical goniometer (**B**), the horizontal low-power microscope (**C**) and attached TV camera (**D**)

ple in the DAC with a stationary horizontal microscope attached to the goniometer (Fig. 1.30). The translational motions of the goniometer positioned the sample on the optic axis of the microscope and the rotational feature inverted the sample on a horizontal axis while maintaining optical alignment (Fig. 1.31). A TV camera and recording instrument were incorporated in the design to permit observations of the ball position over long periods of time. At very high viscosities, for example, observations were made over a period of weeks.

In a practical sense, one significant result of this work is the independent confirmation of the glass transition pressure in 4:1 methanol:ethanol mixture at 10.4 GPa (Fig. 1.32). However, in a more general sense, I think it would be a useful endeavor to explore the use of this method to measure the pressure dependence of explosive liquids, an area of study which has received little attention in the past.

As deficiencies in the ungasketed method such as the large parabolic gradient in pressure, the relatively small volume of sample (a thin layer between the anvil flats) became more appreciated from the line broadening studies, use of a metal gasket became almost universal. As illustrated above in Fig. 1.29, if the gasket confines a powder, pressure gradients are reduced considerably from the case where no gasket is used, but, when the powder is in a liquid pressure-transmitting medium, gradients in pressure are nonexistent until the liquid reaches its freezing pressure or the viscosity of the liquid approaches the glass transition pressure (Fig. 1.33). Thus, it was quickly recognized that if liquids could be confined under pressure in a DAC, then other materials in the form of powders and single crystals could be included and the liquid would act as a hydrostatic pressure-transmitting medium as long as it remained a fluid. Not only did this improve existing x-ray powder diffraction studies by providing a well-characterized pressure environment, but, as we saw earlier, it

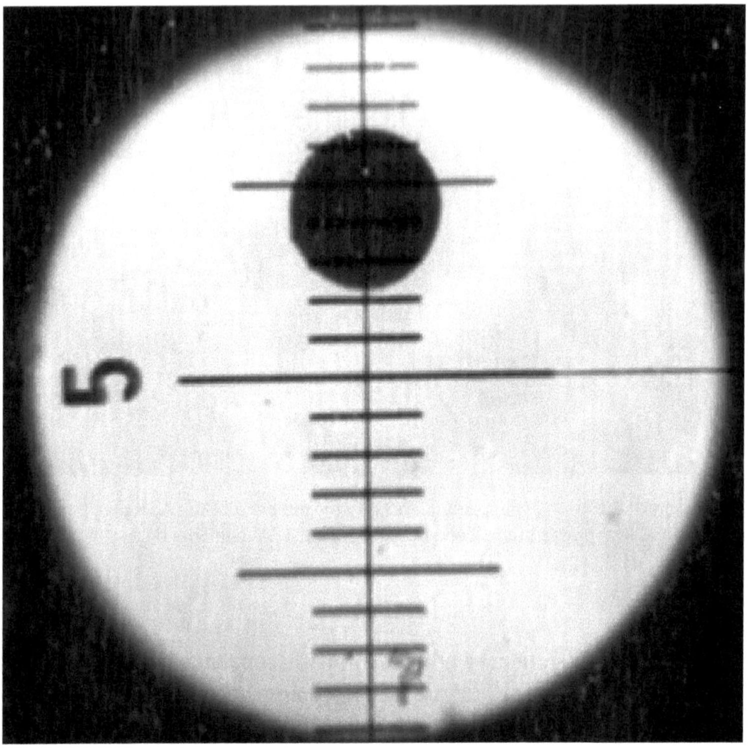

Fig. 1.31 A view of Dow-Corning 200 fluid in the gasket of a DAC, a solid Ni sphere about 0.033 mm in diameter, a very small chip of ruby located at the bottom of the calibrated superimposed scale in the eyepiece of the microscope. The smallest graduation on the scale for the particular optics used is 0.0095 mm. Later, a ruby sphere was used both as the falling ball and the pressure sensor, eliminating fluid flow interference between two objects in the liquid [45]

also led to single-crystal x-ray diffraction under high pressure, a revolutionary development in high-pressure techniques.

To facilitate pressure measurements an optical fluorescence system for rapid routine pressure measurement was developed at NBS in 1973, which permitted pressures to be measured in the DAC with an accuracy and speed hitherto unachieved [42]. The precision of the measurement with that instrumentation with the ruby in a hydrostatic environment is 0.05 GPa. The NBS system, shown in schematic detail in Fig. 1.34, includes an optical polarizing microscope for visual observation of the sample which is also displayed continuously on a color video monitor. In addition to the sample image, the calculated pressures (both FWHM and model values), the measured temperature, the measured and calculated model R-line spectra and the differences between them, and also the time and date are displayed next to the image of the sample on the monitor every 3 s or at longer intervals when greater precision in the pressure measurement is desired. All of this information can be recorded on

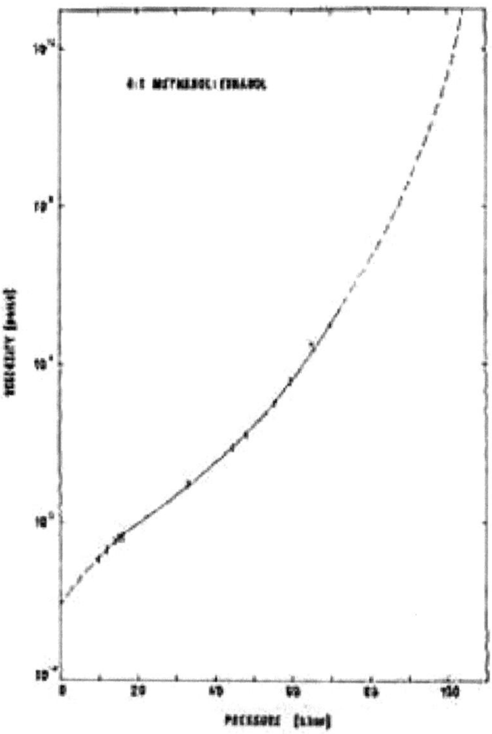

Fig. 1.32 Viscosity as a function of pressure of 4:1 mixture by volume of methanol:ethanol at room temperature corrected for an infinite parallel wall effect. The dashed lines are extrapolations to the glass transition (10^{-13}poise) at about 10.4 GPa (determined by the ruby method) and to the viscosity of pure methanol at 1 atmosphere (6×10^{-3} poise) [45]

video tape as a record of the experiment and can be referred to at a future time with all the relevant information. A photograph of the apparatus is shown in Fig. 1.35.

By 1974, DAC technology had advanced to a high degree of sophistication. With the development of the rapid, accurate, routine, and convenient ruby fluorescence method of pressure measurement at NIST, the problem of pressure calibration and measurement, a major obstacle to the widespread use of the DAC, was removed. As a result, the DAC experienced an unprecedented expansion in its use and acceptance by the high-pressure community as a tool for physical and chemical investigations at high pressure. The ruby fluorescence pressure scale was subsequently extended by various scientists, in 1978 to 1 Mbar, and in 1986 to 5.5 Mbar. No longer was the diamond cell simply a qualitative or semiquantitative instrument, but it became a serious quantitative research tool for carrying out high-pressure studies. Today it is the instrument of choice as a research tool for high-pressure research. Our successes with these two instruments, the lever-arm and hydraulically loaded pressure cells, motivated other research laboratories in the United States and also in other parts of the world, particularly in Japan and Germany, to initiate research programs using

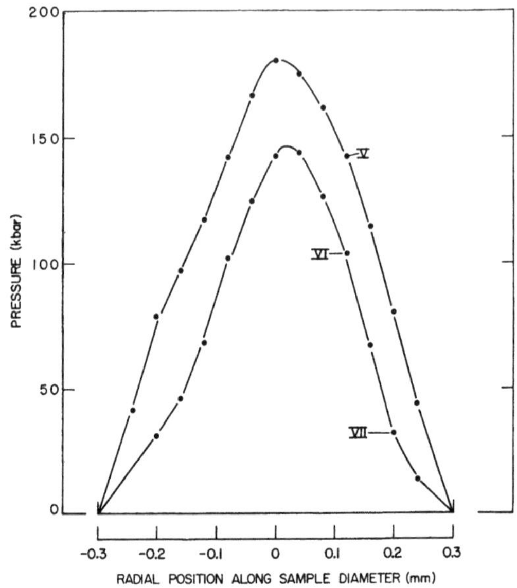

Fig. 1.33 Pressure distribution in an ungasketed sample of powdered NaCl intimately mixed with fine powdered ruby crystals (approximately 10% of ruby by volume) in the diamond anvil cell at two different applied loads (V > VI > VII). The mixture itself through friction acts as its own seal as the sample thins out under load. Because the coordinate scale factors are identical for Figs. 1.29 and 1.33, a comparison between the two figures illustrates clearly that even for a small area within the sample, pressure gradients in the ungasketed system exceed greatly the gradients in gasketed systems [44]

these devices. The following paragraph describes some of those instruments that were developed.

For the purposes of this presentation, I will discuss briefly several well-known pressure cells, each of which has strengths in particular areas of application for which they were specifically designed, and also weaknesses in other areas of application. I do not intend to go into great detail discussing the operation of these instruments. In this connection, the reader is referred to reference [46], for an outstanding review of DACs, discussed in great detail and completeness. The most important component of a DAC is the Bridgman-opposed anvil arrangement of the diamonds, which permits the generation of high pressures on anything placed between them. This opposed-anvil arrangement is a fundamental characteristic of all DACs regardless of their overall mechanical design. The other essential components serve to (1) align the anvils so that their faces are parallel and their areas coaxial, and (2) generate forces on the anvils sufficient to produce the desired pressures on something placed between their faces. Several designs have arisen to fulfill the requirements of (1) and (2), and also to meet specific experimental conditions. Variations in the mechanical design of the DAC have arisen from different ways of generating the applied force and of aligning the diamonds to make them parallel and on axis.

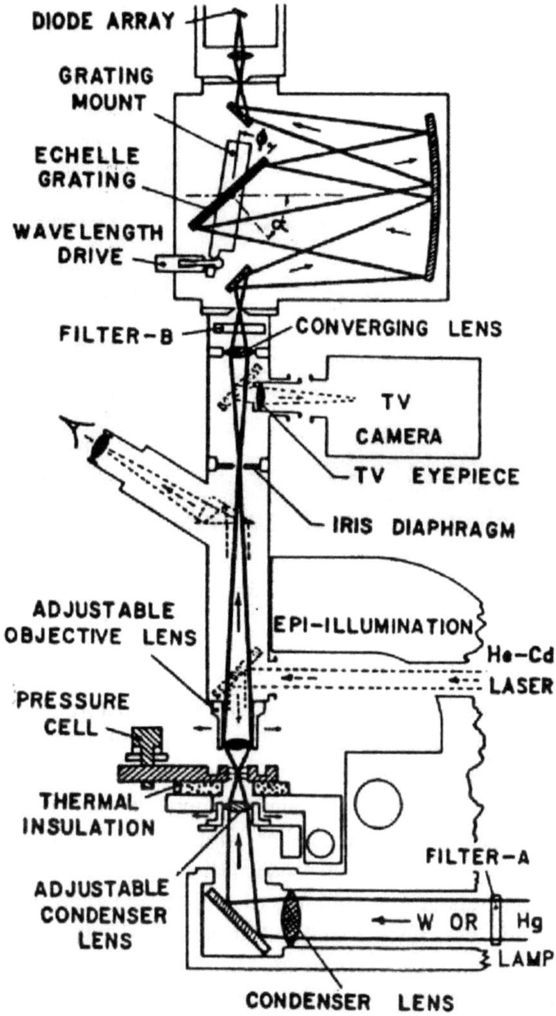

Fig. 1.34 A detailed description of this figure is inappropriate for the purposes of this presentation. Only a brief explanation is given here. A focused He-Cd laser beam is directed via EPI illumination onto the ruby crystal in the pressure cell. The resulting fluorescent emission is directed to the echelle grating and wavelength analyzed to determine its shift relative to the wavelength at atmospheric pressure. The pressure is then calculated and displayed on the monitor utilizing computer programs. A complete description of the system is given in reference [42]

1.7 Some Basic Types of DACs

One such instrument is the NBS cell developed in our laboratory in 1975 and still used quite widely today [47]. Some history behind its development is appropriate to give here because it demonstrates how advances in DAC technology were made in

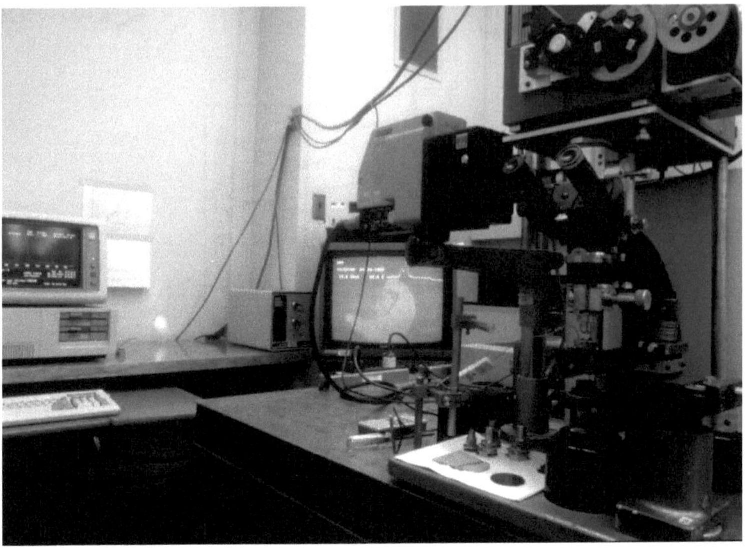

Fig. 1.35 This photograph shows the microscope, mounted DAC, echelle grating, video camera, computer with monitor and television with sample image and relevant data displayed on the screen [42]

those early years. In our early experiments with DACs before 1975, we learned to appreciate how critical it was to achieve and maintain good anvil alignment, because many diamonds were destroyed in the process of testing new mechanical alignment designs. Fortunately for us, we had an abundant supply of diamonds (provided gratis by the US Customs Service, as noted earlier in this chapter), and, although diamond failures were common in those days because of primitive DAC designs, we were not overly concerned about this. Initially, to support and align the anvils we used ordinary steel plates. One plate could be translated and rotated and then fixed in position by three lateral screws 120° apart. The other plate could be tilted with two screws 180° apart. A combination of tilt, rotation, and translation provided the necessary parameters for making the anvil faces parallel and axially aligned. To achieve the latter, one anvil face was intentionally made larger in area than the other. The weakest part of this design was the tilting mechanism, which failed under high loads often resulting in destruction of the diamond, even though, at that time, we had only a rough estimate of the pressure that was reached at the point of failure. The other weakness was the relatively soft metal (unhardened) used to fabricate the plates. We noted that under high loads the table of the diamond indented the seat supporting the anvil, usually unevenly, which contributed to misalignment of the faces. We were literally working in the dark as far as understanding the details of the improvements that were being made was concerned. In this connection, I cannot overstate the importance of our having developed the ruby fluorescence pressure-measurement technique, because, with that new capability, we began to understand in-depth the problems we faced and the solutions that were needed to correct them. For example, with the ability to measure pressure, we were able to determine the maximum pressure achieved before a particular design under study failed. With such information, we proceeded to redesign the alignment and force generation components of the cell in a stepwise fashion. Learning from our earlier failures, we ultimately developed the ball and socket alignment arrangement and force-generating mechanism shown in the diagram below.

In this design, the principle employed to generate force is a spring-loaded lever-arm assembly, where the applied force is produced by the simple rotation of a large screw, which compresses Belleville spring washers, each having a capacity of 272 kgf when fully compressed to the flat position. The spring-lever-arm arrangement generates a uniform and continuously varying force as the screw is rotated, a feature that has proved, in practice, to be highly desirable because it eliminates distortions in the anvil assembly and also undesired large changes in pressure. Use of Belleville spring washers has also proved to be highly useful because they can be stacked either in series, in parallel, or in combinations of both, to alter sensitivity and load characteristics as preferred. It is specifically control of these characteristics which are desired because they can be tailored to the peculiarities of the sample under investigation. The applied load, magnified by 2 through the lever-arm system, is transmitted to the pressure plate which bears against the extended piston containing one of the diamond anvils. The opposing anvil is fixed in its position and acts as an entablature for the piston anvil. The fixed anvil or entablature is supported by a translating diamond mount plate, which permits the two anvil faces to be aligned axially along the direction of the load. The piston anvil is supported in a tilting diamond mount hemisphere and permits parallel adjustment of the opposing faces. With these simple adjustment features, anvil alignment can be accomplished with ease and precision and is also permanently maintained. Both diamond mount supports contain conical cutouts which permit 180° optical access and can be modified to accommodate the particular requirements of the measurement of interest. A loose-fitting removable bifilar-type resistance coil heater sheathed in Inconel alloy and electrically insulated with MgO powder is conveniently inserted in a cavity surrounding the anvil-piston assembly for heating as shown in Fig. 1.36. In effect the sample is located at the center of an externally heated coiled resistance furnace. For low temperatures, the furnace can be replaced with a cooling coil for the passage of liquid nitrogen or other lower-temperature coolants. Personally, I favor the spring-loaded lever-arm design over others as a matter of convenience, static heating with resistance coil heaters and ease in loading with gasketed samples. However, I note, that regardless of the mechanism employed for generating the force, the criteria for producing very high pressures remain the same: (1) the ability to align the anvil faces accurately and (2) maintaining this alignment rigidly under the loads necessary to produce high pressures. The alignment arrangement consists of a hardened (RWC 60) diamond mount plate, which can be translated with the aid of three screws (120° apart) to position the anvils axially. The other anvil is mounted in a similarly hardened steel hemisphere that can be tilted with three screws (120° apart) to make the opposing anvil faces parallel as indicated by the disappearance of interference fringes in the gap between the assembled anvil faces. In 1975, we were

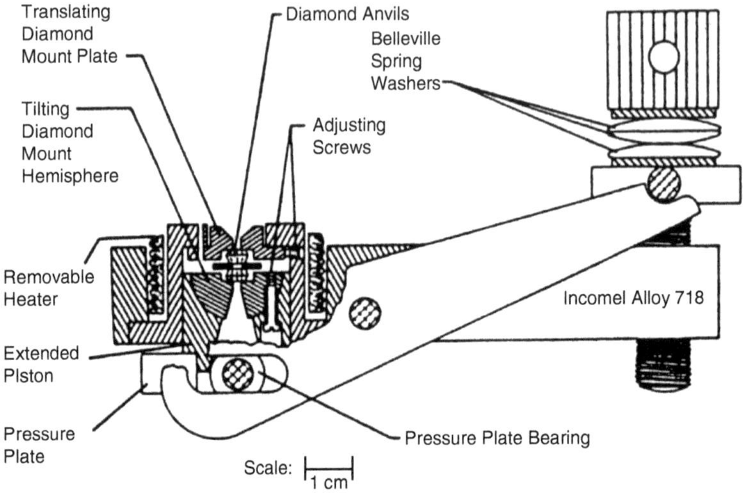

Fig. 1.36 The NIST diamond anvil cell with temperature capability [47]. This cut-away cross section shows the essential components of the NIST DAC including the anvil support alignment design, lever-arm assembly and spring washer loading system. For high-temperature experiments the entire cell, except for the spring washers, is fabricated from Inconel 718, a high-temperature high-strength superalloy which permits sample temperatures as high as 1,073 K to be maintained routinely [58]

able to achieve the highest pressures ever measured by the ruby method (in excess of 60 GPa) in a DAC. Indeed, the famous 19th century physicist, William Thomson (better known to us as Lord Kelvin) had great insight when he commented about the importance of being able to measure a quantity to increase our knowledge and understanding of that quantity. Seventeen years before the founding of NBS, Lord Kelvin, in one of his many lectures [48], said:

> *When you can measure what you are speaking about, and express it in numbers, you know something about it; but when you cannot measure it, when you cannot express it in numbers, your knowledge is of a meager and unsatisfactory kind: it may be the beginning of knowledge, but you have scarcely, in your thoughts, advanced to the stage of science.*

Because of the amazing success of that work, increasing numbers of scientists became interested in learning how to use a DAC especially because one could now routinely measure the pressure on a sample in the cell using an internal pressure sensor (Fig. 1.37). In that paper [47], I stated that I would be pleased to provide working drawings of the instrument upon request. Soon after the publication of that work, P. W. Bell of the Geophysical Laboratory in Washington, DC visited me along with his machinist to obtain the drawings, and also to get my opinion on what I thought would improve the performance of the instrument based on my experience using it. If I were to fabricate another cell, I said I would make the piston and cylinder much longer to improve and maintain alignment at loads higher than I had attained in our present experiments. I thought that was the weakest part of our design. In

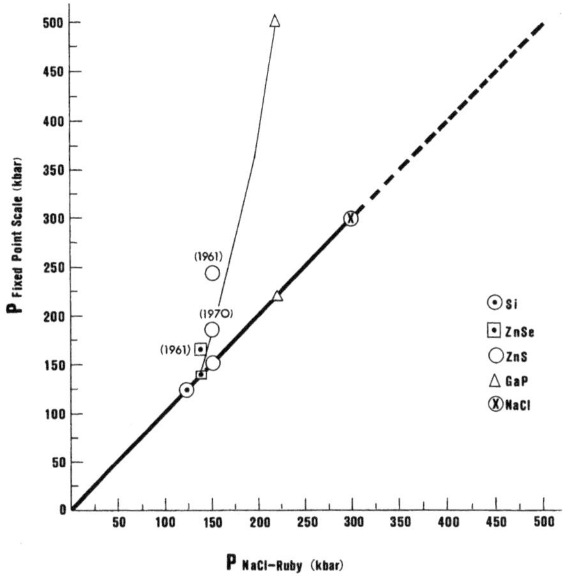

Fig. 1.37 Transition pressures for various materials used as fixed-points on the Fixed Point Pressure Scale (FPPS). The graph compares FPPS with the ruby-NaCl scale. The points identified with 1961 refer to the original FPPS; those identified with 1970 refer 1970 revised scale [47]

1977, Mao and Bell came out with the DAC design shown below with the astonishing announcement of having successfully attained the unprecedented pressures of over a megabar [49]. It took just 2 years to go from 60 GPa to over 100 GPa, an accomplishment creating quite a stir in the high-pressure community, particularly for geophysicists, who were interested in studying phenomena occurring in the interior of the earth. In addition, studies with the ultrahigh-pressure DAC resulted in a revision of the fixed-point pressure scale commonly used to calibrate large presses at that time. The values of the calibration points based on the ruby pressure scale were lowered by a factor of 2 in the range above 15 GPa as illustrated in the graph below.

Like the NIST cell, the Mao-Bell DAC uses a Belleville spring-loaded lever-arm mechanism for generating the force, but is considerably larger and operates in an inverted configuration (Fig. 1.38). For stable anvil alignment, this cell uses a long 60–70 mm detachable piston/cylinder assembly. Anvil alignment involves translating and tilting two hardened half-cylinder rockers. Basically a scaled-up version of the NIST DAC, this cell facilitates the generation of very large forces to produce pressures in the megabar range originally intended for the study of materials of geological interest. During the past several years, cells of this type have continued to break records in achieving the highest static pressures in the laboratory. It has also been used extensively to study the condensed noble gases as well as other gases, which are loaded in a large cryostat at low temperatures. However, its relatively

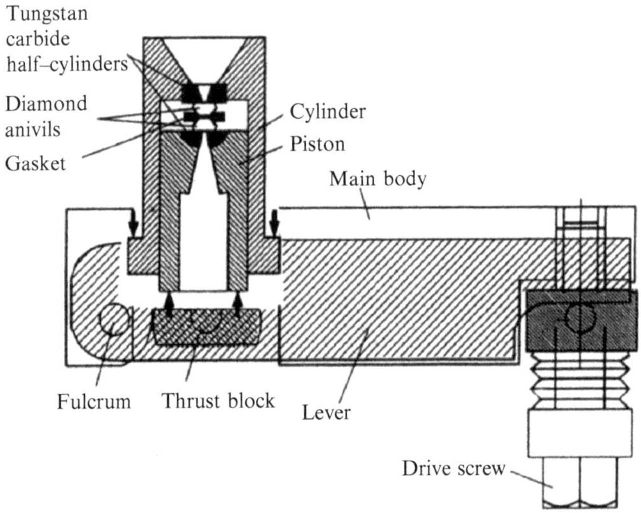

Fig. 1.38 The Mao-Bell diamond anvil cell

large size makes it inconvenient to adapt to measurement techniques where limited space is available.

Another type of DAC designed for x-ray powder diffraction studies is the Bassett cell [50]. The force-generating mechanism in this cell consists of a driving nut in a threaded stainless steel cylinder with relatively massive walls to provide strength and rigidity. The anvil assembly consists of a stationary anvil whose support is threaded from one end of the cylinder and a movable anvil in a sliding piston arrangement is driven from the opposite end of the cylinder as shown in Fig. 1.39. Rotation of the driving nut advances the piston pushing it against the stationary entablature.

Anvil alignment is achieved with two half-cylinder rocker anvil supports whose axes are rotated by 90° in the plane normal to the load direction. Forces generated by this mechanism tend to be discontinuous and in relatively large steps rather than continuous, as is the case provided by spring washers, because the pitch of the thread determines the magnitude of the load generated for a given fraction of a rotation. In cells of this kind, the thread pitch is necessarily small in order to sustain the large forces needed to produce high pressures particularly because there is no mechanical magnification factor as in the case of the lever-arm system. Another disadvantage with this design is in the inconvenience of providing static heating capability using a resistance coil furnace. This cell has been used extensively for high-pressure x-ray powder diffraction studies and with laser heating of samples to very high temperatures for phase diagram studies. Pressures as high as 40 GPa have been reported with samples in the ungasketed configuration using 0.3 mm anvil flats.

The Holzapfel cell [51] is a design based on a knee-action mechanism. Two parallel-threaded rods, which connect the two larger levers shown in Fig. 1.40, are synchronously rotated with a special gear-set wrench pulling the lower ends of the

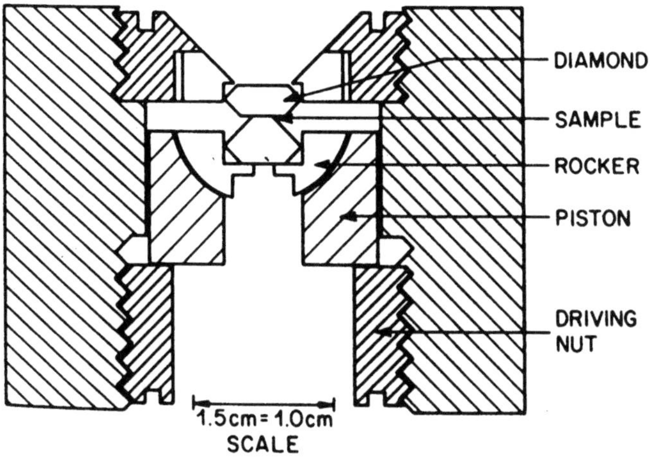

Fig. 1.39 Cross-section drawing of the Bassett cell showing the hemi-cylinder hardened rocker supports for anvil alignment and the very simple force-generating mechanism of a driving nut pushing against a movable piston in a guiding cylinder

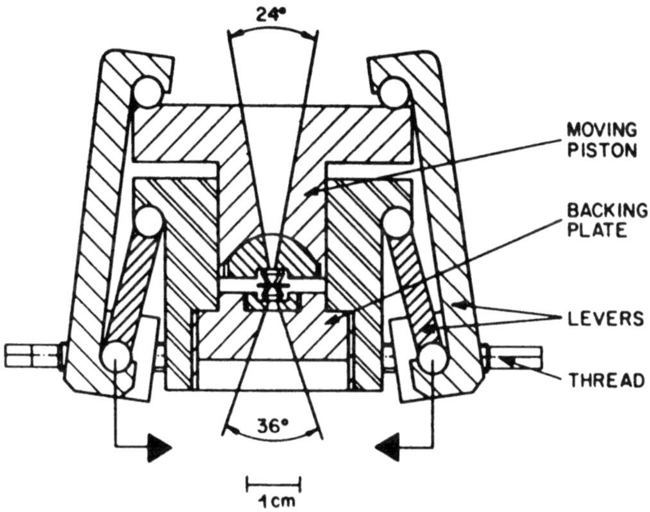

Fig. 1.40 Cross-section of the Holzapfel DAC. The thrust for the piston is generated by moving the opposed knee-type lever arms closer together when the threads are advanced synchronously without distortion. A tilting hemispherical diamond mount and a translating plate are used to align the diamonds similar to the NIST design

levers together. A force is exerted on the moving piston, which generates the pressure. This knee mechanism results in a large force multiplication factor, which permits pressures in the 50 GPa range to be reached. This cell uses the NIST mechanism of a tilting hemisphere and a translation plate for anvil alignment. The cell has been adapted for x-ray powder and single-crystal diffraction, Raman, Brillouin, and other optical studies at high pressure. The special geometry of the knee mechanism including the levers and threaded rods require precision-machined components which

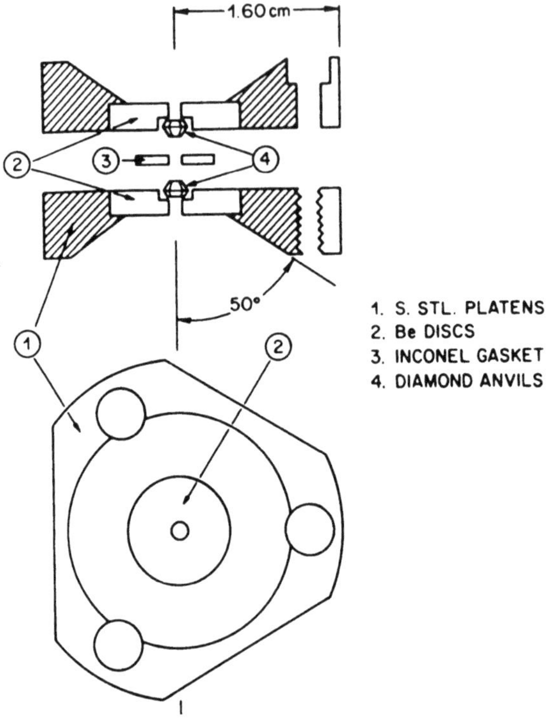

Fig. 1.41 Miniature Merrill-Bassett cell designed originally for single-crystal x-ray diffraction studies, but now used for other kinds of experiments also, consists of Be anvil support plates for high radiation transmission. Thrust is applied by tightening three symmetrically positioned pulling screws. Maintaining anvil alignment may be an issue if the screws are not rotated at the same rate

must all fit together with great accuracy, otherwise the thrust will not act strictly parallel to the cylindrical axis of the instrument and problems may arise in maintaining alignment under applied loads.

The Merrill-Bassett DAC is the simplest of all designs [52] (Fig.1.41). Three screws pull two platens together. When tightened synchronously with a special gear-set wrench, diamond alignment is maintained and pressures are generated. Of all the DACs, this device is the most compact, rendering it useful for single-crystal x-ray diffraction measurements because it can be mounted directly on the circle of an automated x-ray goniometer without additional supporting mechanisms. For this application Be supports are required for the diamond anvils because they are relatively transparent to x-rays and provide a wide-angle window for the incident x-ray beam and also for the diffracted beams. Unlike steel, however, Be is relatively weak and limits the support it can provide to reach very high pressures. Another problem is the lack of positive guidance for the diamond anvils to help maintain alignment. Because of these disadvantages, pressures in excess of 15 GPa are rarely reached with this DAC.

Today, the sample is generally prepared in the gasketed configuration. Metal gaskets, in addition to prolonging the life of the anvils and extending their pressure range, permit the sample to be encapsulated in a fluid pressure-transmitting medium, thus providing a truly hydrostatic environment for the sample. The hydrostatic (zero shear stress) state is highly desirable because it eliminates the presence of shear stresses of unknown magnitude which often lead to serious difficulties in the interpretation of the desired measurement. Thus, the maintenance of the hydrostatic environment is a prime consideration for achieving error-free results in any high-pressure experiment.

In most of the earlier work with the DAC, sample pressures were estimated from the applied load. Such estimates yielded an average pressure over the sample in the ungasketed case, but are even less reliable using a gasketed sample, because the gasket absorbs an unknown amount of the load. Losses in load due to friction both in the sample and in the instrument were also unknown. The lack of a rapid reliable pressure measurement procedure seriously hampered research using the DAC, because one could not prescribe and then produce a desired pressure condition and thus could not give meaningful in situ direction to an experimental study.

1.8 Optical Polarizing Microscopy

Optical polarizing microscopy coupled with a DAC is a very simple technique (Fig. 1.42), yet its power as a research tool is awesome. The DAC can be mounted on a rigid supporting base to prevent detectable motion under magnification as the pressure screw is rotated as shown in the schematic diagram below.

To demonstrate the power of this simple technique the four photos below (Fig. 1.43) illustrate the ZnS transition at 15 GPa and room temperature in a

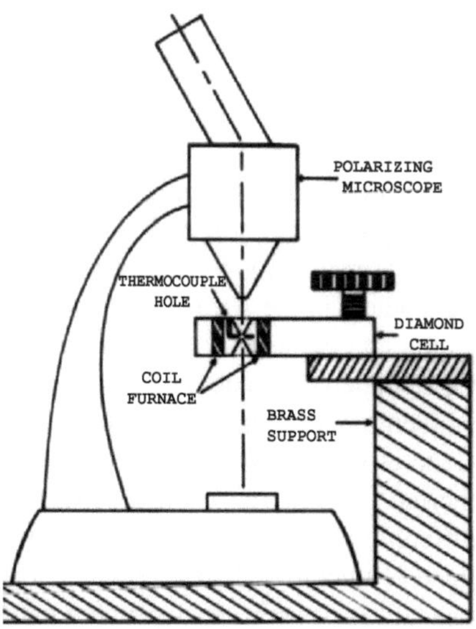

Fig. 1.42 A simple schematic diagram of the optical polarizing microscope technique with thc DAC

Fig. 1.43 Pressure-induced transition in zinc sulfide at about 15 GPa to an opaque semimetal

methanol:ethanol:water mixture (16:3:1 by volume, respectively) in a gasketed DAC as viewed through the polarizing microscope [53, 54]. The clear triangular-shaped crystal in the photo on the left is ZnS and the irregular-shaped crystal above it is the ruby pressure sensor. The range of pressures for these four photos is 14.9 GPa on the left to 15.1 GPa on the right. The transition to a semimetal can be arrested as it progresses and is very sharp and an excellent fixed point for a pressure scale. The transition can be detected by electrical resistance measurements because it involves a sharp decrease in electrical resistance to an opaque semimetal.

1.9 High P–T Properties of Explosives and Propellants

The thermophysical and thermochemical properties of energetic materials at elevated pressures are of interest because many of these compounds are extensively used as propellants and explosives. During detonation, the shock wave produced by these materials may reach 50 GPa and temperatures up to 5,500 K, resulting in polymorphic transformations and the initiation of chemical reactions, including thermal decomposition [55]. Therefore, it is important to understand the reactions involving chemical decomposition in these materials, including their pressure dependence. Given the complexity of the reactions and the observed temperature dependencies in many nitramine explosives and monopropellants such as RDX (1,3,5-trinitrohexahydro-1,3,5-triazine), HNIW (hexanitrohexaaza-isowurtzitane), HMX (octahydro-1,3,5,7-tetranitro-1,3,5,7-tetrazocine), ADN (ammonium dinitramide), PNA (p-nitroaniline), and NM (nitromethane), an understanding of the pressure effect is necessary for accurate modeling of combustion and explosive behavior. Parameters in the Arrhenius equation, characterizing the reaction rate, such as activation energy, usually are derived from ambient pressure studies and subsequently extrapolated to the detonation regime. Published results [56–58], indicate that these ambient pressure kinetic parameters are not applicable to the high-pressure regime. Generally, investigations of the reaction kinetics were experimentally performed by thermo-gravimetric or differential scanning calorimetry. However, in such studies different results are obtained when the experiments are carried out in either a confined or unconfined state. The discrepancies are thought to be due either to the autocatalytic or pressure-dependent nature of the reaction mechanism. Consequently, applicable high-pressure data is critically needed to increase our understanding of the reaction mechanism under detonation conditions, which can improve the accuracy of the mathematical modeling procedures.

A combination of DAC techniques, all utilizing the ruby fluorescence method of pressure measurement, can be used to measure the kinetics of decomposition reactions. These include: (1) Fourier transform infrared spectroscopy for the kinetic measurements, (2) energy dispersive x-ray powder diffraction (EDXD) for crystallographic identification of the observed polymorphic forms and also compression measurements, (3) optical polarizing light microscopy (OPLM) for delineating the stability fields in the equilibrium pressure–temperature phase diagram, as well as for confirming observed phase transitions, and (4) Raman scattering (RS) for measuring the pressure dependence of the shifts in the vibrational modes. To demonstrate the power of the DAC in studying energetic materials as a function of static pressure and temperature, I have selected two examples from the literature, RDX (1,3,5-trinitrohexahydro-1,3,5-triazine) [59] and nitromethane [60,61], which represent the pioneering work done in this area at NIST. I describe how these four experimental techniques were used to obtain data on polymorphism (phase diagrams), chemical reactivity, and rates of thermal decomposition, including reaction mechanisms and Arrhenius activation energies and volumes. References giving similar information and experimental results on the P–T phase diagrams, thermal decomposition reactions, the nitramine compounds, HMX, HNIW, ADN, and PNA and other energetic materials such as ammonium perchlorate (AP) and pentaerythritrol (PETN), can be found in references [62–68]. Also, a comprehensive review relevant to this area of study, including experimental, theoretical, and computational methods characterizing decomposition, combustion, and detonation of energetic materials, is recommended [69].

1.9.1 P–T Equilibrium Phase Diagrams for RDX and NM

1.9.1.1 RDX

In the original study characterizing the equilibrium P–T phase diagram of RDX for the first time, OPLM observations of single crystals in a hydrostatic environment were used to identify three solid polymorphs, α, β, and γ [59] (Fig. 1.44). Two solid polymorphs, known from earlier work [70,71], were originally named I and II, but, following universally accepted phase equilibria convention, we renamed them α- and β-RDX in the present study. The α–γ phase boundary was easily detected by observing with the aid of OPLM discontinuous changes in birefringence in α-phase single crystals. The transition, independent of temperature, is rapid, reversible and involves no breakup of the single crystal as long as the pressure-transmitting environment is hydrostatic. The in situ energy dispersive x-ray results give a small volume decrease of about 1.6% and are in agreement with published data for γ-RDX [72].

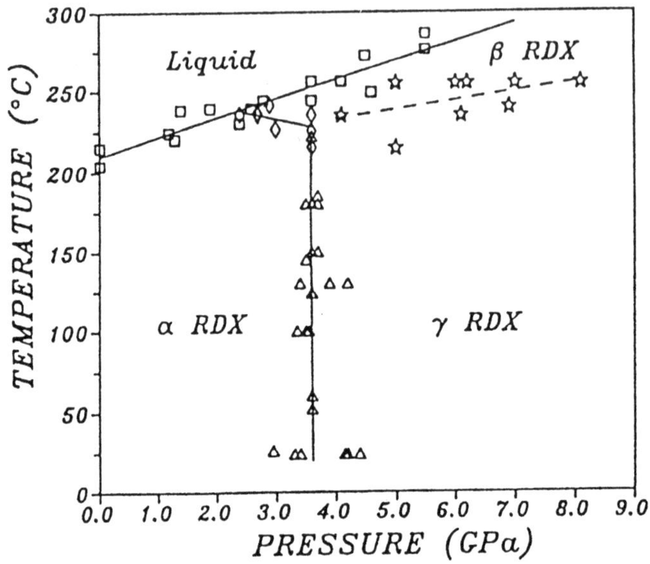

Fig. 1.44 The equilibrium phase diagram of RDX as a function of pressure and temperature as determined with the aid of a DAC [72]

Above about 488 K, a new solid phase, β, exists and is the phase in equilibrium with liquid. Unlike the γ-phase, the β-phase cannot be detected by visual observation under OPLM, but was deduced from changes in both IR and x-ray patterns. (Energy dispersive x-ray diffraction measurements of RDX powders in the same hydrostatic environment confirmed the structural changes associated with the three different polymorphs, even though the β-phase has not been identified crystallographically. Also, observed changes in the FTIR spectra obtained for the three solid phases provided additional confirmation for the phase transitions.) When γ-phase is brought back to RT and nearly to atmospheric pressure, it reverts to the α-phase. Thus, the β-phase remains metastable to almost atmospheric pressure, where it rapidly reverts to α-phase. The liquidus, as expected, increases linearly to 7.0 GPa. The increased scatter in the data points of the liquidus is a reflection of the decrease in accuracy of the ruby fluorescence measurements at higher temperatures. The scatter in the data points at lower temperatures, e.g., RT, is due to sample strain arising from the shear stresses produced in the Fluoroinert liquid used as a chemically inert pressure-transmitting medium. The glass transition in Fluoroinert is about 4.2 GPa at RT and shear stresses produced in the medium at these pressures will affect the transition pressure in RDX. At higher temperatures, the viscosity of Fluoroinert decreases significantly and shear stress plays almost no role in initiating the α–γ transition, resulting in less scatter in the data points.

It is known that RDX has one metastable conformer, β-RDX, in addition to the room temperature stable orthorhombic form, α RDX (space group Pbca). The structure of the α-phase has been determined by single-crystal neutron diffraction, but the β-form has not been solved unequivocally, because of the difficulty in growing and preserving quality single crystals suitable for analysis. The high-pressure phase is in the same crystal system (orthorhombic) as the ambient α-RDX, but a definite structure was not reported. Gamma RDX was identified earlier from volume compression data derived from x-ray diffraction measurements, although it was never assigned a label or name. The volume compression data showed a 1.6% decrease in volume near 4.0 GPa indicating a first-order phase transition. No change in the orthorhombic-type unit cell was found. The pressure dependences of the orthorhombic lattice parameters decrease monotonically with increasing pressure. However, near 4 GPa both *a* and *c* continue to decrease, but *b* abruptly increases resulting in a small decrease in volume of about 1.6%. Above about 488 K, a new solid phase, β, exists which, at higher temperatures, is the phase in equilibrium with liquid. As was noted earlier, β-phase has not been identified crystallographically. It also cannot be detected by visual observation under polarized light, unlike the γ–phase [72] (Fig. 1.45).

The infrared spectra of α and γ-RDX indicate that their molecular structures must be similar, because only slight changes between the two spectra are observed (Fig. 1.48). The γ-to-β transformation, however, results in a significant change in the molecular spectra of the two phases (Fig. 1.46). In particular, the bands at $\approx 1,060\,\text{cm}^{-1}$ and at $\approx 1,400\,\text{cm}^{-1}$ that are observed in the α and γ phases are not seen in the β-phase. The $1,060\,\text{cm}^{-1}$ bands have been assigned to cyclic vibrations coupled to deformations of the NO_2 groups through the C–N–N bond angle. The $1,400\,\text{cm}^{-1}$ bond results from a CH_2 wagging motion. In addition, two prominent

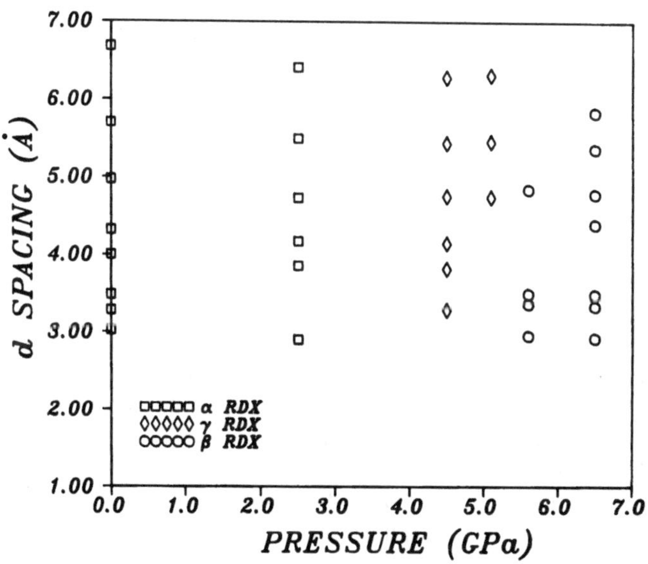

Fig. 1.45 Crystallographic *d*-spacings for three phases of RDX (α, β, and γ) as a function of pressure at room temperature. For α-RDX, the normal orthorhombic pattern is shown at 2.6 GPa (□). At 4.5 and 5.1 GPa, another orthorhombic pattern is shown (◊, for γ phase) and the circles (○) indicate "*d*" spacings for the β phase at 5.6 and 6.7 GPa [72]

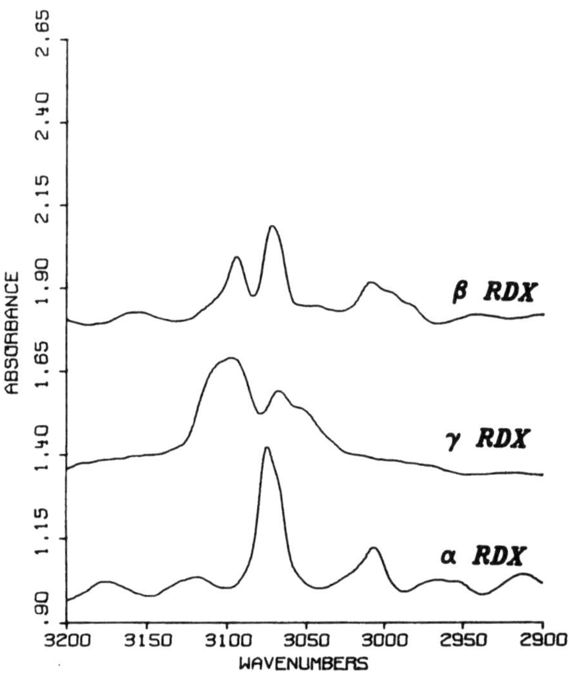

Fig. 1.46 Infrared Spectra for α-, β-, and γ-RDX in the CH stretching frequency range. These spectra are shown to illustrate the significant differences among the three phases of RDX, making them easily distinguishable [72]

bands are observed in α-phase in the CH stretching region, $\approx 3,100\,\text{cm}^{-1}$. However, in the β-phase at least five bands are observed in this region. A detailed assignment of the vibrational spectra for α-RDX and its isotopes has been reported. A detailed analysis of the vibrational spectra of β-RDX found in the present study is not available at present because it requires additional isotopic study and structural information. However, the crystal structure of γ-RDX was investigated recently utilizing a DAC by FTIR spectroscopy and powder x-ray diffraction measurements followed by Rietveld refinements of the structure [72]. Although γ- and α-phases were found to belong to the same space group, *Pbca*, as was noted earlier, they exhibited a different crystal packing, attributable to rotations and translations of molecules. While the molecular structures of the α- and γ-phases have the same conformation, the N–NO_2 torsion angles changed slightly (Fig. 1.47).

It is of interest to note that infrared spectra reported for a metastable RDX phase obtained by the evaporation of a thymol-RDX solution [72], appear to be very similar to the spectra obtained here for the β-phase under different conditions of P and T. Because of the great similarity in these IR spectra, it seems reasonable to assume that the two phases are the same, indicating that the stability field of this metastable phase is now established.

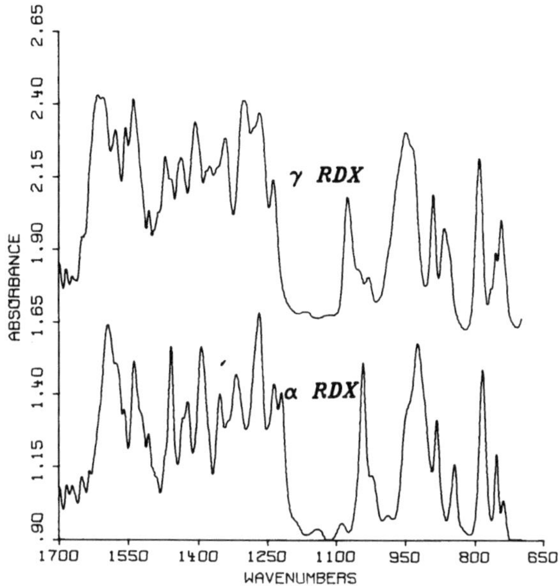

Fig. 1.47 Infrared absorbance spectra of α- and γ-RDX at 1.4 and 4.1 GPa, respectively, both at RT. Because the two spectra are very similar, their molecular structures are assumed to be also similar [72]

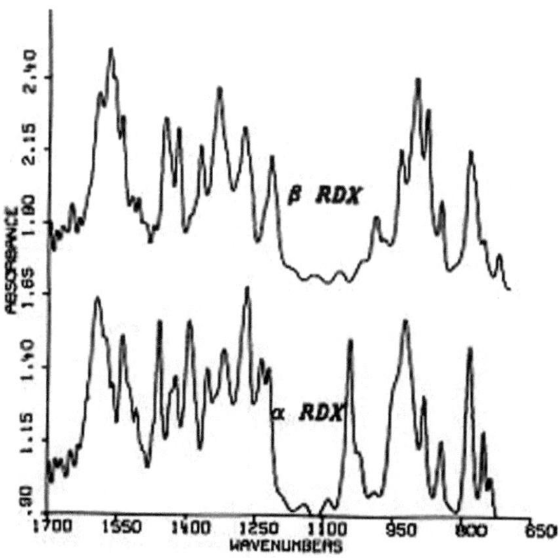

Fig. 1.48 Infrared absorbance spectra of α- and β-RDX at 2.4 GPa and 508 K. In the wavenumber range measured, significant differences are indicated

1.9.1.2 NM

Similar measurements by the same experimental techniques have been published for nitromethane, the simplest of the aliphatic nitro compounds. The melting point [60] (Fig. 1.49), and more recently several solid phases including a partial phase diagram [73] (Fig. 1.50), have been reported. Also, thermal decomposition parameters and kinetics of decomposition have been determined [74–79].

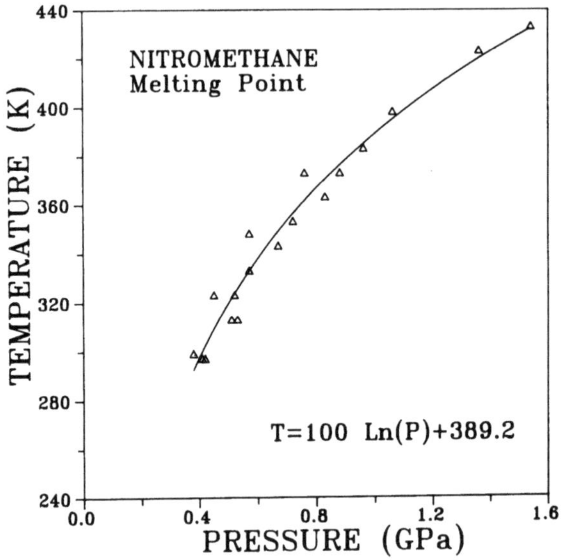

Fig. 1.49 Equilibrium melting point of nitromethane as a function of P and T. The data points were fitted to the logarithmic function shown in the figure. Above 1.54 GPa and 433 K, the curve was not defined because rapid thermal decomposition occurs [60]

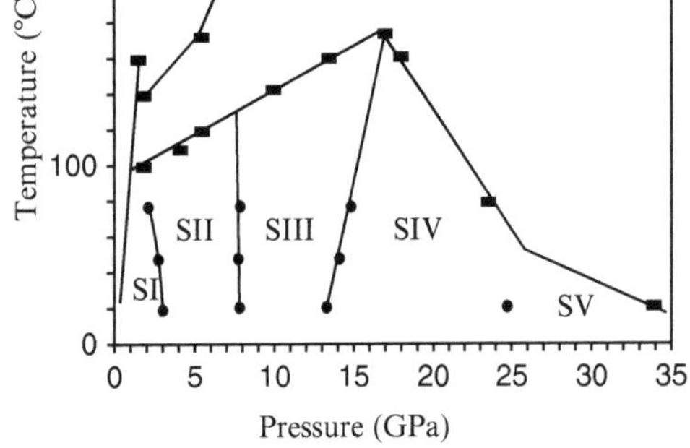

Fig. 1.50 The P–T phase diagram for nitromethane taken from [73]

To again illustrate the power of the OPLM method in conjunction with a DAC, the pressure dependence of the melting point of nitromethane is discussed because utilizing this method permitted delineation of the equilibrium solid–liquid phase boundary to the high-pressure regime for the first time. The two-phase condition, crystal and liquid, can be maintained in a gasketed DAC as a function of T by a corresponding appropriate change in P, thereby ensuring that the measurements are truly at equilibrium conditions. In this manner, the equilibrium phase boundary between any solid and its liquid can be defined over a desired P-T range. In the case of nitromethane, the P-T range present study extends to 1.54 GPa and 433 K, at which point thermal decomposition initiates. Decomposition is readily observed by a darkening of both the liquid and the crystal, ultimately producing a brown solid residue, liquid, and gases, similar to what is shown in Fig. 1.52 for the decomposition products of RDX. Upon release of P, only the solid residue remains in the gasket and it can be retrieved and analyzed by mass spectrometry to determine its chemical constituents. Using the logarithmic function in Fig. 1.49, which describes the melting point of nitromethane at 298 K, the calculated freezing pressure is 0.40 GPa. The liquid phase is easily superpressed to about 2 GPa, where the metastable liquid rapidly crystallizes to a polycrystalline phase, the same phase (orthorhombic, space group, $P2_12_12_1$) that crystallizes at low temperatures and 1 atmosphere [60].

Other studies on polymorphism in nitromethane have been published, in which a solid–solid transition corresponding to the locking of the methyl group rotation was reported [74]. More recent work discloses three new solid phases, called III, IV, and V, discovered by Raman scattering measurements in the P–T range of 0–35 GPa and 293–623 K [72]. Their stability fields in the P–T phase diagram were defined by breaks or discontinuities in observed shifts in different Raman modes as a function of P and T. Four solid–solid transitions and two irreversible transformations were determined in that study (Fig. 1.50).

The I–II transition at 293 K is at (3.0 ± 0.2) GPa, detected by breaks or discontinuities in line-widths in the observed Raman modes. The I–II transition was reported earlier at 3.5 GPa and ambient T [55]. The II–III transition, independent of T, is at (7.5 ± 0.5) GPa at ambient T. It was detected by the appearance of new Raman bands and changes in peak line-widths. The transition is probably first order and thought to be the result of an increase in the number of molecules per cell. The III–IV transition is at (13.2 ± 1.0) GPa at ambient T. It was also detected by breaks or discontinuities in the slope of the different Raman modes with pressure. This transition is P- and T-dependent. The IV–V transition, at (25 ± 1) GPa and ambient T, was not studied at higher T because it transforms at about 328 K. The transformation of nitromethane to the CI compound is irreversible and slow. Because its Raman peak could not be distinguishable from ground, the authors concluded that CI is amorphous. Similarly, the CI–CII is irreversible, but more rapid than the nitromethane–CI transition. CII did not give a Raman signal. This work is an example of identifying phases and delineating phase boundaries to establish the P–T phase diagram based on the measurement of the more intense vibrational modes of solid phases of nitromethane up to 35 GPa and 623 K. The specific behavior of nitromethane under these conditions is unusual as well as interesting for such a basic compound, often considered as a model for energetic nitro compounds. What is needed now is to study nitromethane and its various polymorphs by other experimental measurement techniques in combination with the DAC to detect and confirm the existence of these phases and chemical reactions and to carry out detailed analyses of the results obtained to further our understanding of the behavior of this very fundamental energetic compound.

1.9.2 Pressure Dependence of Thermal Decomposition Rates

In the pressure regime above 1 GPa, few published methods are currently available for kinetic measurements particularly for decomposition reactions of energetic materials where micro volume samples are necessary for the safety of the investigators. The following narrative describes FTIR absorption experiments to obtain relevant kinetic data for thermal decomposition reactions of two important materials, RDX and nitromethane, as a function of pressure and temperature. Through a combination of the measured pressure effects on the thermal decomposition kinetics, and, in the case of NM, the identification of decomposition products, possible reaction mechanisms are proposed for the thermal decomposition process in NM at high pressures. The chemical reactivity and phase stability of RDX and NM at high pressures and elevated temperatures were characterized in equilibrium P–T phase diagrams.

Innovative advances in the measurement of the pressure dependence of decomposition rates of propellants and explosives were made at NIST during the 1980s. The experiments, utilizing time-resolved FTIR technology, were seminal in nature because such measurements had never been made before at high pressures. Because this area of study has not received much attention in the energetic materials community, I shall discuss this subject in detail in the hope that I can rekindle some interest in it.

In 1984, a Fourier transform infrared microscopic method for measuring kinetic effects and thermal decomposition rates for energetic materials at high temperatures and high pressures in combination with a DAC with heating capability was developed at NIST, in collaboration with the Naval Surface Warfare Center at White Oak, MD. Because commercial IR microscopes were unavailable at that time, an FTIR spectrometer modified with an on-axis Cassegrain-type beam condenser was employed. The experimental configuration for the DAC coupled to the Cassegrain-type focusing system is shown in Fig. 1.51. With this system, time-dependent FTIR absorption spectra were obtained for the first time on RDX at high pressures and high temperatures [58].

This Cassegrain optical system with the mounted DAC was small enough to fit directly on an XYZ positioning device located in the sample chamber of the IR instrument [58]. The sample in the DAC could be positioned readily at the focal point of the optical system to achieve maximum IR throughput. Fortunately, infrared microscopes are available today and can be used in combination with a DAC, so that these measurements are much easier to carry out.

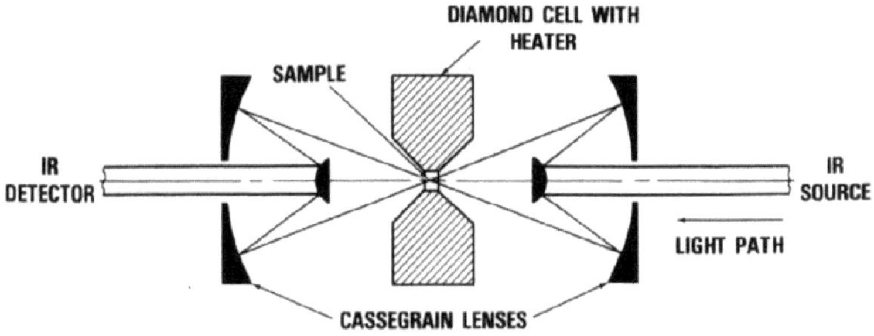

Fig. 1.51 Schematic diagram of Cassegrain optics used with the DAC [58]

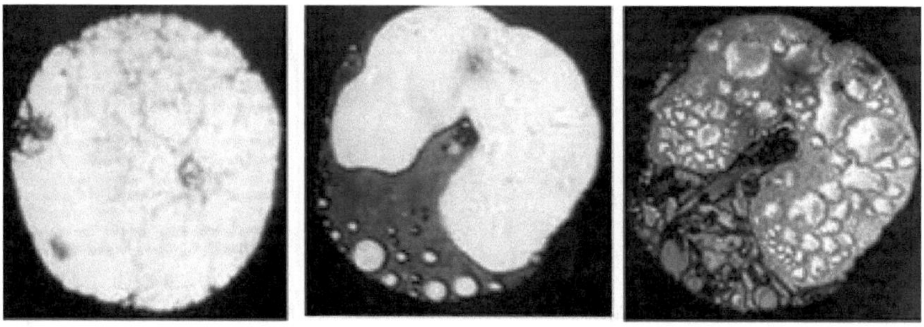

Fig. 1.52 Left: Thin film starting sample of RDX at 2 GPa; Center: RDX at 0.9 GPa and RT after decomposition was initiated at 4 GPa and 516 K. A brownish liquid is present with dispersed gaseous bubbles. Right: P lowered from 0.9 GPa to ambient P and T. Brown residue plus liquid remains in the gasket [58]

For an FTIR experiment, a sample of RDX in a gasketed DAC at an average pressure of 2 GPa and at ambient temperature is shown in Fig. 1.52. To prepare this sample, RDX powder was packed into a Pt-Rh gasket (≈ 0.457 mm diameter and 150 μm thick) along with two small ruby chips (one located near the center of the sample and the other at the far left center). The ruby chips, two small dark regions in a homogeneous field, act as pressure sensors. In this sample configuration, the pressure is quite nonhydrostatic as indicated by line broadening of the R-lines. Consequently, a large uncertainty is associated with the overall sample pressure. This large uncertainty can be reduced appreciably by including a thin film of powdered NaCl in the sample configuration. This is done by first compressing powdered NaCl between the ungasketed anvils to form a thin film (0.25 μm in thickness) and then assembling the gasket containing the powdered sample to be studied as was done in our earlier experiments. The presence of NaCl reduces the inhomogeneity in the stresses significantly, but does not entirely eliminate it. The decomposed products are also shown in Fig. 1.52.

1.9.2.1 RDX

Thermal decomposition in RDX has been determined in α and β phases only (Figs. 1.53 and 1.54). No decomposition in γ phase has been reported in the pressure–temperature regime, 1.4 < GPa > 6.9 and 478 < K > 508, because it always transforms to the β-phase before decomposition initiates.

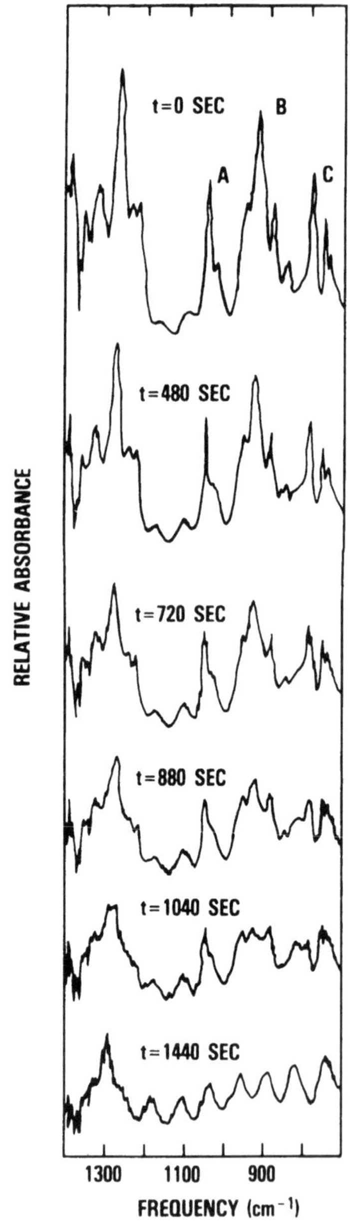

Fig. 1.53 Thermal decomposition of α-RDX. A series of six time-dependent infrared absorption spectra of α-RDX taken at 4.4 GPa and 515 K in a DAC. The sample thickness is approximately 15 μm [58]

The measured thermal decomposition rate for α-RDX increases with increasing pressure between 1.4 and 2.4 GPa. On the other hand, β-RDX decomposes with a very different pressure dependence. Like that observed for β-HMX [62], the rate decreases with increasing pressure and decomposition products are more complex with some solid residue present after pressure reduction to 1 atmosphere.

Decomposition in β-RDX takes place at temperatures well below the melting point (Fig. 1.56). At 488 K and 5 GPa, for example, decomposition at a very slow rate was observed which increases as the temperature is increased. The products of decomposition were found to exhibit interesting behavior and require further discussion. After decomposition at 4 GPa and 516 K, the pressure cell containing decomposed β-RDX was cooled to room temperature and the pressure measured 0.9 GPa indicating a large decrease in pressure. Both solid and liquid are observed whose IR spectra indicate the presence of H_2O, CO_2, and N_2O chemical species. As the pressure is lowered, gases evolve from the liquid at a pressure of about 0.3 GPa. At atmospheric pressure, the remaining solid residue gave no measurable IR absorption indicating the IR active materials escaped as gases as the pressure seal was broken. At present the solid residue is unidentified.

Decomposition rates can be estimated because the concentration of RDX remaining in the DAC is proportional to the measured absorbance. The fraction decomposed can be obtained by taking ratios of peak RDX absorbencies with time, normalized to absorbency at $t = 0$ (Fig. 1.55). For decomposition at a given P and T, the ratios are combined to give an average value, which when subtracted from unity, yields α, a term proportional to the mole fraction of decomposed RDX. Then, α can be plotted as a function of time for the various pressures and temperatures studied.

It is interesting to note that at constant temperature, increasing pressure increases the rate of thermal decomposition for α-RDX, but decreases the rate for β-RDX, indicating that two different chemical mechanisms or rate-controlling steps are involved for the two distinct structures. One would assume that the α-RDX decomposition mechanism is at least bimolecular in nature with a negative volume of activation, while the β-RDX decomposition mechanism would be unimolecular with a positive volume of activation [59].

Many kinetic expressions, derived for determining rate constants, k, from α–time curves for thermal decomposition reactions in the solid state, have been reviewed extensively [79] (Fig. 1.57). For previously reported kinetic measurements on RDX, the Prout-Tompkins rate equation, $\ln[\alpha/(1-\alpha)] = kt + C$ was used to derive the overall rate constant for thermal decomposition. This equation represents an autocatalytic, branching chain-type reaction giving sigmoid-type α–time curves. Another rate equation which linearizes α–time curves is $-\ln(1-\alpha)=kt + \mathrm{C}$, which represents a diffusion limited reaction [55] (Fig. 1.58). Both equations are oversimplifications of complex rate expressions involving rate constants for nucleation, growth, and termination of reactions in a single solid. Even though the true physical meaning may be unknown, these equations do permit one to obtain mathematical expressions for extracting the rate constant for a given measured reaction.

If one assumes an accurate measure of the rate constant, despite the above caveat, for the overall decomposition reaction for α-RDX, one can determine a meaningful

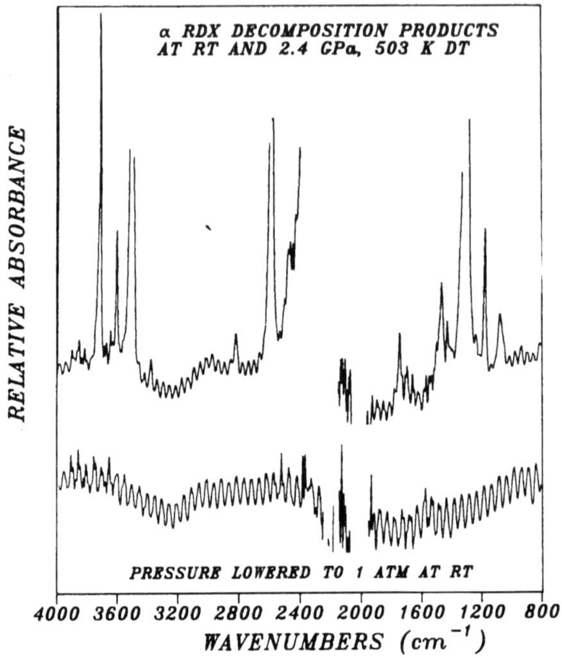

Fig. 1.54 IR absorbance spectra of α-RDX. Top: decomposition products at 2.4 GPa and 503 K. Bottom: decomposition products after the pressure was lowered to atmospheric and the temperature to ambient. Under these conditions, the decomposition products appear to be few and uncomplicated as shown in the FTIR pattern in Fig. 1.54. The products appear to be mainly H_2O, CO_2, and N_2O in the form of liquids, which all transform to gases when the pressure is lowered to approximately 1.2 GPa

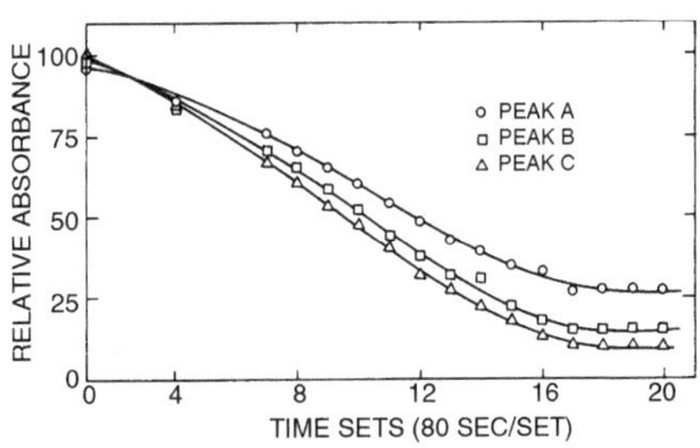

Fig. 1.55 Absorbance vs. time plots of three RDX absorption peaks labeled A, B, and C in Fig. 1.53. These curves have been normalized to an absorbance of 100 at time zero, prior to thermal decomposition at 515 K and 4.4 GPa [58]

dependence of k on pressure and temperature (Figs. 1.59 and 1.60). From the total differential of the natural logarithm of k, it can be shown that the partial derivatives are equal to

$$(d \ln k/d(1/T))_P = -\Delta H^*/R$$

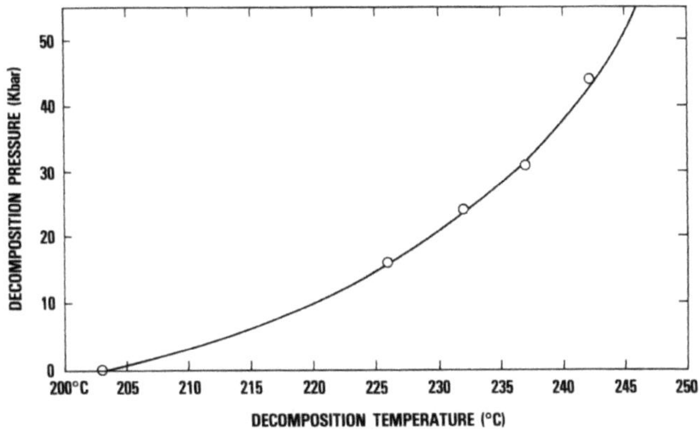

Fig. 1.56 Pressure dependence of the decomposition temperature of RDX indicated by the onset of a decrease in intensity of an infrared absorption peak. Experimental evidence suggests that the melting of RDX is coincident with thermal decomposition [58]

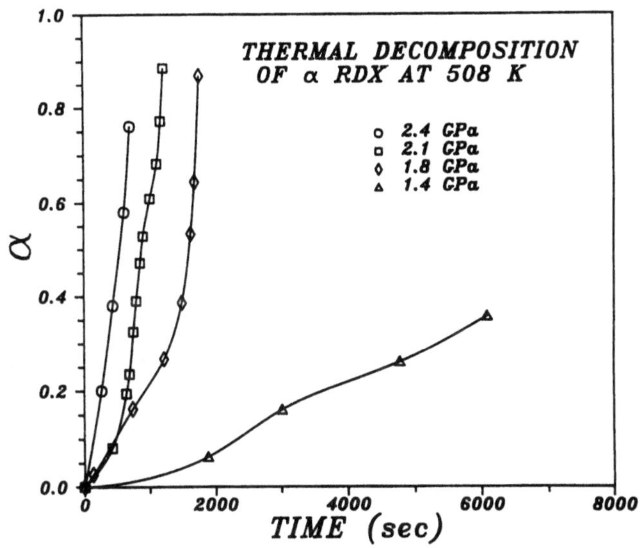

Fig. 1.57 Typical α-time curves for α-RDX thermal decomposition at 508 K at the pressures indicated. α is the fraction of RDX decomposed. Owing to slight changes in the pressure as the decomposition reaction proceeds, the curves show large scatter in the data points. The curves, however, can be identified with the initial and intermediate stages of a sigmoid or s-shaped type curve typical of decomposition reactions. The shape of these curves are related to physical and chemical mechanisms involved in the decomposition reaction [59]

and

$$(d \ln k/dP)_T = -\Delta V^*/RT$$

where ΔH^* and ΔV^* are the experimental enthalpy and volume of activation, respectively [58, 59, 62]. For more complex multiple-step reaction mechanisms ΔH^* and

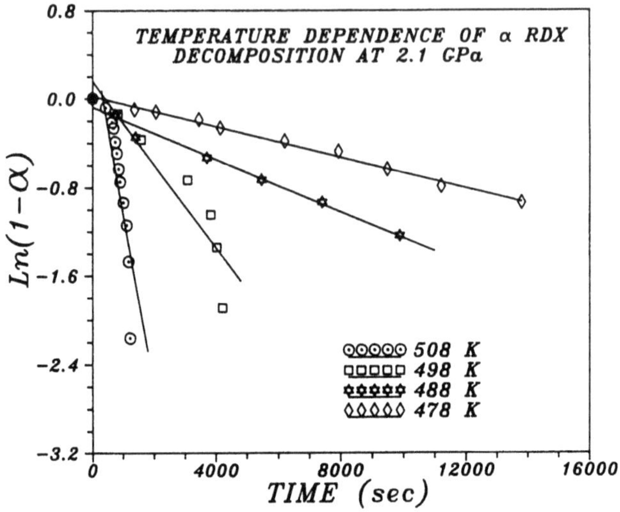

Fig. 1.58 Linearization of the α-time curves for α-RDX thermal decomposition at 2.1 GPa and the temperatures indicated using a diffusion limited model for the reaction mechanism. It should be noted that at higher pressures, the Prout-Tompkin expression gave good linearization, suggesting a change to an autocatalytic-type decomposition mechanism [59]

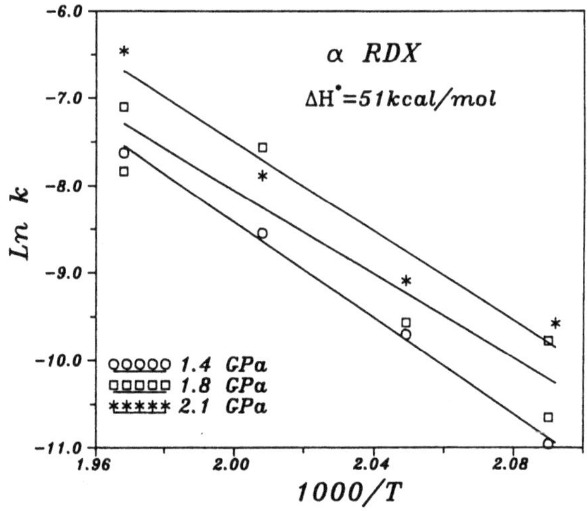

Fig. 1.59 Temperature dependence of the rate constant, k, of α-RDX thermal decomposition for the pressures indicated. The slope is proportional to the activation energy, ΔH^*. Because the lines have almost equal slopes, ΔH^* appears to be independent of pressure and has a positive average value of approximately 51 kcal/mol

ΔV^* are weighted means of the processes with the rate-controlling step providing the main contribution.

In the pressure and temperature range studied in this work, it appears that within the experimental error of the measurements, ΔH^* is independent of pressure and

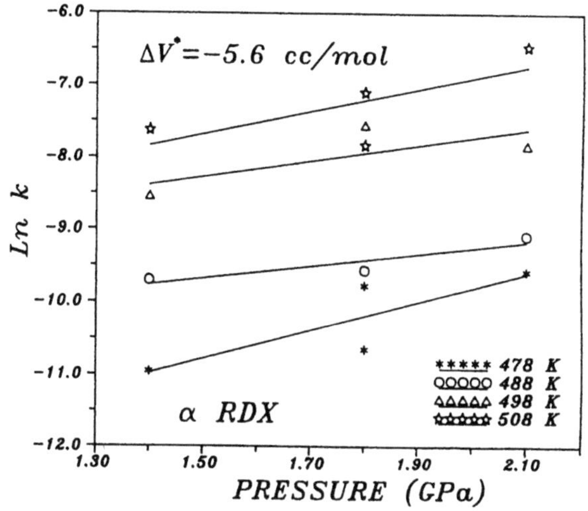

Fig. 1.60 Pressure dependence of the rate constant, k, of α-RDX thermal decomposition for the temperatures indicated. The slope is proportional to the activation volume, ΔV^*. Because the lines have almost equal slopes, ΔV^* appears to be independent of pressure and has a negative average value of approximately $-5.6\,\mathrm{cm}^3/\mathrm{mol}$

ΔV^* is independent of temperature. Having a knowledge of the rate constant, k, ΔH^*, and ΔV^*, the rate constant can be expressed in thermodynamic terms by use of transition state theory,

$$k = (KT/h)e^{-\Delta G^*/RT}$$

and

$$k = (KT/h) \quad e^{-\Delta S^*/RT - \Delta H^*/RT}$$

permitting evaluation of ΔG^* and ΔS^*, the free energy and entropy of the activated state. The importance of these measurements for energetic materials can be emphasized by pointing out the effect of pressure on the Arrhenius chemical-kinetic parameters, often used to model explosions. The Arrhenius activation energy, E^{Arr}, the volume of activation, V^*_{act} and the internal energy of activation, E_{act}, have been related to the reaction rate, k:

$$E^{\mathrm{Arr}} = (R\, d\, \ln\, k)/d(1/T)$$

and

$$E^{\mathrm{Arr=Eact}} + V^*_{act}[P - T(\mathrm{d}P/\mathrm{d}T)].$$

Thus, at very high pressures, the term, $[P - T\ (\mathrm{d}P/\mathrm{d}T)]$, can play a dominant role in the decomposition of energetic materials [59].

1.9.2.2 Nitromethane

The chemical reactivity of NM as a function of pressure and temperature was observed optically with the aid of a polarizing light microscope [60, 61]. The instrumentation permitting these observations is described in detail earlier in this chapter. Above 1.54 GPa and 433 K, thermal decomposition initiates and is readily observed by darkening of the liquid and crystal producing ultimately brown solid residue, liquid, and gases. Upon release of pressure, only the solid residue remains, which has been analyzed by mass spectrometry in order to determine its chemical constituents. The major products of decomposition up to 7 GPa are ammonia, formic acid or its salt and water. The residue was originally reported to be the oxalate salt based on the similarity of the IR absorption spectra, but that result had never been confirmed (Fig. 1.61).

Pressure tends to increase the chemical reactivity of NM as well as the rate of thermal decomposition. It was observed, quite accidentally, that a pressure-induced

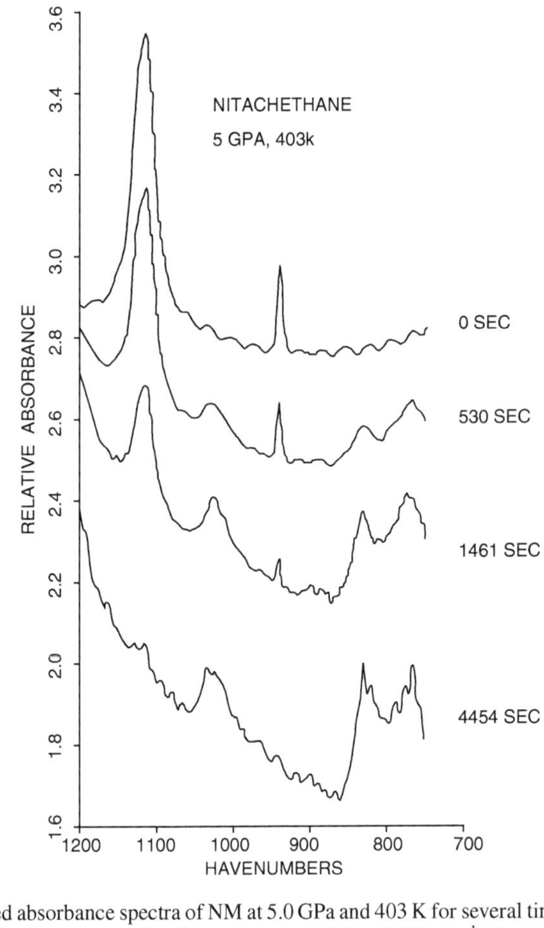

Fig. 1.61 Infrared absorbance spectra of NM at 5.0 GPa and 403 K for several times during thermal decomposition. Only the two peaks shown here at 1,100 and 925 cm^{-1}, due solely to vibrational frequencies of modes in crystalline NM, were used in the analysis of the data. Time $t = 0$ indicates when the sample of NM reached decomposition temperature. The spectrum at 4,454 s shows no NM bands and only those from the decomposition products

spontaneous explosion of single crystals of NM at room temperature can occur. Further study revealed that single crystals grown from the liquid with the 111 and either the 001 or the 100 crystal faces perpendicular to the applied load direction in the DAC, if pressed rapidly to over 3 GPa, explode instantaneously accompanied by an audible snapping sound [60]. The normally transparent sample becomes opaque instantly. Visual examination of the residue revealed a dark brown solid, which was stable when heated to over 300°C. Subsequent x-ray analysis showed the material to be amorphous. Mass spectral analysis of the residue was inconclusive because no well-defined spectra were observed. The fact that most of the sample is recovered as solid residue after the explosion and is stable to over 300°C suggests that the material may be amorphous carbon. This stress-induced explosion occurs only in protonated NM because similar attempts on the deuterated form did not result in explosion [60]. Shock experiments on oriented pentaerythritol (PETN) crystals have shown similar type behavior. In this case, it was suggested that the sensitivity of shock pressures to crystal orientation is the result of the availability of slip planes or system of planes in the crystal to absorb the shock, thereby increasing the threshold to explosion [60]. A similar explanation may be applicable to the NM crystals as well. The deuteration effect must play a role in the initiation chemistry. An isotope effect has been observed previously in the sensitivity of HMX and RDX to shock and thermal conditions [60].

Previously reported time-to-explosion measurements on NM up to 5 GPa indicate that pressure decreases the time required to achieve an explosion [80]. As a result of this behavior, pressure is expected to have a similar effect on the thermal decomposition rate in NM.

Kinetic measurements, similar to those described above for RDX, were made on NM over the ranges $2.0 \leq \text{GPa} \geq 7.1$ and $393 \leq \text{K} \geq 453$ with the following results. As was the case for RDX, the α vs. t curves can be identified with the initial and intermediate stages of a sigmoid (s-shaped) type curve characteristic of thermal decomposition of a single solid in an autocatalytic-type reaction (Fig. 1.62). With increasing temperature the s-character diminishes to where the curve is almost linear at 423 K. A plot of $\ln[\alpha/(1-\alpha)]$ vs. t gives linear dependence which supports a decomposition mechanism consistent with nuclei formation with branching interference. The slopes of these lines give typical Arrhenius temperature behavior with the overall rate increasing with temperature.

Linear fits of $\ln[\alpha/(1-\alpha)]$ vs. time for data obtained at 2.0 GPa for the decomposition temperatures indicated that the overall decomposition rate increases with increasing temperature, typical of Arrhenius behavior. This type of dependence was found for the thermal decomposition of HMX in an earlier work [62], where the results were more consistent than in the case of NM. Unlike the HMX case, NM undergoes significant changes in its decomposition mechanism because the α vs. time curves change above 413 K. Moreover, at pressures greater than 5 GPa, a third change in the reaction mechanism is indicated by a further change in the shape of the α vs. time curves. Because only two absorbance bands with large signal-to-noise error were used to calculate the NM concentration term, α, and because multiple reaction mechanisms appear to overlap in the P–T regime studied, only a qualitative analysis of the data was possible in that work. What is certain is that the

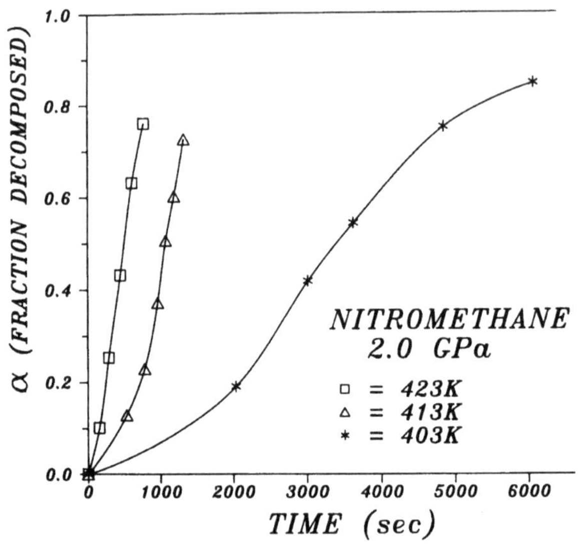

Fig. 1.62 α vs. time sigmoid-type curves obtained for the pressures and temperatures indicated. α is the fraction of decomposed NM at time, t, during thermal decomposition. Curves are logarithmic fits to the data points

decomposition rate in NM was found to increase with increasing pressure, but the overall trend in the data was inconsistent. However, there are some qualitative observations that can be made on the basis of reaction times at the various P–T conditions measured and these are as follows:

1. To 5 GPa at least two different reaction mechanisms are involved in the decomposition and both have a positive pressure dependence on the overall rate of decomposition.
2. Mechanism (1) is dominant below 4 GPa and $T \leq 130°C$.
3. Mechanism (2) operates at all measured temperatures at 5 GPa and at $T \geq 140°C$ at 4 GPa.
4. The mechanism crossover temperature appears to be 140°C. At $T \leq 140°C$ at 2 GPa mechanism (1) is operative, but at $T \geq 140°C$ either mechanism (2) or a mixture of both are operative.
5. Mechanism (1) is linearly dependent in $\ln [\alpha/(1-\alpha)]$ vs. time, similar to that found earlier for β-HMX [62], while mechanism (2) is linear in $\ln(1-\alpha)$ vs. time similar to that reported for RDX [58,59]. The type of predominant chemical mechanism of the decomposition reaction determines the mathematical functionality of α with respect to time [60]. The fact that two different time dependencies were reported indicates the presence of two separate and distinct kinetic processes. It is known that the aci or enol form of NM can play an important role in the thermal decomposition of NM by acting as a catalyst [60,61]. The results obtained for NM cannot rule out the possibility of this form affecting the observed kinetic data.
6. Finally, a third mechanism is reported to operate at pressures greater than 5 GPa.

These results illustrate the complexity of the NM thermal decomposition reaction. More than one kinetic mechanism and/or catalytic effect may be present. However, the results show definitely a positive pressure enhancement of the reaction rate. As described in reference [62], from the expression:

$$d\ln k = \Delta H^*/RT^2 dT - \Delta V^*/RTdP,$$

for small increases in pressure with all other terms being equal, only a negative ΔV^* will cause increases in ln k. Therefore, the thermal decomposition rate of NM must have a negative volume of activation in the P–T regime studied in these experiments. This result directly implies that the NM decomposition reaction in the solid state is at least bimolecular in character, contradicting published mass spectra studies of molecular reactions in the gas phase [60], where the reaction was reported to be unimolecular. This disagreement is not surprising because different decomposition mechanisms can be expected between the crystalline and gaseous phases of NM. The intermolecular collision probability, for example, is much higher in the condensed phase.

1.9.3 Chemical Mechanism of NM Decomposition

Previously reported time-to-explosion measurements on NM up to 5 GPa show that pressure decreases the time to explosion [80]. Therefore, pressure can be expected to have a similar effect on the thermal decomposition rate. If that assumption is correct, then the model for the decomposition reaction mechanism should be at least bimolecular. Considerable work has been reported on the possible chemical reaction pathways for both pyrolysis and detonation processes in NM [60,72]. Most of these proposed mechanisms are based upon a unimolecular process. As far as can be determined, only this work and one other show experimentally that the reaction under pressure results from a bimolecular process, although the reported results were derived from the effect of pressure on time-to-explosion measurements [72].

Based on theoretical electronic orbital calculations, a bimolecular process has been proposed as follows:

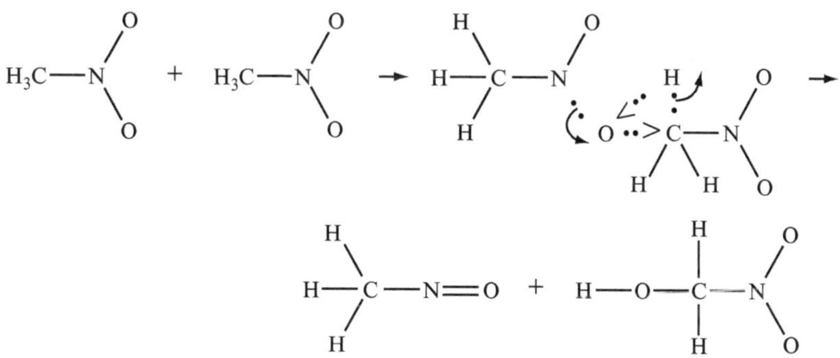

(1.1)

The quantum mechanical calculations show this to be a three-center reaction complex, where, under high pressure, the potential surface contacts result in a simultaneous motion of atoms, indicated by the arrows in the chemical equation, to form the resultant products. The nitrosomethane rapidly reacts to form HCN and finally NH_3 and HCOOH as follows:

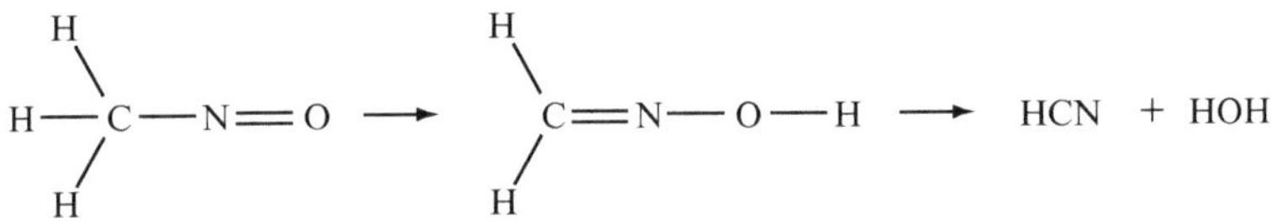

(1.2)

At elevated P and T, HCN rapidly hydrolyzes:

$$HCN + 2HOH \rightarrow NH_3 + HCOOH. \qquad (1.3)$$

The remaining product from reaction (1) rapidly decomposes (reaction 4 not illustrated) into the following possible gaseous compounds: H_2O, N_2O, CO_2, CO, NO, H_2, and solid C. Reaction (1) is rate controlling with a calculated activation energy of 32.5 kcal mol^{-1}. Reactions (2), (3), and (4) are relatively fast compared with reaction (1). Reactions (3) and (4) provide the pathways for the production of NH_3, HCOOH, and H_2O via chemical reaction with HCN which is known to be a major pyrolysis product of NM [60, 72]. These studies confirm the presence of NH_3, HCOOH, and H_2O along with volatile gases as the major products of thermal decomposition of NM under high pressures.

This proposed reaction scheme for NM decomposition at elevated pressures suggests a strong intermolecular interaction of NM in the condensed phase. If this is indeed the case, then it is important to investigate the effect of pressure on the vibrational bands in NM and to analyze the observed shifts in light of intermolecular coupling between the $-NO_2$ and H_3C- groups of interacting molecules. The results from that study support the bimolecular nature of the mechanism for the thermal decomposition of NM under pressure [60, 72].

1.9.4 Compressibility of Explosive Liquids

In the remaining part of this chapter, I discuss two articles appearing in the scientific literature related to energetic materials studies at static high pressures, demonstrating novel applications of the DAC not discussed previously. Where applicable, I suggest new areas to investigate to further advance our scientific knowledge, particularly with respect to energetic liquids where little high-pressure work has been done.

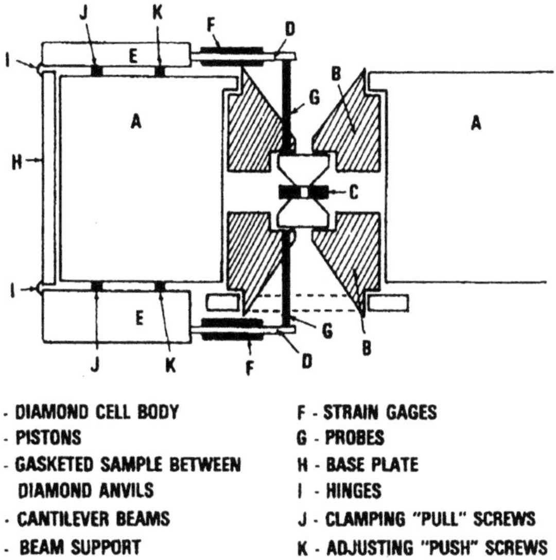

Fig. 1.63 A schematic diagram of the volume measurement apparatus, making use of a strain-gage technique to measure sample volume [81]

One innovative technique [81] I want to mention is a novel method for determining the compressibility of explosive liquids, a measurement that has challenged experimentalists for many years (particularly in the very high-pressure regime) because it requires a determination of the volume pressure dependence of the liquid. The method utilizes a lever-arm type DAC, which incorporates a strain-gage transducer system to measure the thickness of a regular shaped liquid sample as it is compressed in a hydrostatic environment. As shown in Fig. 1.63, the strain-gage transducer functions as an electronic caliper to measure changes in the thickness of the sample (gasket thickness) as it is compressed. Planimeter measurements on enlarged photomicrographs of the gasket hole determine area changes. The ruby technique provides the pressure measurement. Thus, the technique permits direct volume change measurements in the DAC by monitoring the gasket hole area and its thickness. Accuracies of 1% or better were reported (Fig. 1.64). Volume measurements on explosive liquids and liquids in general cannot be determined with desired accuracy by x-ray methods appropriate for powders and single crystals. Thus, this unique method has the potential to provide experimental PVT data on liquids for developing equation of state models (Fig. 1.65).

This strain-gage transducer method represents an unusual application of the DAC, which has direct relevance to the study of energetic liquids, an area having received little attention because of the complexity involved in measuring volume compression of liquids at very high pressures. This technique, although difficult to carry out, provides a means of obtaining PVT data so essential for developing equation of state models for explosive liquids. In this connection, I refer to a recent publication concerning classical MD simulations of nitromethane (NM) under high compression and high-temperature conditions [82]. In that article theoretical results,

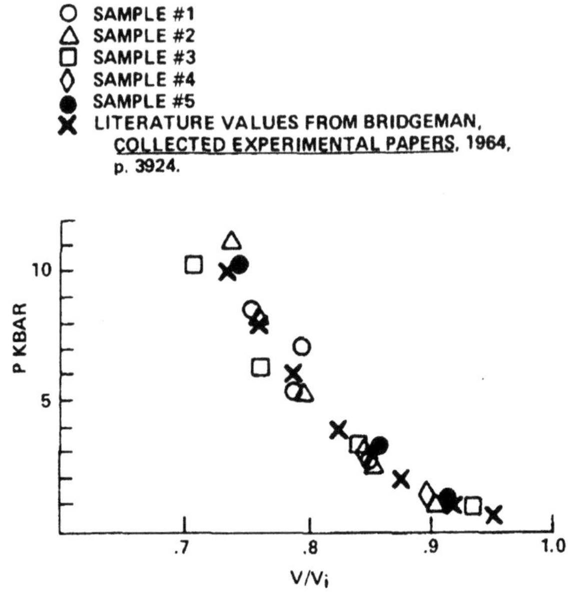

Fig. 1.64 Compressibility data from five samples of n-heptane at 298 K are shown compared to literature values [82]. The agreement is quite satisfactory with a mean square error for a single observation of ±0.013 and for the average of all observations was ±0.003 [81]

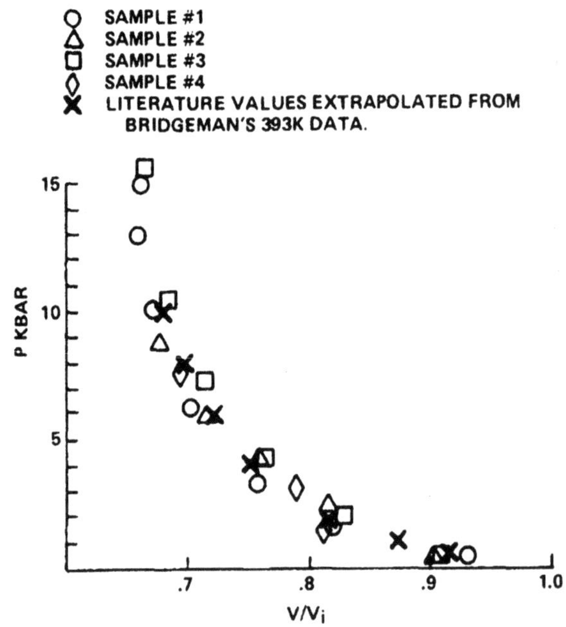

Fig. 1.65 For the measurements at 373 K the results are similar to those obtained at room temperature. A mean square error for a single observation was calculated to be ±0.016 and for the average of all observations was ±0.004 [81]

e.g., thermodynamic functions were compared with experimental compressibility data (density) on NM that were published in 1973, almost 35 years earlier when the accuracy of the pressure scale then in use was not well-characterized. (There was considerable uncertainty in the accuracy of the fixed-point pressure scale used ubiquitously prior to the introduction of the ruby scale.) In 1975, the fixed-point scale was reduced by a factor of 2 in the range above 15 GPa [47]. For some thermodynamic functions, e.g., shock velocity and particle velocity, the classical MD simulation was found to be close to the experimental results. However, the calculated Hugoniot pressure was too high in the high-pressure range, even when other simulation models were applied. There appears to be a deficiency of compressibility data in the scientific literature, not only for explosive liquids, but for liquids in general, probably because of the great difficulty in performing such experiments. While this strain-gage device is, indeed, not easy to use, there is definitely a need to provide more accurate compression data on explosive liquids. The rewards are great for someone adept enough with a DAC (and there are many such individuals now, worldwide) to accomplish such measurements.

To illustrate the usefulness of this technique, I refer to a paper reporting the effects of pressure on the vibrational spectra of liquid NM which uses data on the effect of pressure on the specific volume of NM at temperatures of 298 and 373 K [61] (Fig. 1.66). The strain-gage transducer system described above was used

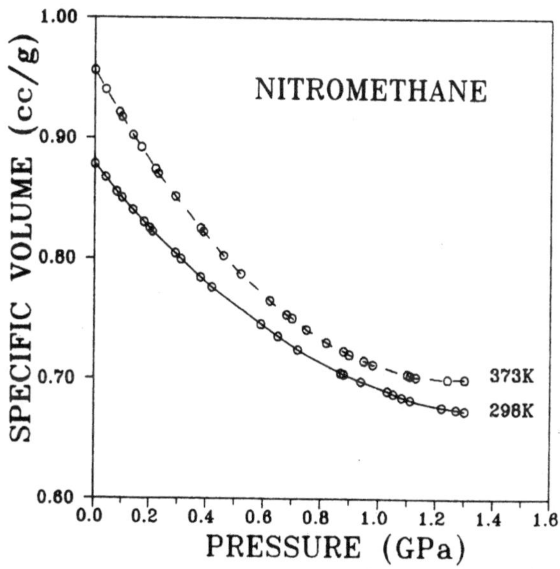

Fig. 1.66 Pressure dependence of the specific volume of NM at 298 and 373 K [61]. The uncertainty associated with the specific volume measurement is $\pm 0.015\,\mathrm{cm^3 g^{-1}}$ and with pressure is ± 0.025 GPa. The curves were drawn through the center of each data point. At 298 K, the specific volume of NM appears linear with pressure to about 0.4 GPa. At higher pressures, where the superpressed liquid exists, the curve bends rapidly, becoming almost flat due to increasing repulsive forces at those higher densities

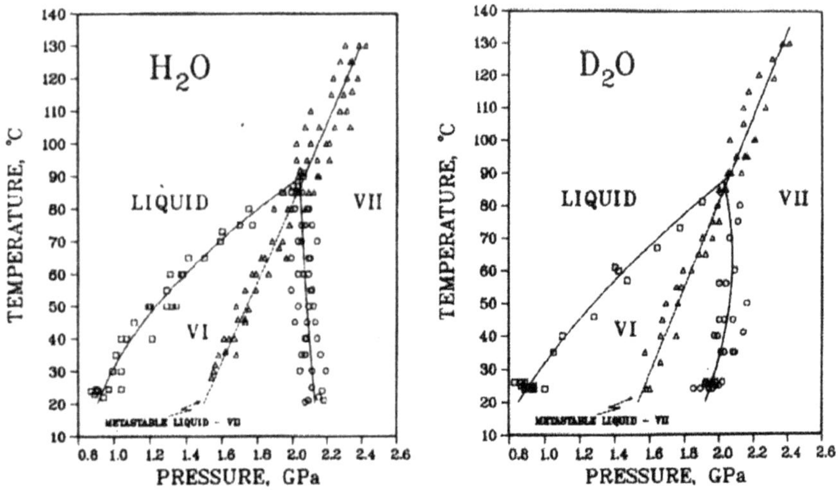

Fig. 1.67 Equilibrium pressure/temperature phase diagrams for H_2O and D_2O illustrating the superpressed state of liquid and its behavior as a function of P and T [85]

to obtain the compression data required to compare the observed frequencies of the NO_2 stretching modes to the calculated frequency shifts due to intermolecular coupling. To make this comparison it was necessary to estimate from specific volume data the decrease in the intermolecular distance with pressure between interacting pairs of molecules.

Another interesting and unique modification is the development of a dynamic diamond anvil cell (dDAC) which permits measurements of pressure/time-dependent phase transformation pathways [83] (Fig. 1.67). The fundamental design of the dDAC consists of a traditional DAC with integrated electromechanical piezoactuators to control the load on the sample. Because of the known anomalous superpressed nature of H_2O and D_2O in the stability regions of phases VI and VII [84, 85], the dDAC instrument was demonstrated on H_2O by observing time-resolved pressure-induced crystallizations of ice VI and ice VII. The authors of this article were able to evaluate the interfacial free energy and were also able to determine that the value for supercompressed water (SW/iceVII) is less than that for (SW/iceVI), indicating that the local order of the supercompressed phase is more similar to crystalline iceVII than crystalline ice VI. This result is consistent with recent studies suggesting that the local order in high density water is bcc-like as in iceVII. The ramifications of this technique when applied to explosive liquids deserve more attention by the energetic materials community. One area to explore further, for example, is the superpressed state of NM. At 298 K, NM freezes to a crystalline solid at 0.40 GPa [61]. The structure of this high-pressure phase is the same as the phase that crystallizes at low temperature and 1 atmosphere (orthorhombic space group $P2_12_12_1$) [73, 74, 86]. It is also known, however, that liquid NM easily superpresses at RT with the metastable liquid existing to about 2 GPa, where it rapidly crystallizes to a polycrystalline phase at 298 K [61]. The metastable liquid exists over a considerable range of pressure (1.6 GPa) before it crystallizes, where it could be studied by various techniques to learn more about this superpressed state. Furthermore, at elevated temperatures, the superpressed state of NM liquid would exist even at higher pressures, a typical behavior of the superpressed liquid state, in general.

The two phase diagrams shown above illustrate the anomalous behavior of H_2O and D_2O, i.e., the metastable extension of the liquid–VII phase boundary into the stability field of solid phase VI. Note that at higher temperatures the superpressed liquid exists at higher pressures for both H_2O and D_2O. To determine these phase diagrams polarizing light microscopy was used in conjunction with a DAC equipped with a miniature resistance coil-heating element. The ruby technique was used to measure pressure. This is yet another example of the power of OPLM in determining phase transitions, phase stability fields and phase diagrams. A wealth of valuable scientific knowledge concerning the superpressed state of nitromethane could be obtained by this experimental method of study. Finally, much to my surprise, my recent literature search on α-2,4,6 trinitrotoluene (TNT), which has a low melting point of 80.1°C (probably existing as a stable liquid over a significant range of pressure and temperature before decomposition), yielded little information on the P/T equilibrium phase diagram of that important explosive. This compound, in my opinion, would be an ideal candidate for investigation by OPLM in a DAC. Such an investigation could lead to a wealth of knowledge concerning the behavior of TNT both in the solid and liquid states, providing motivation to further study the material by other measurement techniques, e.g., x-ray diffraction, IR and Raman spectroscopy, etc.

1.10 Summary

I have attempted to give the reader an overview of the various DAC technologies that were developed in the early period of its growth soon after its invention at the NIST/NBS laboratory and to stress the seminal nature of the advances made, as many of them have been refined by others and are still in use today. In summary, besides the invention of the DAC itself (including high and low temperature capability), other applications to IR spectroscopy, x-ray diffraction (both to powder and single-crystal studies), polarizing light microscopy, Raman spectroscopy, FTIR spectroscopy (including time-dependent studies), were all developments at the NIST/NBS laboratory. Ancillary technologies, including the discovery of hydrostatic pressure-transmitting media, the simple Stokes falling ball methodology for the measurement of viscosity in liquids, and, of course, the all important development of the ruby fluorescence method of pressure measurement, so essential to the success of the of DAC as a quantitative scientific tool, were also seminal advances made at the NIST/NBS laboratory. All of these techniques were discussed in the preceding pages of this chapter. Certainly numerous improvements over these early methods have been made during the past decades and a discussion of these improvements would be impractical to include in this chapter simply because there are too many of them currently practiced in high-pressure research. All of these methods, in conjunction with a DAC, have proved to be exceedingly useful in the study of energetic materials (as well as in many other research areas), with the result that numerous advances have been made in our scientific understanding of these vital materials.

Acknowledgments The author wishes to thank Alan D. Mighell, NIST Research Associate, for his careful and thorough reading of the manuscript and also for his persistent encouragement in writing this chapter. The author also wishes to acknowledge the Ceramics Division of NIST for providing me with a Research Associateship under the sponsorship of Winnie Wong-Ng to support the preparation of this book.

Notes

◊Certain companies, commercial equipment, instruments, or materials are identified in this document. Such identification neither implies recommendation or endorsement by the National Institute of Standards and Technology, nor does it imply that the products identified are necessarily the best available for the purpose.

✦The use of "bar", "poise," and "angstrom (Å)" throughout the text and in some of the tables follow the common practice of workers in the field at the time the research was performed. For the uninitiated, I note that 1 bar $= 10^5$ N/m^2 (or 10^5 pascal) $= 10^6$ dyn/cm^2 $= 1.0197$ kgf/cm^2. The international standard (SI) unit of pressure is the pascal, or newton per square meter. The SI unit for viscosity is the pascal second (Pa·s), where 1 poise $= 0.1$ Pa·s. The SI unit for wavelength is nanometer (nm), where 1 Å $= 0.1$ nm.

References

1. C. E. Weir, E. R. Lippincott, A. Van Valkenburg, E. N. Bunting, (1959) Infrared studies in the 1-micron to 15-micron region to 30,000 atmospheres. J Res Nat Bur Stand, Sect A 63(1), 55–62.
2. G. J. Piermarini, C. E. Weir, (1962) A diamond cell for x-ray diffraction studies at high pressures. J Res Nat Bur Stand, Sect A 66, 325–331.
3. C. E. Weir, (1945) Compression of sole leather. J Res Nat Bur Stand 35(4), 257–271.
4. C. E. Weir, (1948) Effect of temperature on the volume of leather and collagen in water. J Res Nat Bur Stand 41(4), 279–285.
5. C. E. Weir, (1950) High-pressure apparatus for compressibility studies and its application to measurements on leather and collagen. J Res Nat Bur Stand 45(6), 468–476.
6. C. E. Weir, (1954a) Compressibilites of crystalline and glassy modifications of selenium and glucose. J Res Nat Bur Stand 52(5), 247–249.
7. C. E. Weir, (1954b) Temperature dependence of compression of linear high polymers at high pressures. J Res Nat Bur Stand 53(4), 245–252.
8. C. E. Weir, L. Shartsis, (1955) Compressibility of binary alkali borate and silicate glasses at high pressures. J Am Ceram Soc 38(9), 299–306.

9. C. E. Weir, J. D. Hoffman, (1955) Compressibilities of long-chain normal hydrocarbons. J Res Nat Bur Stand 55(6), 307–310.
10. L. H. Adams (Consultant to the Director) (1958) Survey of current high pressure research program at National Bureau of Standards and recommendations regarding future needs in this area. Revised November 19, 1958, NARA RG, Astin File, Box 15.
11. R. A. Paquin, E. Gregory (1963) Modification and calibration of a tetrahedral anvil apparatus, in Giardini AA and Lloyd EC (eds) High pressure measurement. Butterworths, Washington, DC, pp 274–285.
12. E. R. Lippincott, A. Van Valkenburg, C. E. Weir, E. N. Bunting, (1958) Infrared studies on polymorphs of silicon dioxide and germanium dioxide. J Res Nat Bur Stand 61(1), 61–70.
13. J. C. Jamieson, (1957) Introductory studies of high-pressure polymorphism to 24,000 bars by x-ray diffraction with some comments on calcite II. J Geol 65, 334–343.
14. P. W. Bridgman, (1952) The resistance of 72 elements, alloys and compounds to 100,000 kg/cm^2. Proc Am Acad Arts Sci 81, 167–251.
15. P. W. Bridgman, I. Simon, (1953) Effects of very high pressures on glass. J Appl Phys 24, 405–413.
16. E. N. Bunting, A. Van Valkenburg, (1958) Some properties of diamond. Am Mineral 43, 102–106.
17. J. C. Jamieson, A. W. Lawson, N. D. Nachtrieb, (1959) New device for obtaining x-ray diffraction patterns from substances exposed to high pressure. Rev Sci Instrum 30(11), 1016–1019.
18. J. C. Jamieson, A. W. Lawson (1962) Debye-Scherrer x-ray techniques for very high pressure studies in Wentorf RH (ed) Modern very high pressure techniques. Butterworths, Washington, DC, pp 70–91.
19. E. R. Lippincott, C. E. Weir, A. Van Valkenburg, (1960) Infrared studies of dense forms of ice. J Chem Phys 32(2), 612–614.
20. E. R. Lippincott, C. E. Weir, A. Van Valkenburg, E. N. Bunting, (1960) Studies of infrared absorption spectra of solids at high pressures. Spectrochim Acta 16, 58–73.
21. G. J. Piermarini, C. E. Weir, (1962) High-pressure transition in RbF. J Chem Phys 37(8), 1887–1888.
22. C. E. Weir, A. Van Valkenburg, E. R. Lippincott (1962) Optical studies at high pressures using diamond anvils in R. H. Wentorf (ed) Modern very high pressure techniques. Butterworths, Washington, DC, pp 51–69.
23. Alvin Van Valkenburg Biographical File. Archives Collection, Information Services Division. National Institute of Standards and Technology. Gaithersburg, MD.
24. R.J. Schneller Jr (2005) Breaking the color barrier: The U.S. naval academy's first black midshipman and the struggle for racial equality, New York University Press, New York, pp 65–67.
25. Charles Edward Weir Biographical File. Archives Collection, Information Services Division. National Institute of Standards and Technology. Gaithersburg, MD.
26. Elmer Newman Bunting Biographical File. Archives Collection, Information Services Division. National Institute of Standards and Technology. Gaithersburg, MD.
27. Obituary of Ellis Ridgeway Lippincott. *The Washington Post*, December 28, 1974.
28. A. Van Valkenburg A (1963) High-pressure microscopy, in Giardini AA, Lloyd EC (eds) High pressure measurement. Butterworths, Washington, DC, pp 87–94.
29. A. Van Valkenburg, (1970) High pressure optics. Appl Opt 9(1), 1–4.
30. S. Block, C. E. Weir, G. J. Piermarini, (1965) High-pressure single-crystal studies of ice VI. Science 148, 947–948.
31. G. J. Piermarini, C. E. Weir, (1964) Allotropy in some rare earth metals at high pressures. Science 144, (3614) 69–71.
32. C. E. Weir, G. J. Piermarini, (1964) Lattice parameters and lattice energies of high- pressure polymorphs of some alkali halides. J Res Nat Bur Stand, Sect A 68(1), 105–111.
33. C. Weir, G. Piermarini, S. Block, (1965) Single-crystal x-ray diffraction at high pressures. J Res Nat Bur Stand, Sect C 69(4), 275–281.
34. S. Block, C. E. Weir, G. J. Piermarini, (1965) High-pressure single crystal studies of ice VI. Science 148(3672), 947–948.

35. A. Santoro, C. E. Weir, S. Block, G. J. Piermarini, (1968) Absorption corrections in complex cases. Application to single crystal diffraction studies at high pressure. J Appl Crystallogr 1, 101–107.
36. G. J. Piermarini, A. D. Mighell, C. E. Weir, S. Block, (1969) Crystal structure of benzene II at 25 kbar. Science 165, 1250–1255.
37. G. J. Piermarini, A. B. Braun, (1973) Crystal and molecular structure of CCl_4-III: A high pressure polymorph at 10 kbar. J Chem Phys 58(5), 1974–1982.
38. C. E. Weir, G. J. Piermarini, S. Block, (1969) Instrumentation for single crystal x-ray diffraction at high pressures. Rev Sci Instrum 40(9), 1133–1136.
39. C. E. Weir, G. J. Piermarini, S. Block, (1969) Crystallography of some high-pressure forms of C_6H_6, CS_2, Br_2, CCl_4, and KNO_3. J Chem Phys 50(5), 2089–2093.
40. C. E. Weir, S. Block, G. J. Piermarini, (1970) Compressibility of inorganic azides. J Chem Phys 53(11), 4265–4269.
41. R. A. Forman, G. J. Piermarini, J. D. Barnett, S. Block, (1972) Pressure measurement made by the utilization of ruby sharp-line luminescence. Science 176, 284–285.
42. J. D. Barnett, S. Block, G. J. Piermarini, (1973) Optical fluorescence system for quantitative pressure measurement in the diamond-anvil cell. Rev Sci Instrum 44(1), 1–9.
43. G. J. Piermarini, S. Block, J. D. Barnett, R. A. Forman, (1975) Calibration of the pressure dependence of the R_1 ruby fluorescence line to 195 kbar. J Appl Phys 46, 2774–2780.
44. G. J. Piermarini, S. Block, J. D. Barnett, (1973) Hydrostatic limits in liquids and solids to 100 kbar. J Appl Phys 44, 5377–5382.
45. G. J. Piermarini, R. A. Forman, S. Block, (1978) Viscosity measurements in the diamond anvil pressure cell. Rev Sci Instrum 49, 1061–1066.
46. M. I. Eremets (1996) Chapter 3, The diamond anvil cell. in High pressure experimental methods, Oxford University Press, New York, pp 49–92.
47. G. J. Piermarini, S. Block, (1975) Ultrahigh pressure diamond-anvil cell and several semiconductor phase transition pressures in relation to the fixed point pressure scale. Rev Sci Instrum 46, 973–979.
48. W. Thomson (1891) Lecture to the Institution of Civil Engineers, May 3, 1883, Popular Lectures and Addresses by Sir William Thomson (London/New York, Macmillan) vol. 1, p 80.
49. H. K. Mao, P. M. Bell, (1978) High pressure physics: Sustained static generation of 1.36 to 1.72 Megabars. Science 200, 1145–1147.
50. W. A. Bassett, T. Takahashi, P. W. Stook, (1967) X-ray diffraction and optical observations on crystalline solids up to 300 kbar. Rev Sci Instrum 38, 37–42.
51. G. Huber, K. Syassen, W. B. Holzapfel, (1977) Pressure dependence of 4f levels in euopium pentaphosphate up to 400 kbars. Phys Rev B 15, 5123–5128.
52. L. Merrill, W. A. Bassett, (1974) Miniature diamond anvil pressure cell for single crystal x-ray diffraction studies. Rev Sci Instrum 45, 290–294.
53. S. Block, G. Piermarini, (1976) The diamond cell stimulates high-pressure research. Phys Today 29(9), 44–55.
54. I. Fujishiro, G. J. Piermarini, S. Block, R. G. Munro (1982) Viscosities and glass transition pressures in the methanol-ethanol-water system, in C. M. Backman, T. Johannisson, L. Tegner (eds) High pressure in research and industry. 8th AIRAPT Conference Proceedings, Arkitektkopia. Uppsala, Sweden, Vol. II, pp 608–611.
55. Y. B. Zel'dovich, Y.P. Raiser (1966) Physics of Shockwaves and High Temperature Hydrodynamic Phenomena, Academic Press, New York.
56. R. D. Bardo, T. N. Hall, M. J. Kamlet, (1982) Energies and volumes of activation for condensed detonating explosives. J Chem Phys 77(11), 5858–5859.
57. R. D. Bardo, T. N. Hall, M. J. Kamlet, (1979) Volumes of activation in the shock initiation of explosives. Combust Flame 35(3), 259–265.
58. P. J. Miller, G. J. Piermarini, S. Block, (1984) An FT-IR microscopic method for kinetic measurements at high temperatures and high pressures. Appl Spectrosc 38, 680–686.
59. P. J. Miller, S. Block, G. J. Piermarini, (1991) Effects of pressure on the thermal decomposition kinetics, chemical reactivity and phase behavior of RDX. Combust Flame 83, 174–184.

60. G. J. Piermarini, S. Block, P. J. Miller, (1989) Effects of pressure on the thermal decomposition kinetics and chemical reactivity of nitromethane. J Phys Chem 93, 457–462.
61. P. J. Miller, S. Block, G. J. Piermarini, (1989) Effects of pressure on the vibrational spectra of liquid nitromethane. J Phys Chem 93, 462–466.
62. G. J. Piermarini, S. Block, P. J. Miller, (1987) Effects of pressure and temperature on the thermal decomposition rate and reaction mechanism of β-Octahydro-1,3,5,7-tetranitro-1,3,5,7-tetrazocine. J Phys Chem 91(14), 3872–3878.
63. T. P. Russell, P. J. Miller, G. J. Piermarini, S. Block (1993) Pressure/temperature/reaction phase diagrams for several nitramine compounds, in L. H. Liebenberg, R. W. Armstrong, J. J. Gilman (eds) Structure and properties of energetic materials. Materials Research Society Symposium Proceedings, Pittsburgh, PA, pp 199–213.
64. T. P. Russell, P. J. Miller, G. J. Piermarini, S. Block, (1992) High-pressure phase transition in γ-hexanitrohexaazaisowurtzitane. J Phys Chem 96(13), 5509–5512.
65. T. P. Russell, P. J. Miller, G. J. Piermarini, S. Block, (1993) Pressure/temperature phase diagram of Hexanitrohexaazaisowurtzitane. J Phys Chem 97(9), 1993–1997.
66. G. J. Piermarini, S. Block, R. Damavarapu, S. Iyer, (1991) 1,4-Dinitrocubane and cubane under high pressure. Propellants Explos Pytotech 16, 188–193.
67. T. P. Russell, G. J. Piermarini, P. J. Miller, (1997) Pressure/temperature and reaction phase diagram for dinitro azetidinium dinitramide. J Phys Chem B 101, 3566–3570.
68. T. P. Russell, G. J. Piermarini, S. Block, P. J. Miller, (1996) Pressure, temperature reaction phase diagram for ammonium dinitramide. J Phys Chem 100, 3248–3251.
69. R. W. Shaw, T. B. Brill, D. L. Thompson (2005) In Overviews of recent research on energetic materials, Advanced Series in Physical Chemistry, World Publishing, Singapore, Vol. 16.
70. W. C. McCrone, (1950) RDX (Cyclotrimethylenetrinitramine). Anal Chem 22(7), 954–955.
71. R. J. Karpowicz, S. T. Sergio, T. B. Brill, (1983) Beta-polymorph of hexahydro-1,3,5-trinitro-s-triazine. A Fourier transform infrared spectroscopy study of an energetic material. Ind Eng Chem Prod Res Dev 22(2), 363–365.
72. G. J. Piermarini, S. Block, P. J. Miller (1990) Effects of pressure on the thermal decomposition rates, chemical reactivity, and phase behavior of HMX, RDX and nitromethane, in Bulusu SN (ed) Chemistry and physics of energetic materials. Kluwer, Dordrecht, pp 391–412.
73. S. Courtecuisse, F. Cansell, D. Fabre, J. P. Petitet (1995) A Raman spectroscopic study of nitromethane up to 350°C and 35 GPa. J Phys IV (Paris) 5, C4–359–363.
74. D. T. Cromer, R. R. Ryan, D. Schiferl, (1985) The structure of nitromethane at pressures of 0.3 to 6.0 GPa. J Phys Chem 89, 2315–2318.
75. J. W. Brasch, (1980) Irreversible reaction of nitromethane at elevated pressure and temperature. J Phys Chem 84, 2084–2085.
76. R. Ouillon, J. P. Pinan-Lacarré, P. Ranson, (2002) Low-temperature Raman spectra of nitromethane single crystals. Lattice dynamics and Davydov splittings. J Chem Phys 116, (11) 4611–4625.
77. J. P. Pinan-Lacarre, R. Ouillon, B. Canny, P. Pruzan, P. Ranson, (2003) Pressure effect at room temperature on the low-energy Raman spectra of nitromethane-h(3) and –d(3) up to 45 GPa. J Raman Spectrosc 34, 819–825.
78. S. F. Rice, M. F. Foltz, (1991) Very high pressure combustion: Reaction propagation rates of nitromethane within a diamond anvil cell. Combust Flame 87(2), 109–122.
79. W. E. Brown, D. Dollimore, A. K. Galwey (1980) Reactions in the Solid State in C. H. Bamford, C. H. F. Tipper (eds) Comprehensive Chemical Kinetics, Vol 22, Elsevier, Amsterdam, Chap. 3, pp 0–340.
80. E. L. Lee, R. H. Sanborn, H. D. Stromberg (1970) Thermal decomposition of high explosives at static pressures to 50 Kbar. Proc 5th Symp (Int) Detonation, pp 331–337.
81. J. W. Brasch, (1980) Techniques for compressibility measurements on explosive materials using an opposed diamond-anvsil optical cell. Rev Sci Instrum 51, 1358–1362.82. Bridgman PW (1932) Volume-temperature-pressure relations for several non-volatile liquids. Proc Am Acad Arts Sci 67, 1–27.
82. H. Liu, J. Zhao, G. Ji, Z. Gong, D. Wei, (2006) Compressibility of liquid nitromethane in the high-pressure regime. Phys B 382, 334–339.

83. G. W. Lee, W. J. Evans, C. S. Yoo, (2006) Crystallization of water in a dynamic diamond-anvil cell: Evidence for ice VII-like local order in supercompressed water. Phys Rev B 74, (134112) 1–6.
84. K. Yamamoto, (1980) Supercooling of the coexisting state of Ice VII and water within Ice VI region observed in diamond-anvil pressure cells. Jpn J Appl Phys 19(10), 1841–1845.
85. G. J. Piermarini, R. G. Munro, S. Block(1984) Metastability in the H_2O and D_2O systems at high pressure. Mat Res Soc Symp Proc, vol 22, Elsevier Science, pp 25–28.
86. S. F. Trevino, E. Prince, C. R. Hubbard (1980) Refinement of the structure of solid nitromethane. J Chem Phys **73**(6), 2996–3000.

Chapter 2
Synthesis of High-Nitrogen Energetic Material

Mikhail I. Eremets, Ivan A. Trojan, Alexander G. Gavriliuk, and Sergey A. Medvedev

2.1 Introduction

Pure nitrogen can be considered as a material with optimized storage of chemical energy because of the huge difference in energy between triply bonded di-nitrogen and singly bonded nitrogen. $N \equiv N$ bond is one of the strongest chemical bonds known, containing 4.94 eV atom^{-1} while the N–N bond is much weaker with -0.83 eV atom^{-1} [1]. Therefore when transforming from singly bonded nitrogen to diatomic triply bonded molecular nitrogen, a large amount of energy should be released: about 2.3 eV atom^{-1}. Or, in other words, this chemical energy can be ideally stored by transforming a triple bond into three single bonds. Thus, nitrogen may form a high-energy density material with energy content higher than that of any known nonnuclear material. The greatest utility of fully single-bonded nitrogen would be as high explosives. Here, a tenfold improvement in detonation pressure over HMX (one the more powerful high explosives) seems possible [2].

A chemical approach for synthesis of nitrogen-energetic materials is creating large all-nitrogen molecules or clusters bound by single (N–N) or single and double ($N=N$) bonds. Calculations predict different polynitrogen molecules or clusters [3] such as, for instance, N_4, N_8, N_{20}, or even nitrogen fullerene N_{60} (see for review Refs [3–5]) with high-energy-storage capacity. However, none of them has yet been synthesized with exception of $N_4(TdN_4)$ [6], albeit with a very short lifetime of $\sim 1\,\mu s$ [6]. Synthesis of compounds having several nitrogen atoms consecutively is difficult because the single bond in nitrogen is relatively weak. It has been achieved only in compounds with other atoms. For instance, HN_3 and other azides with linear-N_3 radical have been synthesized by Curtius in 1890 [7]. Only recently $N_5{}^+$ was synthesized by Christe and coworkers [8]. On the basis of the $N_3{}^-$ and $N_5{}^+$ species nearly all nitrogen compounds were synthesized by attaching these radicals to a central atom of Te, B, and P such as $N_5P(N_3)_6, N_5B(N_3)_4, Te(N_3)_4$, and others (see for review Refs [2, 9–11]).

S.M. Peiris, G.J. Piermarini (eds.), *Static Compression of Energetic Materials*,
Shock Wave and High Pressure Phenomena,

Presented here is an ultimate case of pure single-bonded nitrogen which can be synthesized at high pressures and temperatures from common triply bonded molecular nitrogen. At high pressures molecules of nitrogen approach each other and start interacting so strongly that intramolecular interaction becomes comparable with intermolecular interaction, and therefore di-nitrogen molecules can dissociate to atomic nitrogen. Atoms of nitrogen with three directed covalent bonds can further create a three-dimensional network popularly termed polymeric nitrogen. Such transformations have been predicted long time ago [12, 13] to happen at high pressures of $\approx$50 GPa, but it was found experimentally at much high pressures of $\sim$200 GPa [14] at low temperatures, and then at a slightly lower pressures of 150 GPa and room temperature [15]. This apparently nonmolecular nitrogen has a disordered network of single- and double-bonded atoms as follows from calculations [16, 17].

Next important step in the preparation of polymeric nitrogen was the theoretical prediction of single-bonded crystalline nitrogen where each nitrogen atom is connected with three other with single bonds [18]. It has a cubic symmetry structure but with an unusual arrangement of atoms: the cubic gauche structure (referred to as cg-N or CG) of the $I2_13$ space group. This structure was recently fully confirmed experimentally [19–22]. Theoretical calculations also agree that the cg-N structure is the most stable among possible proposed single-bonded nitrogen networks [23–29]. This polymeric nitrogen is a unique material. First of all, it has very high accumulated energy: about 1.2–1.5 eV atom^{-1} releases during the transformation from cg-N to N_2 [16–18, 30]. This is 5.5 times higher (energy/mass ratio) when compared to the powerful explosive hexogen [17]. At the same time cg-N is similar to diamond in terms of values of lattice parameters and low compressibility. Thus it could be a hard (or superhard) material with ultimate high-energy capacitance. Importantly, polymeric nitrogen is predicted to be metastable at ambient pressure [18], and an amorphous polymeric nitrogen indeed has been recovered to ambient pressure at low temperatures [14]. Cg-N was proved experimentally to be metastable at least down to 25 GPa. This chapter describes the synthesis and properties of polymeric nitrogen in more detail, and also discusses other polymeric structures which are predicted to be competitive with the cg-N phase in some pressure range [23, 25–28, 31–33].

Finally we mention attempts to decrease the pressure of polymerization by using precursors different from pure nitrogen. Instead of compressing diatomic nitrogen, molecular N_3^- anions in the lattice of NaN_3 were examined [34,35]. The N_3^- anion is a straight chain of three nitrogen atoms linked essentially with double bonds. It can be considered as a molecule which is more weakly bonded than the diatomic triple-bonded nitrogen. Therefore, it could be expected that the N_3^- molecules will create polymeric single-covalent bond networks easier than diatomic nitrogen. New structures that are distinct from linear N_3^--ion-based molecular phases were found near ambient pressure after laser heating [34]. In Ref. [35] the photolytic decomposition of sodium azide under UV pulse irradiation was studied. The resultant product, speculated to be N_7^-, was found to be unstable at room pressure and temperature.

2.2 Polymeric Nitrogen

In this section we will present our results on synthesis and properties of polymeric single-bonded nitrogen. When nitrogen atoms are connected with *single bonds* into a polymeric network it could be a high-energy density material (HEDM). Moreover, polymeric nitrogen should be environmentally clean because the final product of the transformation is nitrogen. This polymeric nitrogen material could also be an ideal rocket fuel or propellant.

A way to create an atomic solid with a single-bonded crystalline structure is application of high pressures, as predicted in 1985 by McMahan and LeSar [12]. Simple cubic or distorted sc structures similar to other group-V elements were calculated. This transformation is expected to accompany with a large drop of volume and considerable hysteresis. This new material can be called "polymeric nitrogen": the starting dimer is the molecular nitrogen while the final product is a crystal built with single covalent bonds. We note here that nitrogen behaves uniquely under pressure. If we compare it with other molecular solids such as H_2, O_2, I_2, and others [36], they are double-covalent bonded and at high pressure transform to metals while nitrogen first transforms to an insulating covalent crystal. Nitrogen also is different from other elements of group V. For instance, the stable form of phosphorus at ambient pressure is a P_4 molecule where atoms are connected to each other by single bonds. The reason of this difference is simple: atoms of phosphorous are significantly larger than nitrogen atoms and therefore longer single bonds are preferable. At high pressure nitrogen also does not follow the sequence of phase transformations typical for group-V elements.

Mailhiot et al. [18] found that nitrogen behaves uniquely under pressure: molecular nitrogen transforms into a lattice of single-bonded atoms with cubic gauche structure (cg-N). This unusual structure naturally follows from the directed covalent bonds: the dihedral angle in a N–N single bond energetically prefers a *gauche* conformation [18] as can be seen, for instance, in the molecule of hydrazine. As a result this structure theoretically is much more preferable (Fig. 2.1): it has significantly lower ($0.86\,\mathrm{eV\ atom}^{-1}$) total energy than previously considered phases. This has been confirmed in numerous calculations [17, 23–28, 30, 32, 37–44].

Mailhiot et al. [18] calculated the volume, bulk modulus, and other properties of cg-N. This transformation should happen at $\approx 50\,\mathrm{GPa}$ and be accompanied with a dramatic drop of volume ($\sim 20\%$) and complete change of the phonon spectrum. The cg-N phase was predicted to be metastable at ambient pressure. All this gave clear information for experimental proof. However, the experiments that followed did not reveal any transition in this pressure range while a darkening [45] and nearly opaque state [46] has been observed at the highest pressures up to 180 GPa [46].

Transformation to nonmolecular nitrogen finally came with the dramatic disappearance of the molecular vibron [14] at 190 GPa at low temperatures from transparent to an opaque phase. The same transition was also seen in measurements at room and elevated temperatures [15, 47, 48]. It was found that this transformation has a huge hysteresis [14]. Therefore, the value of equilibrium pressure is about 100 GPa [14] – much lower than the pressure of direct transformation, and closer

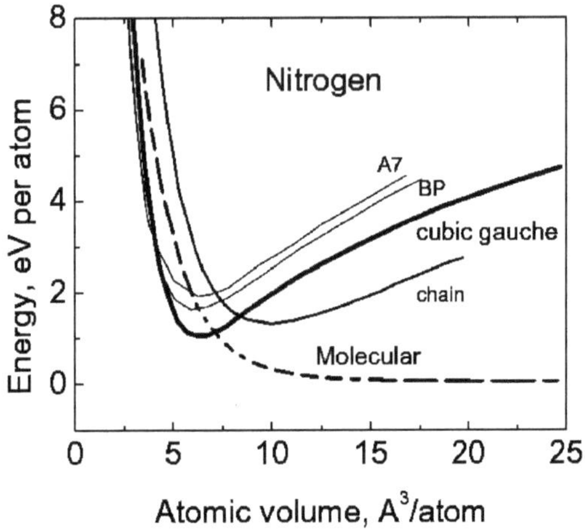

Fig. 2.1 Polymeric phases of nitrogen. Total energy per atom calculated for the threefold-coordinated arsenic (A7), black-phosphorus (BP), and cubic gauche (cg) phases of nitrogen, and for a twofold-coordinated chainlike (ch) phase of nitrogen, as a function of atomic volume (after Mailhiot et al. [18])

to the theoretical [18] calculations. Moreover, the hysteresis was so large that it allowed the nonmolecular nitrogen to be recovered at low temperatures [14]. Though the nonmolecular nitrogen has properties close to those predicted for the polymeric nitrogen [18], the material was a narrow-gap semiconductor [14, 16, 47], which is in contradiction to the predicted dielectric cg-N. First-principle simulations elucidated the structure of the opaque phase: an amorphous or disordered network of single- and double-bonded nitrogen atoms [16, 17] as shown in Fig. 2.2a. However, there is no direct experimental evidence that the nonmolecular phase creates a polymeric network. Available optical data only indicate the amorphous nature for this material [14, 16, 47]. This can be seen, for instance, from Fig. 2.2b where only very broad bands are present in the Raman spectrum.

Next decisive step towards polymeric nitrogen was done with laser heating of molecular nitrogen above 100 GPa [19]. Here a direct transformation of molecular nitrogen to a new transparent phase, which turned out to be the long sought after polymeric nitrogen was found. Polymeric nitrogen was synthesized at $T > 2{,}000$ K at $P > 110$ GPa. These record parameters were achieved from a novel arrangement of the diamond anvil cell (Fig. 2.3). First, a gasket compacted from powdered cubic boron nitride (cBN) mixed with epoxy was used [49]. This gasket is 1.5–2 times thicker (8–10 μm at 60 GPa) than typical hard gasket materials such as Re or W at the same pressure. This allows larger sample volumes. Also, this increases the gap between the boron laser radiation absorber (that heats the sample) and the anvils (Fig. 2.3), a precondition to achieving very high temperatures at megabar pressures. Second, the cell was specifically designed for use in with challenging x-ray diffraction measurements. X-ray diffraction measurements of a nitrogen sample (which is a light element and a weak scatterer) at pressures up to 170 GPa were performed

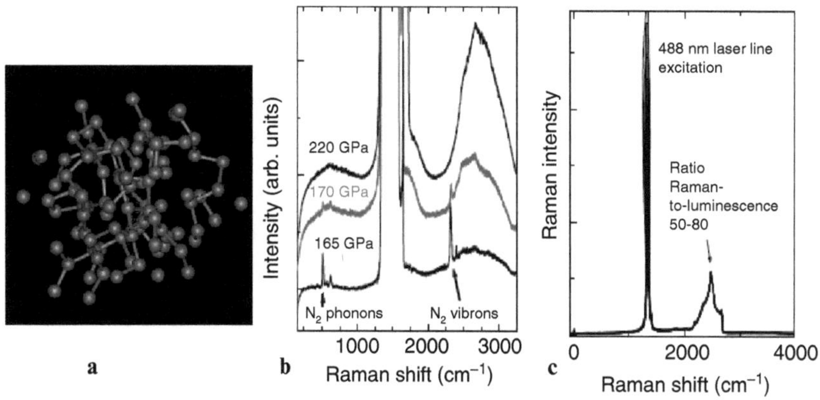

Fig. 2.2 Amorphous polymeric nitrogen. (**a**) Structure of low-coordinated ($Z = 3$–4) nitrogen at ambient pressure at 100 K produced by simulations showing atom positions and bonds between them (from Ref. [16]). (**b**) Raman spectra of nitrogen at megabar pressures at 300 K taken through diamond anvils having a negligible own luminescence as can be seen from (**c**) where Raman spectra from 15 anvils are presented which perfectly coincide: only first- and second-order Raman peaks from diamond are in the spectra

at the ID9 beam line at ESRF [50]. The cBN gasket is extremely advantageous in x-ray diffraction studies: it produces only a few weak peaks comparable with nitrogen in intensity. Also, boron as an absorber of the 1.064 μm laser radiation was used (Fig. 2.3a). Importantly, boron is a very weak scatterer of x-rays and a signal produced by a 1 μm thick plate does not interfere with the diffraction pattern from the nitrogen sample. Moreover, the weak diffraction peaks from boron are uniformly distributed over a wide spectrum without strong peaks [51]. The boron plate was formed by pressing nanoparticles of amorphous boron of 99.99% purity. Boron is initially a semiconductor and transforms to a poor metal at megabar pressures [49]. It is a good absorber of the 1.06 mm laser radiation, however it violently reacts with nitrogen at $T > 1{,}800$ K and pressures <15 GPa [52]. At megabar pressures this reaction is nearly suppressed. We observed small color changes at the center of the boron plate at the highest temperatures that are likely due to formation of cBN, as evident from well-recognized peaks of cBN in the diffraction pattern (Fig. 2.3). However, the thin boron/cBN plate remained opaque, and heating was very stable to the highest temperatures allowing reliable temperature measurements [53]. Note, that at pressure below ∼50 GPa reaction of boron with nitrogen becomes marked.

Heating to high temperature released stress inside the sample, resulted in sharp x-ray diffraction peaks (Fig. 2.3), and dramatically increased the accuracy of the measurements. For pressure measurements ruby luminescence was used [54]. For that a micrometer-sized piece of ruby was placed at the edge of the gasket hole. To study the possible reaction of ruby with nitrogen, ruby was omitted in another experiment, yielding the same result. The pressure in this experiment was determined from the sharp high-frequency edge of the diamond anvil [55]. The diamond anvil cell of original construction was described elsewhere [56]. Nitrogen gas compressed to 2,500 bars was loaded at room temperature. Very low-luminescence synthetic

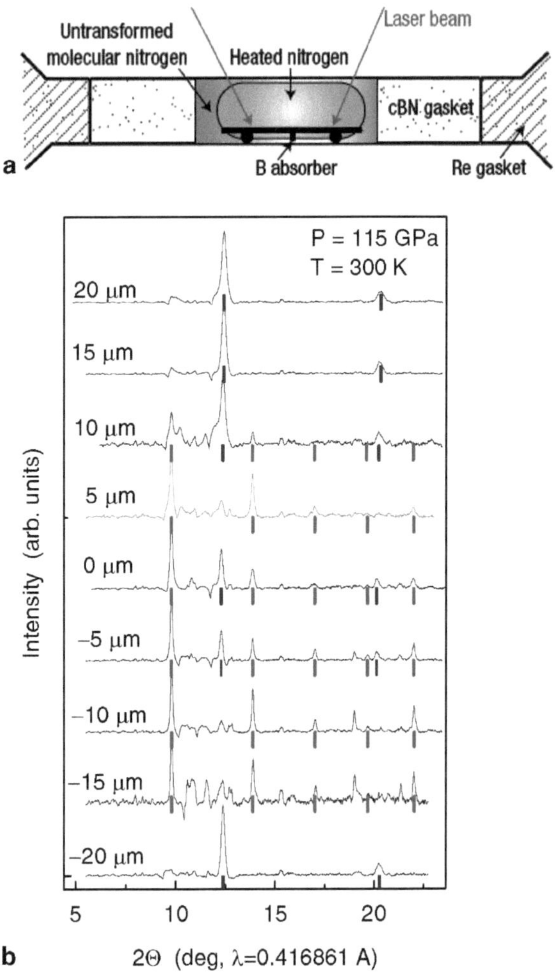

Fig. 2.3 X-ray diffraction measurements of nitrogen after heating to 2,600 K at 140 GPa. (**a**) A cross section of the sample arrangement. The heating ∼1 μm-thick boron plate (absorber of laser radiation; black) rests on cBN pieces that thermally insulate the plate from the bottom anvil. The sample squeezed between the anvils is surrounded by the cBN/epoxy gasket followed by the metallic (Re) supporting ring. The culet of diamond anvil is of diameter 90 μm. X-ray diffraction was measured with a monochromatic X-ray beam with 0.4168 Å wavelength focused to a spot of diameter ∼5 μm. (**b**) X-ray patterns taken across the culet of diamond anvil to distinguish between the input from the gasket and the different phases of the sample. Spectra taken from the most heated part (±5 μm) were attributed to the new phase of nitrogen. At the gasket only pure cBN patterns are present. At the edge of the nitrogen sample ($r = -15$ μm) there is a significant input from the molecular phase at the 10°–14° range. The black and grey vertical lines indicated calculated positions of reflections corresponding to cg-N and cBN, respectively

diamonds (Fig. 2.2) with flats of diameter 95 μm beveled at 10° to 380 μm culet and 60–90 μm flats were used in other runs. Raman spectra before and after heating were recorded with a single grating monochromator, notch-filters, cooled CCD and argon and Ti-sapphire lasers at the 458–800 nm spectral range.

In a typical run, the pressure was increased up to 140 GPa at 300 K until the sample started to darken. At this pressure, the sample was heated in steps with intermediate recordings of Raman spectra at room temperature. After the first heating to 980 K, the transformation accelerated. The sample darkened further, but it remained in the molecular state – the vibrons in the Raman spectra of nitrogen did not change. The high-frequency edge from the diamond anvils sharpened, indicating uniform pressure distribution, which dropped to 120 GPa. No further changes were observed after heating up to 1,400 K. After heating to 1,700 K, the sample further darkened and the intensity of the vibron of nitrogen significantly decreased.

At 1,980 K, the vibron peaks became indistinguishable from the background, all wide bands in the low-frequency part of spectrum disappeared, and luminescence measured from the nitrogen sample significantly increased. A ring around the absorber appeared, likely due to melting of the sample. Instead of further darkening, unexpectedly, the sample became transparent (the absorber plate became clearly visible in reflected light). After successive heating up to 2,500 and 2,600 K, the transparent colorless part spread out over a larger area and could be seen in transmitted light. The nitrogen at the edge of the gasket hole was not heated to the high temperatures and remained dark. The pressure in the center of the sample reduced to 115 GPa.

Thus, a dramatic transition happened at 115 GPa at temperatures above $\sim$2,000 K. X-ray diffraction patterns were accumulated from different parts of the sample across the culet in order to distinguish between the contribution from the gasket and the different phases of the sample (Fig. 2.3). In the center of the sample there are up to five reflections of the new phase of nitrogen and two reflections from cBN. These reflections are clearly different: the nitrogen rings are spotty while the cBN ring is not. In addition, the intensity of the cBN peaks varies from point to point of the scan, and at some places the reflections almost disappear.

We fitted the measured x-ray diffraction spectra by the cubic gauche structure of nitrogen (cg-N) and cBN. cg-N has the following parameters: space group $I2_13$, $a_0 = 3.4542$ Å, sites 8**a**, $x = 0.067$. We tried also to fit the spectra to different proposed polymeric structures [12,13,17,25,37] as well as boron from the absorber [51] instead of the cg-N structure, but without satisfactory results. The cg-N structure is schematically shown in Fig. 2.4. All nitrogen atoms are threefold coordinated and bond-lengths are the same for all pairs of bonded atoms. At a pressure of 115 GPa the bond length is 1.346 Å and the angle between bonds is 108.87°. We also determined the cg-N structure upon decreasing pressure down to 42 GPa (Fig. 2.4c). We fitted the experimental points measured in the 42–134 GPa range with Birch-Murnaghan equation of states (EOS) extrapolated to zero pressure with the following parameters: bulk modulus $B_0 = 298$ GPa, $B' = 4.0, V_0 = 6.592$ Å [3]. The cg-N is a stiff solid as follows from the high value of B_0 which is characteristic for strong covalent solids ($B_0 = 365$ GPa for cBN [57]). The bulk modulus determined from different types of EOSs lies in the range of 300–340 GPa. The determined value of atomic volume V_0 is close to 6.67 Å [3] predicted in Ref. [18].

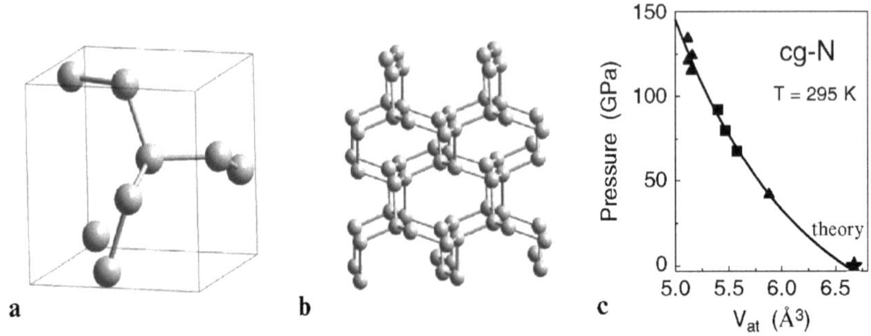

Fig. 2.4 Cubic gauche (cg-N) structure. Each atom of nitrogen is connected to three neighbors with three single covalent bonds. **(a)** The primitive cell, **(b)** an extended structure of the polymeric nitrogen. **(c)** The EOS of cg-N structure. Extrapolation of this EOS to zero pressure gives a volume of cg-N structure about 6.6 Å [3] – in excellent agreement with theoretical predictions of 6.67 Å [3] (Ref. [18])

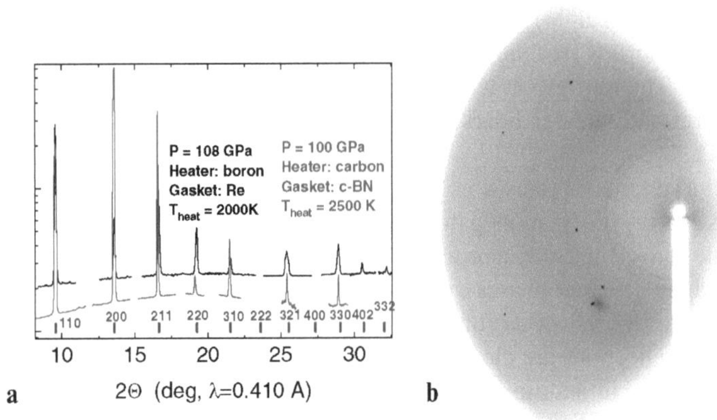

Fig. 2.5 Diffraction from cg-N synthesized at 110 GPa in diamond anvil cell with laser heating. **(a)** Diffraction peaks from two experiments. Grey lines relates to the arrangement with boron absorber heated to 2,000 K and rhenium gasket. Black lines: CVD diamond plate heated to 2,500 K, gasket was prepared from cBN powder mixed with small amount epoxy. Numbers and rods are indexes and position of reflections calculated according to the cubic gauche structure. The peaks are obtained by integration of spots at diffraction pattern shown in **(b)** for the case of a cell with side-input for x-ray beam and combination of boron heater and Re gasket

The cg-N structure has been further confirmed in many experiments [20–22]. For instance, more x-ray reflections were obtained to check cg-N structure [58]. For that a diamond anvil cell, which allows collection of diffraction at higher angles has been built. It has an additional side-input at 15° to the axes of the cell. If the x-ray goes through this window the diffraction rays can be observed up to 35°. An example of the x-ray diffraction for this side-input geometry is shown in Fig. 2.5. Thus two new 321 and 330 reflections of cg-N structure were found. In other experiments with this geometry but with different arrangement of the gasket and the heater 402 and 332 reflections in addition were obtained. Altogether nine reflections from the sample were collected. These reflections were observed also because instead of a fine powder of cg-N larger crystallites were grown which can give strong reflections at large angles. For that, the synthesis temperature was kept as low as possible at a given pressure, and the heating time was increased to minutes. Thus conditions of equilibrium growth were approached.

An absorber of the laser radiation gives an additional complication for analysis and interpretation of the experiment. Therefore, cg-N also has been attempted to synthesize with laser heating but without any absorber. Nitrogen at pressures above ~130 GPa significantly absorbs itself and it was successfully heated directly with YAG laser radiation resulting x-ray diffraction and Raman spectra of cg-N [20–22].

To summarize, we found the cg-structure which was considered theoretically as the most energetically favorable single-bonded structure [17, 23–28, 30, 32, 37–44] and originally proposed by Mailhiot et al. [18]. This is a new allotrope of nitrogen where all atoms are connected with single covalent bonds. To our knowledge, this unusual cubic phase has not been observed previously in any element.

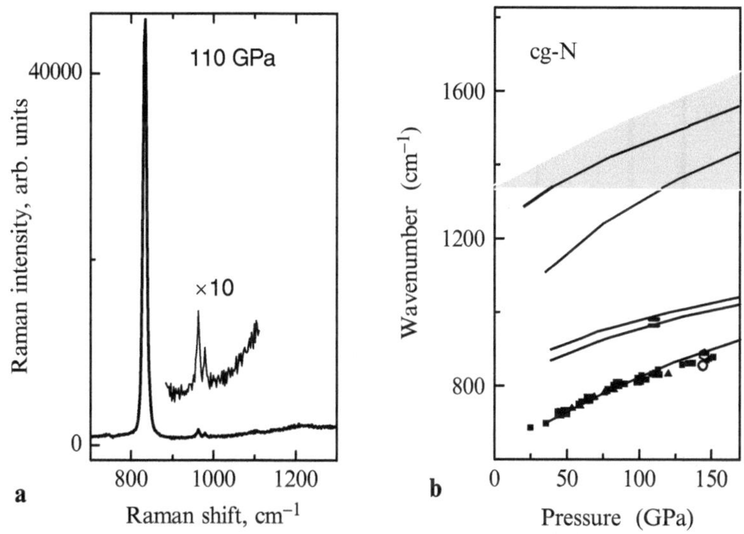

Fig. 2.6 Raman scattering from the cubic gauche polymeric nitrogen. **(a)** Raman spectrum from a large sample of cg-N at 120 GPa. A small intensity doublet at 963 cm^{-1} has been magnified for clarity. **(b)** Pressure dependence of Raman modes of the cg-N phase. Points are taken from different experiments. Lines are theoretical calculations [24]. The shadow area indicates width of Raman band from the stressed diamond anvils which covers the high-energy Raman lines of cg-N

2.2.1 Raman Spectra of cg-N

Transformation to cg-N phase was accompanied with a dramatic change of vibration spectra: after disappearing of the molecular vibron a new Raman peak in the cg-N phase was observed (Fig. 2.6). This is a fingerprint of transformation. The new peak is easily detected and it is strong in some samples. The position of this peak and its pressure dependence well follows calculations for cg-N phase [39] (Fig. 2.6). However, puzzlingly only one peak was observed in all experiments while four peaks are expected from the symmetry $I2_13$ of cg-structure. Only recently this problem has been resolved. From theoretical side, intensities of Raman peaks along with their positions have been calculated in Ref. [24] and it was found that only one peak dominates, others should be much weaker. Experimentally this was confirmed by obtaining a strong Raman signal from a large sample. It was grown in a diamond cell with anvil culets of the toroidal shape [22] (Fig. 2.7). This shape dramatically

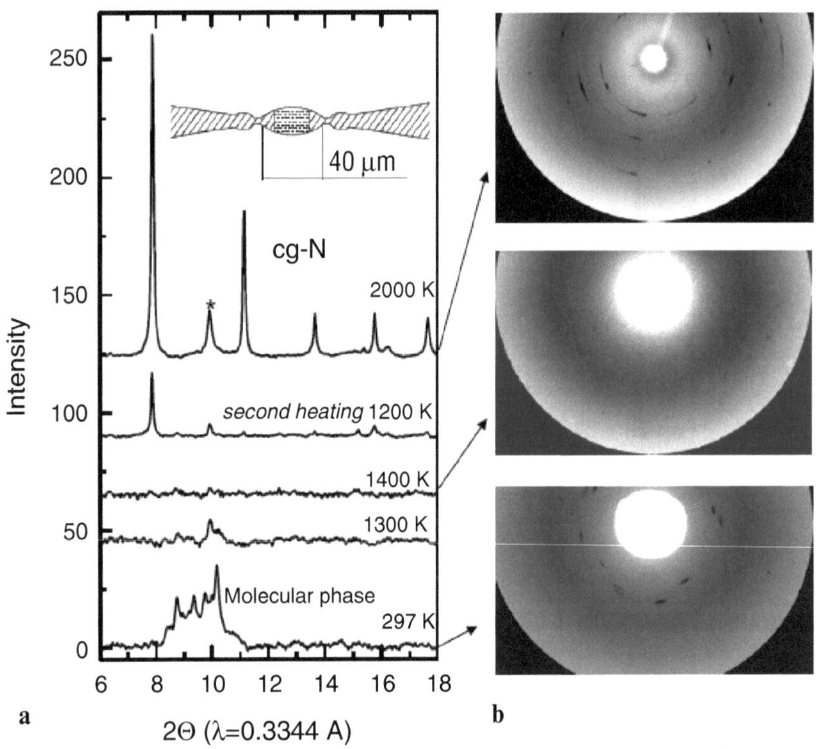

Fig. 2.7 X-ray diffraction measurements of nitrogen at 140 GPa measured in situ at different temperatures. (**a**) The X-ray Diffraction spectra were obtained by integrating the CCD image patterns (shown at (**b**)) and extracting the background. Star indicates the cBN peak. X-ray beam with 0.3344 Å wavelength was focused to a spot of diameter ≈ 5 μm and collimated by a 20 μm pinhole. The radiation from the YAG laser was focused in the same place. Toroidal profile of diamond anvil culets is shown in inset (**a**)

increases the high-pressure volume and allows large samples to be obtained and consequently to have intensive x-ray diffraction (Fig. 2.7) and Raman spectra (Fig. 2.6). Thus we were able to analyze the Raman spectrum of cg-N in detail and to find one more additional weak split peak (Fig. 2.6). Its position as well as the intensities (which are ~100 times smaller than that of the dominated Raman peak) coincides well with the calculations [24, 39]. The remaining two weak peaks owing to cg-N structure [24, 39] can be hidden under the strong background of the Raman band from the stressed diamond anvils (the dashed area at Fig. 2.6).

2.2.2 Control Experiments

In the experiments performed at high temperatures $T > 2{,}000\,K$, reaction of hot nitrogen with heater or surrounding gasket or diamonds is not excluded. Therefore, we had done numerous checking experiments to insure that the observed Raman and x-ray diffraction signals belong to cg-N, not to a product of interaction of nitrogen with gasket, absorber, or diamond. Different combinations of absorber and gasket materials (boron–cBN, boron–rhenium, -TiN–rhenium, beryllium–cBN, hBN–rhenium) results in the same x-ray diffraction pattern and Raman spectrum manifesting synthesis of cg-N. The following two arrangements are mostly evident: (a) absorber made of CVD (diamond grown by chemical vapor deposition) diamond plate and cBN/epoxy gasket, and (b) boron absorber and Re gasket. In both cases we obtained only x-ray reflections corresponding to the cg-N structure. This means that the observed diffraction pattern does not relate to compounds of nitrogen with rhenium or boron. Reaction with carbon is also excluded because the CVD plate heater apparently does not react with nitrogen at temperatures of up to 2,500 K: diffraction from the heated polycrystalline CVD plate presents only diamond rings which have same intensity as the unheated plate. We also did not observe any Raman peaks in the range of 200–3,500 cm^{-1} except of the cg-N peak at $\approx 830\,cm^{-1}$ and the vibron of a small intensity – a reminder of incomplete transformation of molecular nitrogen. Reaction with diamond anvils is also excluded for the following reason. We know that molecular nitrogen nearly completely transforms at the new phase (the molecular vibron disappeared). The thickness of the nitrogen sample in our experiments was $\sim 5\,\mu m$. This means that diamond anvils should be etched at $>1\,\mu m$ in depth to create a hypothetical carbonitride compound. We did not see any changes at the surface of the anvils after the experiments while submicroscopic changes can be easily seen with the interference microscope. Indeed, in rare cases occasionally we damaged the diamond surface with sharply focused laser radiation but these spots were visually easy to identify. In addition, Raman spectra from the damaged areas revealed peaks related to carbonitrides but not the cg-N peak. Finally, diamonds are warmed to less than 100° during laser heating, as we measured with a Pt thermometer (Pt foil touched to an anvil). Therefore, reaction of nitrogen with diamonds is unlikely.

2.3 Transformation from Molecular to Polymeric Nitrogen

In this section we will discuss details of transformation of molecular to polymeric nitrogen. First, we estimate a minimum pressure of the phase transformation. For that, we heated a sample of nitrogen at successively increasing pressures. At 20 GPa and 800 K and at 60 GPa and 2,500 K no changes in the Raman spectra were observed. At 82 GPa we saw no significant changes in the Raman spectra below 2,400 K. After heating at this temperature, a peak at 2,392 cm^{-1} from a new molecular nitrogen phase was observed below the main vibron line of molecular nitrogen at 2,410 cm^{-1}, indicating reduced bond order of the N–N bond and charge redistribution due to enhanced interaction between molecules [41]. At 95 GPa, this phase disappeared after heating to 1,500 K and only the previously known molecular phase was observed after heating up to 2,590 K. After increasing pressure to 110 GPa, the sample behaved similar to other experiments on synthesis of cg-N (as shown, for instance, in Fig. 2.3). Luminescence strongly increases at temperatures above 800 K (Fig. 2.8). This is apparently due to accumulation of defects in the crystalline lattice. After heating to higher temperature of 1,520 K luminescence decreases but nitrogen obviously remains in the molecular form – intensity of the vibron remains the same

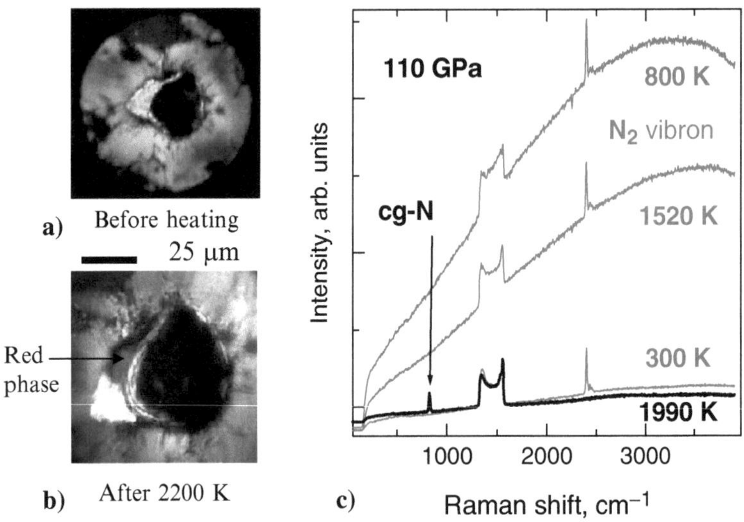

Fig. 2.8 Laser heating of molecular nitrogen at 110 GPa. **(a)** View of the sample through diamond anvils. Nitrogen is contained in a hole in rhenium gasket where a 1 μm thick boron plate black is placed as an absorber of heating radiation of YAG laser. Nitrogen is over the plate and seen as transparent at the left corner of the hole of the Re gasket which surrounds the sample. **(b)** After laser heating up to 1,990 K nitrogen transforms to cg-N which can be seen around the boron absorber as yellow or colorless substance. A transient red phase is at periphery of the heated area. Nitrogen remained in molecular form in the unheated corner of the sample at the left. **(c)** Raman spectra of nitrogen measured at room temperature after heating to the mentioned temperatures. In result of heating to 1,990 K vibron disappeared and a peak corresponding to cg-N phase appeared

and no other peaks appear. Drastic spontaneous transformation happens at heating to ~2,000 K: nitrogen transforms to a nonmolecular form (the vibron disappears) and a cg-N single-bonded nitrogen forms with a characteristic Raman peak at $820\,cm^{-1}$. During laser heating at the transformation, the temperature drops for several hundred degrees so that additional power is needed to maintain the temperature. Upon the transformation, pressure significantly drops on 5–20 GPa (depending on particular arrangement of the sample in the chamber) indicating a significant reduction of the volume of the sample. The phases of the transformation can be seen in Fig. 2.8. The cg-N looks colorless or yellow over the boron absorber and at its edge. A transient high luminescence amorphous phase is formed at the colder part of sample. It is red and is well separated from the cg-N phase. The colorless molecular nitrogen remains unchanged in the unheated part of the sample.

In many other experiments we observed that nitrogen remains in molecular form at pressures lower than 110 GPa, even being heated to 3,000 K. At higher pressures, the needed temperature is lower: for instance, at 140 GPa cg-N can be created at $T \sim 1{,}400$ K.

For further clarification of transformation of molecular to polymeric nitrogen the starting structure - knowledge of the crystal structure of molecular nitrogen is required. However, it is available only below approximately 50 GPa where the diatomic solid nitrogen exhibits a rich phase diagram [59, 60]. At higher pressures of ~60 GPa at room temperature, Raman and infrared absorption data indicate a transformation from the ε-N_2 rhombohedral (*R*3*c*) [61–63] to a ζ-N_2 phase [60, 64, 65]. This phase persists in the molecular state up to further transformation to the nonmolecular state at 150–180 GPa. The crystal structure of ζ-N_2 has not yet been firmly determined. It was proposed as *R*3*c* [65, 66], but later it was found that this structure is not consistent with Raman and IR data performed at low temperatures [60, 64]. A low-symmetry (orthorhombic or monoclinic) structure with two sites for atoms was proposed [64]. In x-ray studies of nitrogen up to 65 GPa at room temperature, Jephcoat et al. [67] observed a transition at ~60 GPa, but the signal was weak and the pressure was not high enough to separate the new phase and determine its structure. New molecular phases have also been synthesized at high pressures and temperatures (70–90 GPa and 600–1,000 K); although x-ray diffraction data are sufficient to identify them as new phases, the crystal structures have not yet been determined [48].

We performed extensive x-ray diffraction measurements of molecular nitrogen up to 170 GPa and tried to identify the structure of ζ-N_2 phase. A majority of the diffraction patterns of the molecular phases were collected at the Advanced Photon Source (APS, HPCAT at Sector 16), while the cg-N-phase and some patterns of molecular nitrogen were collected at the European Synchrotron Radiation Facility (ESRF, beamlines ID-9 and ID30). In all cases, we used an x-ray beam focused down to a spot of ~5 μm, and angle dispersive diffraction techniques [19].

Diffraction patterns of nitrogen in the 60–150 GPa pressure range are shown in Fig. 2.9. At 60 GPa, several diffraction peaks of the previous ε-phase widen. At 69 GPa, these peaks are split but new peaks do not appear, which suggests that the ε-N2 $\rightarrow$ ζ-N2 transformation is not accompanied by a large lattice distortion. Above

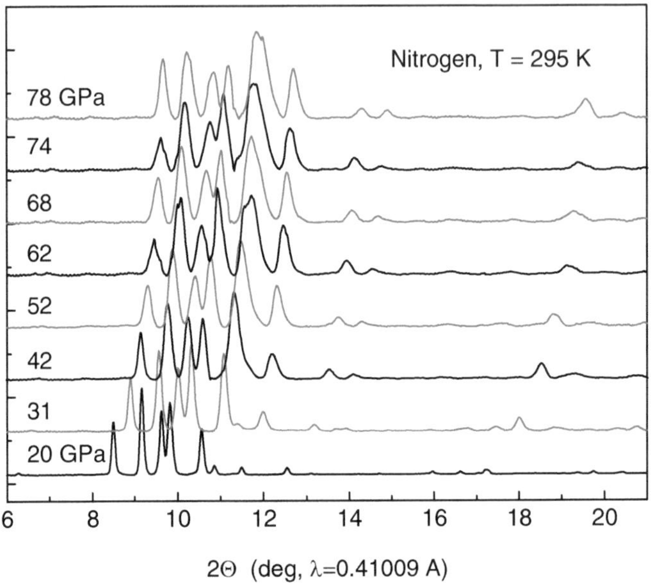

Fig. 2.9 X-ray diffraction spectra of molecular nitrogen taken at different pressures

80 GPa, diffraction patterns demonstrate only one phase which does not change up to 150 GPa, with the exception of the appearance of an amorphous halo above 100 GPa and some redistribution of the intensity. We analyzed the diffraction pattern of the ζ-phase obtained at 80 GPa. We tried the hexagonal indexing to verify the $R3c$ structure. The structure was found not to be a rhombohedral structure; only hexagonal indices fit the data, verifying that the ζ-phase does not have the $R3c$ structure. To identify this low-symmetry structure, the best indexing was found with an orthorhombic unit cell. Three space groups: $P222_1$, $P2_12_12$, and $P2_12_12_1$ satisfied the diffraction spectra. The $P222_1$ group has been selected because of its proximity to the space group $I2_13$ (cg-N structure). However, its comparison with the cg-N structure may not be valid because a gradual martensitic transformation between these structures is unlikely. Such transition requires very high pressures, which is where the equations of state (EOS) of ζ-N2 and cg-N would intersect (Fig. 2.10). Instead, the final transition to the cg-N structure is accompanied by a drastic change in volume, approximately 22% at 110 GPa (Fig. 2.10). Note, that at this pressure, the atomic volume of the molecular phase (ζ-N_2) approaches that for cg-N at zero pressure (Fig. 2.10). The $P222_1$ group structure is under debate [68] and further efforts on determination of apparently low-symmetry structure of ζ-N2 are needed.

Molecular–polymeric transformation in nitrogen remarkably resembles phase transformations in carbon and boron nitride in terms of compressibility and volume change (Fig. 2.10). There may also be similarities in the microscopic mechanisms of these transformations in nitrogen and carbon (and BN). Transitions between the carbon and BN phases are reconstructive diffusion-type phase transformations where

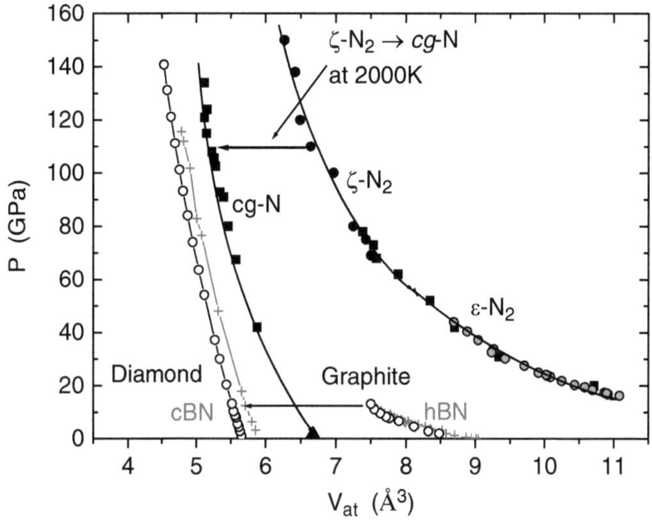

Fig. 2.10 Pressure–volume equation of state (EOS) of nitrogen. Our experimental data for ε-N_2 are shown with open circles. Grey circles are data from Ref. 61. At $P > 60$ GPa ε-phase transforms to ζ-phase which remains stable at pressures up to $P > 150$ GPa, where upon it transitions to a nonmolecular phase with an amorphous-like structure. η-N_2 can be directly transformed to the cubic gauche structure (cg-N) with laser heating above 2,000 K at $P > 110$ GPa (Ref. [20]). The EOS of this phase has been measured with increasing pressure up to 134 GPa, and then on releasing pressure down to 42 GPa where the sample escaped the cell. Extrapolation of this EOS to zero pressure gives a volume of cg-N structure about 6.6 Å [3] in excellent agreement with theoretical predictions of 6.67 Å [3] (Ref. [18]). The EOSs for carbon [71, 72] graphite and diamond) and BN [73, 74] are also presented to show the proximity of these covalent bonded materials with nitrogen

growth of a new phase can be accompanied by the fragmentation of the parent crystal and the creation of a disordered amorphous layer between crystallites, as has been reported for the hBN ↔ cBN transitions [69]. Similarly, it has been proposed recently that such transformations can occur through virtual melting along interfaces in the material [70]. This general picture of fragmentation of phases and creation of amorphous material is consistent with the disordered state of the nonmolecular nitrogen.

To check, if transformation from molecular to cg-N goes through an amorphous phase, or a transient phase [41] we performed for the first time x-ray diffraction in situ with laser heating. This experiment requires a combination of two complex techniques: x-ray diffraction measurements from light elements like nitrogen, and laser heating at megabar pressures. We performed measurements at 13 ID beam line at APS (Chicago). An intensive x-ray beam of 0.3344 Å wavelength was collimated to $\approx 5\,\mu$m spot. This beam produced a luminescence in the sample, therefore, the beam can be visualized with a sensitive CCD camera. This helped us to combine precisely the x-ray and laser-heating beams. The YAG laser beam was focused to a larger spot than the x-ray beam to diameter of $\approx 10\,\mu m$, and therefore diffraction has been collected from nearly uniformly heated area. Temperature was determined by recording the spectrum of thermal radiation and fitting it with the Planck radiation function.

An important experimental improvement is the already mentioned toroidal shape of diamond anvil culet (Fig. 2.7). This allowed us to increase thickness of the sample to ≈ 14–$15\,\mu m$ (with flat anvils it is ≈ 3–$5\,\mu m$) and thus dramatically increase x-ray scattering and decrease accumulation time of a nitrogen sample to seconds.

As squeezed to pressure about 140 GPa, the nitrogen sample becomes partly dark, and no Raman signal could be detected from the dark sample, indicating the transition to the amorphous nonmolecular phase in agreement with previous studies [14, 15, 47] However, the area in the center of the sample remained in molecular state as inferred by Raman spectra typical for molecular ζ-N_2 phase. The x-ray diffraction pattern obtained from this part of sample can also be attributed to the molecular ζ-phase while no diffraction peaks were observed in diffraction pattern from the dark region of the amorphous sample. As the sample was heated above 1,000 K, the intensity of diffraction lines of ζ-phase gradually decreases (Fig. 2.7) indicating the decrease of the amount of this phase. No new diffraction lines indicating the transformation to a new crystalline phase could be detected in the diffraction pattern. With temperature increase, the molecular ζ-phase transforms to the amorphous state. At $T \sim 1{,}450$ K the diffraction lines of ζ-phase disappears completely. This transformation to the amorphous phase is irreversible – the diffraction pattern of the sample quenched to the room temperature does not reveal reflections from ζ-phase. This amorphous phase is nonmolecular as inferred by the absence of vibron lines in the Raman spectra of the quenched sample. In the subsequent heating of this part of the sample containing the amorphous material the well-defined diffraction pattern of newly formed cg-N appeared as the sample was heated above 1,300 K. The amount of the cg-phase was growing with further temperature increase up to a highest temperature of about 2,000 K. No changes of diffraction pattern were observed as the sample was cooled down to room temperature. The same sequence of transformations (molecular→amorphous→cg-nitrogen [cgN]) was reproducibly observed at continuous increases in temperature of unheated parts of the sample with laser. In this manner nearly the whole sample was entirely transformed to the cg-phase revealing a very pronounced diffraction pattern (Fig. 2.7).

Our data evidence that at high pressure–high temperatures molecular nitrogen transforms to the atomic cg-N through an amorphous state. This transformation between solid molecular and cg-nitrogen occurs with continuous increase of temperature at the same pressure; therefore, this is a solid-state transformation without melting. The amorphous state is likely a polymeric network of single-bonded and double-bonded nitrogen atoms as follows from calculations [16, 17]. An amorphous state intermediate between phases is typical for diffusion-type phase transformations. It has been clearly observed in transformations with large change of volume to such as graphite-diamond ($\Delta V \approx 26\%$) [71–73] and hexagonal BN – cBN ($\Delta V \approx 24\%$) transformations [73, 74]. The molecular–cg-N transitions is also diffusion-type first-order transition between very different structures with a large drop of volume ($\Delta V \approx 22\%$).

2.3.1 Metastability of Polymeric Nitrogen and its Recovery

Another aspect of transitions with large change of lattices and volume is a large hysteresis of the phase transformation and metastability – once formed, a high-pressure phase can be recovered to ambient pressure (a well-known example is diamond). From an energetic point of view a reason of metastabilty is a large kinetic barrier separating high- and low-pressure phases. For carbon, the energetic barrier at zero pressure between graphite and diamond is $\Delta E \approx 0.33\,\text{eV atom}^{-1}$ (Fig. 2.11) and $0.38\,\text{eVpair}^{-1}$ for BN [73–75] while the energy difference between diamond and hexagonal graphite is nearly zero [73]. Therefore, diamond (which has slightly higher energy) separated by the large barrier from graphite is practically as stable as graphite.

In difference with cBN or diamond, nitrogen in polymeric state has much higher energy than in the ground state (molecular nitrogen): this difference is $\sim 1.2\text{–}1.3\,\text{eV atom}^{-1}$. However, to transform from cg-N to molecular nitrogen a barrier of $\sim 0.9\,\text{eV}$ [18] should be overcome (Fig. 2.11). The activation energy $E_A = 0.63\,\text{eV}$ is also characteristic for the amorphous polymeric nitrogen (which has a short-range with predominantly threefold coordinated atoms) as calculated in Ref. [16]. These large barriers suggest that the highly energetic cg-N and the amorphous polymeric nitrogen might be metastable even at zero pressure [18, 39]. More careful consideration includes surface. Unsaturated bonds which are intrinsic defects on the surface of cg-N crystal significantly decrease the barrier [16] down to $\sim 0.1\,\text{eV}$ [29]. Still this is a high barrier which implies that cg-N might be stable at ambient pressure at least at low temperatures. Note, that the broken bonds at the

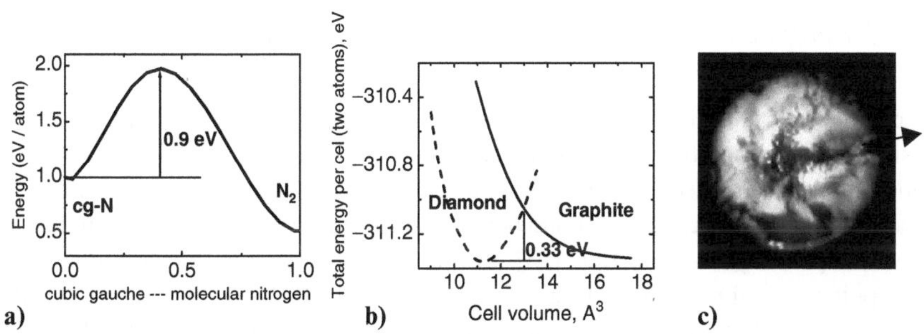

Fig. 2.11 Energy barriers separated a metastable phases from a ground state. **(a)** Calculated total energy per atom as a function of the reaction coordinate for cg $\rightarrow \beta$'-O_2 transformation, where β'-O_2 designates a distorted two-layer variant of the β-O_2 molecular structure. This is approximately zero pressure path, with the starting atomic volumes equal to the equilibrium values for the cg-N phase (6.67Å^3atom) and the final volume the equilibrium value for β-O_2 molecular (19.67Å^3atom^{-1}) The zero of energy corresponds to the diatomic β-O_2 structure (from Ref. [18]). **(b)** Graphite-diamond relation. Total energy per cell (two atoms), as functions of the cell volume for the hydrostatic pressure path. In **(a)** the solid curve is for the graphite and the dashed curve is for the diamond (from Ref. [73]). **(c)** A typical example of the rhenium gasket broken by the cg-N sample escaped from the cell at 70 GPa in direction shown by the arrow

surface which are the nuclei of instability can be saturated by hydrogen atoms [29]. This is similar as in the case of amorphous polymeric nitrogen [16].

The above estimation suggest possibility of recovering of polymeric nitrogen. However, this turned out to be a difficult experimental problem: typically the sample escapes from the diamond anvil cell at pressures of 70–100 GPa by rupturing the gasket. This is a typical occurrence, and obviously this is due to the inherent weakening of the gasket at decreasing pressure [56]: at releasing, load edges of diamond culet elastically recover and do not support the plastically deformed thin gasket anymore. A novel technique must be developed for repetitive recovery of the sample at ambient pressure. Nevertheless, after a number of efforts amorphous nonmolecular nitrogen once has been recovered to ambient pressure at temperatures below 100 K [14]. Cg-N has not been recovered yet – it also escapes at the same pressures at our numerous attempts with different configurations of the gasket and the anvils. At present pressure down to 42 GPa has been achieved occasionally due to a sudden drop of pressure from 115 GPa because of failure of one of the anvils. The sample of cg-N nevertheless survived but partly transformed to molecular nitrogen. At this pressure and 300 K it was apparently unstable – it had transformed to molecular nitrogen (cg-N Raman peak disappeared) under approximately 40 min of focused laser radiation of wavelength of 488 nm and power of $W = 10\,\text{mW}$. This behavior is similar to nonmolecular amorphous nitrogen which transformed back to molecular form at $P < 50\,\text{GPa}$ at 300 K [14]. This indicates that most likely both forms of nonmolecualr nitrogen are polymeric nitrogen in disordered or single-crystalline forms. This also means that cg-N will be likely recovered to ambient pressure similar to the amorphous nitrogen.

2.3.2 Perspectives of Synthesis of New Forms Polymeric Nitrogen

Starting from prediction of polymeric nitrogen [12, 18] theoretical calculations give insight to its properties and stimulated experiment. Recent calculations become more and more precise and can predict amazingly well structures of high-pressure phases. In our experience, we found nearly exact prediction of high-pressure phases in silane [76, 77], see also Ref. [23]. In this respect theoretical exploration of the new polymeric nitrogen structure [16, 23–31, 33, 37, 38, 42–44] is very promising and can guide direction of experimental search. These calculations agree that cg-N structure is the most stable. Besides, many other structures were found which are close in energy to cg-N, for instance, another three-dimensional polymeric structure related to cubic gauche (Fig. 2.7) and less stable by only $49\,\text{meV atom}^{-1}$ [23].

Systematic search of new metastable allotropes of nitrogen has been done by Zahariev et al. [28]. They considered Peierls-like distortions of a reference structure which was taken simple cubic structure (SC) which is to be the natural high-pressure reference structure for group 15 elements. They found many structures with total energy close to cg-N, which clearly has the lowest enthalpy. Two structures were discovered to be mechanically stable at zero pressure. Notably, all the previously considered simple single-bonded phases of polymeric nitrogen (BP, A7, and LB) with the exception of cg-N, were found to be mechanically unstable at all pressures considered.

There is a number of works, predicting new nonmolecular phases compatible to cg-N in the low-pressure region. C*haired web* (CW) structure was found which is thermodynamically more stable than the cubic gauche (cg-N) phase at the ambient pressure: the enthalpy favors CW over CG by approximately 20 meV [31]. The temperature-dependent free-energy and zero-point energy corrections favor cg-N over CW by 54 meV. With the rise in temperature there is a crossover of the two free-energy dependences around 200 K. At ambient temperature the new CW phase is thermodynamically more stable than the cg-N. Metastability of the CW phase was demonstrated by both phonon calculations, and first-principles molecular dynamics simulations. The CW phase is an insulator at low pressure with a calculated band gap of $\sim$5 eV.

Alemany and Martins [37] also predicted a phase which has lower enthalpy than cg-N at $P < 15$ GPa. It has a zigzag chainlike arrangement is thus partially polymeric and each atom has a coordination number of two. This phase is metastable at zero pressure and is metallic. Mattson et al. [25] found a very similar and also metallic chain structure but with the different packing which leads to a significantly lower energy ($\sim$0.18 eVatom^{-1}) [25] making it very close in energy (but slightly higher) to cg-N. A new structure with the zigzag chains in the new phase connected by two atoms with double bonds has been proposed in Ref. [44], but it has higher energy than other chain phases.

Stability of cg-N in the limit of high pressures was studied by Alemnay and Martins [37]. They found that at $P > 200$ GPa BP, an orthorhombic distortion of *sc* is more stable than CG.

Experimentally all these predicted nonmolecular phases were not yet confirmed.

Theoretical calculations gave further insight of unique properties of cg-N. This is a wide energy gap dielectric: the band gap is 4.13 eV at zero pressure and 3.89 eV at 240 GPa [27], and 5.2 eV at 150 GPa in Ref. [38]. The generalized gradient approximation (GGA) calculation of the band gap at 240 GPa is in good agreement with previous work [37] while Monte Carlo calculations [32] give band gap 8.1 eV. Chen et al. [42] found the unique constant-gap behavior ($E_g = 4.2$ eV under pressures 40 and 130 GPa, and it decreases at lower and higher pressures. The $E_g \sim 4$ eV at megabar pressures well correlates with our observations of the colorless transparent phase. Further detailed calculations of compressibility, lattice dynamics [24], dielectric constants $\varepsilon^{\infty} = 4.44$, $\varepsilon^{0} = 4.81$ at P_0 of cg-N and their pressure dependence have been done in Ref. [41].

Chen et al. [42] raised attention to an interesting possibility that cg-N can be also a superhard material. The distance between nitrogen atoms in the cg-N at ambient pressure is 1.466 Å [19] – significantly shorter than that for diamond (1.54 Å). The high strength of cg-N is also indicated by its high bulk modulus: 260–340 GPa [18, 19, 21, 27, 38, 44, 50]. On the other hand, strength of cg-N could be weaker in comparison with diamond because there are only three bonds for each atom for nitrogen instead of four bonds for diamond. Chen et al. [42] calculated the hardness of cg-N as $\approx$83 GPa which is less than diamond (100–130 GPa) but higher than other known superhard materials: (c – BC_2N (76 GPa) and cBN (63 GPa)).

2.4 Conclusions

It is shown here that pure nitrogen can exist in polymeric forms where atoms connect each other by single covalent bonds. At room or lower temperatures nitrogen transforms to an amorphous polymeric form at 150–200 GPa. Under laser heating at T $\sim$ 2,000 K and $P > 110$ GPa nitrogen has been synthesized in a single-bonded crystalline form. It has unusual cubic gauche structure (cg-N) with space group $I2_13$. In the cg-N structure all-nitrogen atoms are threefold coordinated and bond-lengths are the same for all pairs of bonded atoms. The experimentally determined structure is in perfect agreement with the theoretical prediction. Polymeric nitrogen can be metastable at ambient pressure as has been confirmed experimentally for the amorphous phase. The metastable polymeric nitrogen is predicted to be the most powerful high-density energy, because of a uniquely large difference in energy between single-bonded and triple-bonded atoms. In particular, it was calculated that polymeric nitrogen can produce detonation pressure ten times higher than that for HMX. Interestingly, cg-N is also predicted to be a superhard material second after diamond.

It is also shown here that high nitrogen materials can be prepared at lower pressure utilizing starting materials other than pure nitrogen. The challenge remains the recovery of such new high-energy density high-nitrogen or pure-nitrogen materials down at ambient pressures and temperatures.

References

1. Huheey, J. E., Keiter, E. A. & Keiter, R. L. *Inorganic Chemistry: Principles of Structure and Reactivity* (Harper Collins, New York, 1993).
2. Christe, K. O. Recent Advances in the Chemistry of N_5^+, N_5^- and High-Oxygen Compounds. *Propel, Explosiv. Pyrotech.* **32**, 194–204 (2007).
3. Samartzis, P. C. & Wodtke, A. M. All-nitrogen chemistry: how far are we from N_{60}? *Int. Rev. Phys. Chem.* **25** 527–552 (2006).
4. Rice, B. M., Byrd, E. F. C. & Mattson, W. D. Computational aspects of nitrogen-rich HEM. *High Ener Density Mater. Struct. Bond.* **125**, 153–194 (2007).
5. Barlett, R. J. Exploading the mysteries of nitrogen. *Chemi. Indus.* 140–143 (2000).
6. Cacase, F., Petris, G. d. & Troiani, A. Experimental detection of tetranitrogen. *Science* **295**, 480–481 (2002).
7. Curtius, T. *Berichte. Deutsch. Chem. Gesellschaft* **23**, 3023–3033 (1890).
8. Christe, K. O., Wilson, W. W., Sheehy, J. A. & Boatz, J. A. N_5^+: A novel homolepic polynitrogen ion as a high energy density material. *Angew. Chem. Int. Ed.* **38**, 2002–2009 (1999).
9. Klapotke, T. M. in *High Energy Density Materials* (ed. Klapotke, T. M.), pp. 85–121 (Springer, Heidelberg, 2007).
10. Knapp, C. & Passmore, J. On the Way to "Solid Nitrogen" at Normal Temperature and Pressure? Binary Azides of Heavier Group 15 and 16 elements. *Angew. Chem. Int. Ed.* **43**, 2–4 (2004).
11. Glukhovtsev, M. N., Jiao, H. & Schleyer, P. v. R. Besides N_2, what is the most stable molecule composed only of nitrogen atoms? *Inorg. Chem.* **35**, 7124–7133 (1996).
12. McMahan, A. K. & LeSar, R. Pressure dissociation of solid nitrogen under 1 Mbar. *Phys. Rev. Lett.* **54**, 1929–1932 (1985).

13. Martin, R. M. & Needs, R. Theoretical study of the molecular-to-nonmolecular transformation of nitrogen at high pressures. *Phys. Rev. B* **34**, 5082–5092 (1986).
14. Eremets, M. I., Hemley, R. J., Mao, H. K. & Gregoryanz, E. Semiconducting non-molecular nitrogen up to 240 GPa and its low pressure stability. *Nature* **411**, 170–174 (2001).
15. Goncharov, A. F., Gregoryanz, E., Mao, H. K., Liu, Z. & Hhemley, R. J. Optical evidence for nonmolecular phase of nitrogen above 150 GPa. *Phys. Rev. Lett.* **85**, 1262–65 (2000).
16. Nordlund, K., Krasheninnikov, A., Juslin, N., Nord, J. & Albe, K. Structure and stability of non-molecular nitrogen at ambient pressure. *Europ. Lett.* **65**, 400–406 (2004).
17. Mattson, W. D. PhD thesis (University of Illinois at Urbana-Champaign, 2003).
18. Mailhiot, C., Yang, L. H. & McMahan, A. K. Polymeric nitrogen. *Phys. Rev. B* **46**, 14419–14435 (1992).
19. Eremets, M. I., Gavriliuk, A. G., Trojan, I. A., Dzivenko, D. A. & Boehler, R. Single-bonded cubic form of nitrogen. *Nature Mater.* **3**, 558–563 (2004).
20. Gregoryanz, E. et al. High P-T transformations of nitrogen to 170 GPa. *J. Chem. Phys.* **126**, 184505 (2007).
21. Lipp, M. J. et al. Transformation of molecular nitrogen to nonmolecular phases at megabar pressures by direct laser heating. *Phys. Rev. B* **76**, 014113 (2007).
22. Trojan, I. A., Eremets, M. I., Medvedev, S. A., Gavriliuk, A. G. & Prakapenka, V. B. Transformation of molecular to polymeric nitrogen at high pressures and temperatures. In situ X-ray diffraction studies. *Appl. Phys. Lett.*, to be published (2008).
23. Oganov, A. R. & Glass, C. W. Crystal structure prediction using ab initio evolutionary techniques: Principles and applications. *J. Chem. Phys.* **124**, 244704 (2006).
24. Caracas, R. Raman spectra and lattice dynamics of cubic gauche nitrogen. *J. Chem. Phys.* **127**, 144510 (2007).
25. Mattson, W. D., Sanchez-Portal, D., Chiesa, S. & Martin, R. M. Prediction of new phases of nitrogen at high pressure from first-principles simulations. *Phys. Rev. Lett.* **93**, 125501–125504 (2004).
26. Zahariev, F., Hu, A., Hooper, J., Zhang, F. & Woo, T. Layered single-bonded nonmolecular phase of nitrogen from first-principles simulation. *Phys. Rev. B* **72**, 214108 (2005).
27. Yu, H. L. et al. First-principles calculations of the single-bonded cubic phase of nitrogen. *Phys. Rev. B* **73**, 012101 (2006).
28. Zahariev, F., Dudiy, S. V., Hooper, J., Zhang, F. & Woo, T. K. Systematic method to new phases of polymeric nitrogen under high-pressure. *Phys. Rev. Lett.* **97**, 155503–1555034 (2006).
29. Zhang, T., Zhang, S., Chen, Q. & Peng, L.-M. Metastability of single-bonded cubic-gauche structure of N under ambient pressure. *Phys. Rev. B* **73**, 094105–094107 (2006).
30. Uddin, J., Barone, V. N. & Scuceria, G. E. Energy storage capacity of polymeric nitrogen. *Molecul. Phys.* **104**, 745–749 (2006).
31. Zahariev, F., Hooper, J., Alavi, S., Zhang, F. & Woo, T. K. Low-pressure metastable phase of single-bonded polymeric nitrogen from a helical structure motif and first-principles calculations. *Phys. Rev. B* **75**, 140101 (2007).
32. Mitas, L. & Martin, R. M. Quantum Monte Carlo of nitrogen: atom, dimer, atomic, and molecular solids. *Phys. Rev. Lett.* **72**, 2438–2441 (1994).
33. Lewis, S. P. & Cohen, M. L. High-pressure atomic phases of solid nitrogen. *Phys. Rev. B* **46**, 11117–11120 (1992).
34. Eremets, M. I. et al. Polymerization of nitrogen in sodium azide. *J. Chem. Phys.* **120**, 10618–10618 (2004).
35. Peiris, S. M. & Russell, T. P. Photolysis of Compressed Sodium Azide (NaN_3) as a synthetic pathway to nitrogen materials. *J. Phys. Chem. A* **107**, 944–947 (2003).
36. Hemley, R. J. Effects of high pressure on molecules. *Annu. Rev. Phys. Chem.* **51**, 763–800 (2000).
37. Alemany, M. M. G. & Martins, J. L. Density-functional study of nonmolecular phases of nitrogen: metastable phase at low pressure. *Phys. Rev. B. 024110* **68**, 024110 (2003).
38. Zhao, J. First-principles study of atomic nitrogen solid with cubic gauche structure. *Phys. Lett. A* **360**, 645–648 (2007).

39. Barbee, T. W. & III. Metastability of atomic phases of nitrogen. *Phys. Rev. B* **48**, 9327–9330 (1993).
40. Yakub, E. S. Diatomic fluids at high pressures and temperatures: a non-empirical approach. *Physica B* **265**, 31–38 (1999).
41. Caracas, R. & Hemley, R. J. New structures of dense nitrogen: pathways to the polymeric phase. *Chem. Phys. Lett.* **442**, 65–70 (2007).
42. Chen, X. Q., Fu, C. L. & Podloucky, R. Superhard Dense Nitrogen. *Phys. Rev. B* **77**, 064103 (2008).
43. Ross, M. & Rogers, F. Polymerization, shock cooling, and the high-pressure phase diagram of nitrogen. *Phys. Rev. B* **74**, 024103 (2006).
44. Wang, X. L. et al. Prediction of a new layered phase of nitrogen from first-principles simulations. *J. Phys.: Condens. Matter.* **19**, 425226–425229 (2007).
45. Reichlin, R., Schiferl, D., Martin, S., Vanderborgh, C. & Mills, R. L. Optical studies of nitrogen to 130 GPa. *Phys. Rev. Lett.* **55**, 1464–1467 (1985).
46. Bell, P. M., Mao, H. K. & Hemley, R. J. Observations of solid H_2,D_2,N_2 at pressures around 1.5 megabar at 25°C. *Physica B* **139–140**, 16–20 (1986).
47. Gregoryanz, E., Goncharov, A. F., Hemley, R. J. & Mao, H. K. High-pressure amorphous nitrogen. *Phys. Rev. B* **64**, 052103 (2001).
48. Gregoryanz, E. et al. Raman, infrared, and x-ray evidence for new phases of nitrogen at high pressures and temperatures. *Phys. Rev. B* **66**, 224108–5 (2002).
49. Eremets, M. I., Struzhkin, V. V., Mao, H. K. & Hemley, R. J. Superconductivity in boron. *Science* **293**, 272–274 (2001).
50. Eremets, M. I., Gavriliuk, A. G., Trojan, I. A., Dzivenko, D. A. & Boehler, R. in *ESRF Highlights 2004* (ed. Admans, G.), pp. 37–38 (Imprimerie du Pont de Claix, Grenoble, 2005).
51. Sanz, D. N., Loubeyre, P. & Mezouar, M. Equation of state and pressure induced amorphization of β-boron from X-ray measurements up to 100 GPa. *Phys. Rev. Lett.* **89**, 245501 (2002).
52. Yoo, C. S., Akella, J., Cynn, H. & Nicol, M. Direct elementary reactions of boron and nitrogen at high pressures and temperatures. *Phys. Rev. B* **56**, 140–146 (1997).
53. Boehler, R., Bargen, N. v. & Chopelas, A. Melting, thermal expansion, and phase transitions of iron at high pressures. *J. Geophys. Res. B* **95**, 21731–21736 (1990).
54. Mao, H. K., Xu, J. & Bell, P. M. Calibration of the ruby pressure gauge to 800 kbar under quasihydrostatic conditions. *J. Geophys. Res.* **91**, 4673–4676 (1986).
55. Eremets, M. I. Megabar high-pressure cells for Raman measurements. *J. Raman Spectr.* **34**, 515–518 (2003).
56. Eremets, M. I. *High Pressures Experimental Methods* (Oxford University Press, Oxford, 1996).
57. Knittle, E., Wentzcovitch, R. M., Jeanloz, R. & Cohen, M. L. Experimental and theoretical equation of state of cubic boron nitride. *Nature* **337**, 349–352 (1989).
58. Eremets, M. I., Gavriliuk, A. G. & Trojan, I. A. Single-crystalline polymeric nitrogen. *Appl. Phys. Lett.* **90**, 171904 (2007).
59. Manzhelii, V. G. & Freiman, Y. A. (eds.) *Physics of Cryocrystals* (American Institute of Physics, College Park, MD, 1997).
60. Bini, R., Ulivi, L., Kreutz, J. & Jodl, H. J. High-pressure phases of solid nitrogen by Raman and infrared spectroscopy. *J. Chem. Phys.* **112**, 8522–8529 (2000).
61. Mills, R. L., Olinger, B. & Cromer, D. T. Structures and phase diagrams of N_2 and CO to 13 GPa by x-ray diffraction. *J. Chem. Phys.* **84**, 2837–2845 (1986).
62. Olijnyk, H. High pressure x-ray diffraction studies on solid N_2 up to 43.9 GPa. *J. Chem. Phys.* **93**, 8968–8972 (1990).
63. Hanfland, M., Lorenzen, M., Wassilew-Reul, C. & Zontone, F. Structures of molecular nitrogen at high pressure. *Rev. High Pressure Sci. Technol.* **7**, 787–789 (1998).
64. Goncharov, A. F., Gregoryanz, E., Mao, H.-K. & Hemley, R. J. Vibrational dynamics of solid molecular nitrogen to megabar pressures. *Low Temper. Phys.* **27**, 866–869 (2001).
65. Schiferl, D., Buchsbaum, S. & Mills, R. L. Phase transitions in nitrogen observed by Raman spectroscopy from 0.4 to 27.4 GPa at 15 K. *J. Phys. Chem.* **89**, 2324–2330 (1985).

66. LeSar, R. Improved electron-gas model calculations of solid N_2 to 10 GPa. *J. Chem. Phys.* **81**, 5104–5108 (1984).
67. Jephcoat, A. P., Hemley, R. J., Mao, H. K. & Cox, D. E. Pressure-induced structural transitions in solid nitrogen. *Bull. Am. Phys. Soc.* **33**, 522 (1988).
68. Gregoryanz, E. et al. On the epsilon-zeta transition of nitrogen. *J. Chem. Phys.* **124**, 116102 (2006).
69. M. I. Eremets et al. Disordered state in first-order phase transitions: Hexagonal-to-cubic and cubic-to-hexagonal transitions in boron nitride. *Phys. Rev. B* **57**, 5655–5660 (1998).
70. Levitas, V. I., Henson, B. F., Smilowitz, L. B. & Asay, B. W. Solid-solid phase transformation via virtual melting significantly below the melting temperature. *Phys. Rev. Lett.* **92**, 235702-1-4 (2004).
71. Hanfland, M., Beister, H. & Syassen, K. Graphite under pressure: equation of state and first-order Raman modes. *Phys. Rev. B* **39**, 12598–12603 (1989).
72. Occelli, F., Loubeyre, P. & LeToullec, R. Properties of diamond under hydrostatic pressures up to 140 GPa. *Nature Materials* **2**, 151–154 (2003).
73. Furthmueller, J., Hafner, J. & Kresse, G. Ab initio calculation of the structural and electronic properties of carbon and boron nitride using ultrasoft pseudopotentials. *Phys. Rev. B* **50**, 15606–15622 (1994).
74. Albe, K. Theoretical study of boron nitride modifications at hydrostatic pressures. *Phys. Rev. B* **55**, 6203–6210 (1997).
75. Fahy, S., Louie, S. G. & Cohen, M. L. Pseudopotential total-energy study of the transition from rhombohedral graphite to diamond. *Phys. Rev. B* **34**, 1191–1199 (1986).
76. Eremets, M. I., Trojan, I. A., Medvedev, S. A., Tse, J. S. & Yao, Y. Superconductivity in hydrogen dominant materials: silane. *Science* **319**, 1506–1509 (2008).
77. Pickard, C. J. & Needs, R. J. High-pressure phases of silane. *Phys. Rev. Lett.* **97**, 045504 (2006).

Chapter 3
Equations of State and High-Pressure Phases of Explosives

Suhithi M. Peiris and Jared C. Gump

3.1 Introduction

Energetic materials, being the collective name for explosives, propellants, pyrotechnics, and other flash-bang materials, span a wide range of composite chemical formulations. Most militarily used energetics are solids composed of particles of the pure energetic material held together by a binder. Commonly used binders include various oils, waxes, and polymers or plasticizers, and the composite is melt cast, cured, or pressed to achieve the necessary mechanical properties (gels, putties, sheets, solid blocks, etc.) of the final energetic material. Mining, demolition, and other industries use liquid energetics that are similarly composed of an actual energetic material or oxidizer together with a fuel, that is to be mixed and poured for detonation. Pure energetic materials that are commonly used are nitroglycerine, ammonium nitrate, ammonium or sodium perchlorate, trinitrotoluene (TNT), HMX, RDX, and TATB. All of them are molecular materials or molecular ions that when initiated or insulted undergoes rapid decomposition with excessive liberation of heat resulting in the formation of stable final products. When the final products are gases, and they are rapidly produced, the sudden pressure increase creates a shock wave. When decomposition is so rapid that the reaction moves through the explosive faster than the speed of sound in the unreacted explosive, the material is said to detonate. Typically, energetic materials that undergo detonation are known as high explosives (HEs) and energetic materials that burn rapidly or deflagrate are known as low explosives and/or propellants.

The singular character of energetic molecules is that they are very easily initiated, even just by friction caused when twisting of the container cap to open the container. That is, these materials are so sensitive to perturbation that their reaction or decomposition is easily started. This certainly implies that the molecule is inherently unstable or strained at ambient pressure and temperature, or that its heat of formation is very low in comparison to the formation of other phases or products from the same chemical composition. In the case of high explosives, positive values for the heat of

S.M. Peiris, G.J. Piermarini (eds.), *Static Compression of Energetic Materials*,
Shock Wave and High Pressure Phenomena,

formation is not unknown, perhaps explaining the multiple meta-stable crystalline structures or polymorphs discovered at ambient pressure and temperature. Consequently, the phase diagrams of HEs and their high pressure, high-temperature phase transitions are often very complex.

The decomposition reaction or detonation of such high explosives is also complex, being a combination of almost instantaneous chemical, mechanical, and physical changes. It is convenient to approximate or envision detonation in the following sequential steps. When an explosive is initiated (subjected to a stimulus such as a shock wave or an insult such as friction) energy is transferred to the material causing mechanical deformation and heating. The hot and compressed material proceeds to produce chemically excited species and initial bond breaking (considered reaction initiation) takes place. The chemistry of these materials is such that once reaction is initiated, sequences of chemical reactions follow at supersonic speed. This results in the supersonic liberation of chemical energy and production of gaseous reaction products. These product gases support or push a propagating shock wave resulting in pressure–volume work. The utility of a weapon rises from this pressure–volume work known as the continuum response of the explosive, and it is highly advantageous to design this response to address the specific effect desired from a weapon.

The final effect or result of a detonation, being based on such complex phenomena as described in the previous paragraph, is not easy to predict. Yet the design of efficient mission-specific weapons and the limitations of conducting large-scale tests makes the prediction, simulation, and modeling of explosives essential. Such simulations and models in order to make accurate predictions must incorporate sufficient understanding of the complex detonation process. Realistic estimations of high-pressure, high-temperature thermodynamic properties, chemical kinetics, and reaction mechanism are necessary to accurately describe an explosive's response. Unfortunately, the rapidly changing high-pressure, high-temperature, thermodynamic properties and chemical reactions result in non-equilibrium temperature, pressure, and volume conditions. Therefore, well-defined experiments with accurate measurements of thermodynamic parameters and chemical concentrations are extremely challenging. Extrapolations made from ambient-pressure, ambient-temperature data are unreliable and lead to large uncertainties in models of explosive phenomenon. Subsequently, over the past 2 decades, experimental methods that approach the shock pressure and temperature of detonating explosives while still allowing time (static conditions) for detailed study, have been devised. Such experimental methods and data are described in Chapters 3–5, and 6 of this book.

This particular chapter, Chapter 3, focuses on the equations of state (EOS) and phase stability of unreacted solid explosives at static high pressure and temperature. Experimental data described herein obtained mostly by using x-ray diffraction methods, highlight the isothermal compressibility of explosives and elaborate on the lattice structure of the material. Experiments that investigate phase stability using vibrational spectroscopies such at Raman and IR are also described. The dependence of both the EOS and the phase stability on pressure-transmitting media and quasi-hydrostatic to non-hydrostatic compression are discussed. The various meta-stable structures of energetics and the phase transitions that arise from heating or compression are presented together with multiple phase transitions to phases that could also be meta-stable. Further, the effect of the rate of pressure or temperature increase on phase transition pressure and temperature, is also included in the later part of this chapter.

3.2 Equations of State

3.2.1 Background

An "equation of state" relates the thermodynamic states of matter such as pressure (P), volume (V), and temperature (T). In static-pressure studies, the temperature is often held constant and isothermal EOS are obtained.

There are several approaches to deriving EOS. Most approaches use the classical thermodynamic definitions of pressure and temperature.

$$\mathrm{P} = -\left(\frac{\partial U}{\partial V}\right)_S \text{ and } \mathrm{T} = \left(\frac{\partial U}{\partial S}\right)_v$$

Where U is the total internal energy or equilibrium potential energy. The interpretation or derivation of U is usually based on the material whose EOS is to be described. Early theoretical work was based on perfect gases that followed ideal gas relationships. Subsequent attempts are more selectively based on potential energy of finite hard spheres, of harmonic vibrations. The more sophisticated quantum-mechanical potentials account for attractive and repulsive forces, non-spherical distribution of electrons and even hydrogen bonding between molecules. These methods and potentials are well described in Chapters 7 and 8 of this book.

Limiting EOS to isothermal conditions where only mechanical work is performed allows another classical expression for pressure, using the Helmholtz free energy, F:

$$\mathrm{P} = -\left(\frac{\partial F}{\partial V}\right)_T$$

This definition is found useful because when considering elasticity and compression of solids, mechanical work can be expressed in terms of deformation or strain under force. Such an expression for finite, elastic, and isothermal strain was first derived by Murnaghan, and later expanded by Birch [1]. This derivation assumes that there is always a "state of zero strain" for any temperature at which the solid exists. This assumption limits the application of this EOS to "unstrained" crystals as opposed to strained or meta-stable crystalline phases, as, interestingly, most of the HEs might be. However, at least these EOS are not dependent on the many assumptions about electronic interactions, anharmonicities, and binding energies, required for first-principle quantum-mechanical EOS.

In the Birch-Murnaghan (BM) isothermal finite-strain derivations, F is assumed to depend only on the strain components in different directions of an elastic crystalline system. (That is, the Helmholtz free energy is considered the Strain energy, and thermal expansion is not considered as strain.) When strains are small (in the order of 10%) strain components can be defined by a Lagrangian scheme based on initial unstrained coordinates. However, an Eulerian scheme, where the coordinates of the strained state are thought to be independent of the initial unstrained state, was found to be more useful for higher compression [1]. The derivation defines negative Eulerian strain, or positive compression, f as:

$$f = \frac{1}{2}\left\{\left(\frac{V}{V_0}\right)^{-2/3} - 1\right\}$$

where V is the volume at some pressure and V_0 is the volume at zero pressure. Then strain energy (or Helmholtz free energy, F) resulting from hydrostatic compression is represented as a Taylor series in f:

$$F = \frac{9}{2}V_0K_0\left(f^2 + \frac{2}{3}af^3 + bf^4 + \ldots\right)$$

Differentiating gives the Pressure (P):

$$P = 3K_0(1+2f)^{5/2}(f + af^2 + bf^3 + cf^4 + \ldots)$$

where K_0 is the zero-pressure isothermal bulk modulus (the inverse of compressibility). Evaluating this equation with then known compressibility data, the values of a, b, and c, etc. were limited [1,2] to give:

$$\begin{aligned} P &= 3K_0(1+2f)^{5/2} \\ &\times\left\{f + \frac{3}{2}(K_0' - 4)f^2 + \frac{3}{2}\left[K_0K_0'' + K_0'(K_0' - 7) + \frac{143}{9}\right]f^3 - \ldots\right\} \end{aligned}$$

where K_0', K_0'' are the first and second pressure derivatives of the zero-pressure isothermal bulk modulus [2].

This BM EOS formalism has been widely used for understanding the behavior of minerals and materials in the interior of the earth. Experimentally obtained P–V data are "fitted" to the formalism to yield the bulk modulus and its derivatives. These parameters are then used to extrapolate the EOS to higher pressures. As many orders of the series as necessary are used to fit P–V data. Using up to the third-order (equation shown below) is common practice. The application of a value of 4 to K_0' in the third-order equation, results in essentially the second-order equation.

$$P = \frac{3}{2}K_0\left[\left(\frac{V}{V_0}\right)^{-\frac{7}{3}} - \left(\frac{V}{V_0}\right)^{-\frac{5}{3}}\right]\left\{1 + \frac{3}{4}\left[K_0' - 4\right]\left[\left(\frac{V}{V_0}\right)^{-\frac{2}{3}} - 1\right]\right\}$$

Several other derivations of EOS are well known, including one named the "universal equation of state" derived by Vinet et al. [3]. Here internal energy is expressed in terms of the Wigner-Seitz radius scaled by a scaling factor, to yield:

$$P = \frac{3K_0(1-X)}{X^2} e^{\left\{\frac{3}{2}(K_0'-1)(1-X)\right\}} \text{ where } X = \left(\frac{V}{V_0}\right)^{\frac{1}{3}} \left(\frac{V}{V_0}\right)^{\frac{1}{3}}$$

Jeanloz has shown that this EOS is an algebraic approximation of the Birch-Murnaghan EOS within the range $3 < K_0' < 6$ [4]. Since most of the isothermal EOS data on high explosives use the Birch-Murnaghan EOS, and since it can be shown to be an approximation to other EOS formalisms as well, this chapter will focus on the use of the Birch-Murnaghan EOS.

3.2.2 Method

Typical experiments consist of loading diamond anvil cells with the sample and a pressure-transmitting medium. Samples are loaded into holes drilled into metal gaskets that are a couple hundred microns thick. The holes are in the order of tens to hundreds of microns in diameter, so the total amount of sample necessary for each experiment is very small. Samples of explosives, being commercially produced for military use, are usually not 99.99% pure like other high-purity chemicals bought from chemical manufacturers. Typical military standard production of RDX (hexahydro-1,3,5-trinitrotriazine) could include up to 7% HMX (octahydro-l,3,5,7-tetranitrotetrazocine) [5]. Similarly HMX samples could include RDX. With materials like CL-20 (HNIW, hexanitrohexaazaisowurtzitane, 2,4,6,8,10,12-hexanitro 2,4,6,8,10,12-hexaaza tetracyclo [5,5,0,$0^{5,9}$,$0^{3,11}$]dodecane), which has multiple solid-state phases (or crystal structures) that are meta-stable at ambient P–T, samples may consist of mixed phases. More recently, less-sensitive forms of these materials have also been prepared, and structural differences in the conventional material versus the insensitive material are still under investigation [6]. Therefore, it is now necessary to describe the origin and purity of the particular samples used to obtain experimental data when publishing equations of state or similar results.

To utilize powder x-ray diffraction methods, the samples must be finely ground. Grinding these friction- and electrostatic-sensitive materials can be dangerous. Grinding should be limited to very small sample quantities such a 0.1 mg, and precautions, such as static-discharge grounding or wetting down of the working surface, should be taken. Depending on the pressure-measuring technique employed, ruby dust, or other material is often loaded with the sample [7, 8]. Adding pressure-transmitting media to the gasket hole to create hydrostatic or at least quasi-hydrostatic conditions may be complicated by the typical liquid organics used (such as ethanol and/or methanol) reacting with the organic sample material. Therefore, either non-reacting noble gas pressure media should be used, or inorganics such as FX-75 or Dow-Corning 200 fluid can be employed [9].

Explosives, being sensitive to various stimuli, are decomposed or reacted when subjected to certain wavelengths of radiation. Energy-dispersive x-ray (EDX) diffraction using white radiation most often causes sample decomposition. In addition, crystal structures of explosives are of low symmetry and the diffraction resolution obtained from EDX in 2θ or d-space is insufficient to yield good structural fits. Therefore, single-wavelength angle-dispersive diffraction becomes the method of choice. Diamond anvil cell samples being extremely small, synchrotron radiation is necessary for obtaining diffraction patterns in minutes instead of hours.

Usually, samples are loaded, pressed to a known or measured pressure (P), and then exposed to x-rays to obtain diffraction patterns. Often, commercial software packages are used to analyze diffraction patterns and assign miller indices. Such software also calculates unit cell parameters and unit cell volumes (V). It is important to retain the errors associated with the unit cell parameters to propagate errors in the unit cell volumes obtained. These volume errors together with errors associated with the measurement of pressure should be used in the equation of state fits. When utilizing the Birch-Murnaghan EOS formalism, the fit should be weighted with the reciprocal of the error in each pressure measurement and the thermodynamic parameters from the "best" weighted fit should be published. Unfortunately, because of the percentages of error and scatter in the P–V points, not weighting the fit and not employing the best fit can result in various fitted lines that all seem to be "good" fits but yield widely disparate EOS.

In our studies, high temperatures were achieved by using a hydrothermal diamond anvil cell (HDAC), developed by Bassett et al. [10]. Temperature was monitored by thermocouples placed near the culet tip of each diamond. To determine pressure for our high-temperature measurements, a ruby fluorescence measurement was taken after the addition of pressure-transmitting fluid while the cell was still at ambient temperature, to determine initial sample pressure [7, 8]. The sample was then heated to the target temperature. A new ruby fluorescence measurement was taken to determine the position of the ruby peaks at the target temperature. This target-temperature ruby peak position was then used to calculate peak shifts as the sample was compressed at the target temperature. Final pressure was obtained by adding the pressure increase at the target temperature to the initial pressure created by the addition of the pressure medium. This method ensures that pressure and temperature dependence of the ruby lines are treated independently, and the pressures reported here do not include thermal pressure created when heating.

3.2.3 Data

The earliest EOS from ambient-temperature static compression of explosives was obtained by the group of Bart Olinger and Howard Cady at Los Alamos National Lab. Their method for evaluating the bulk modulus and its pressure derivative from their static-pressure P–V data involved using the Hugoniot relations to obtain pseudo-particle (u_p) and pseudo-shock (u_s) velocities.

$$u_s = \sqrt{\frac{PV_0}{\left(1 - \frac{V}{V_0}\right)}}$$

$$u_p = \sqrt{PV_0\left(1 - \frac{V}{V_0}\right)}$$

Then, fitting the u_p, u_s data to the linear relationship $u_s = c_T + s_T u_p$ where c_T is the bulk isothermal sound speed which relates to the bulk modulus and s_T relates to the pressure derivative of the bulk modulus.

$$c_T = \sqrt{K_0 V_0}$$

$$s_T = \frac{(K_0' + 1)}{4}$$

Their initial publication of such data for the two most popular explosives RDX and HMX appears in the Proceedings of a Commissariat a l'Energie Atomique conference in Paris, France [11]. During their compression of α RDX (the polymorph stable at ambient pressure and temperature) they observed a phase transition at about 4 GPa to a phase with similar orthorhombic symmetry. Therefore, they report the bulk modulus of α RDX obtained from data up to 4.4 GPa, to be 13 GPa with a pressure derivative of 6.6. Subsequently, Yoo and Cynn also compressed α RDX [12]. They obtained a bulk modulus of 13.9 GPa with a pressure derivative of 5.8 GPa. They also obtained unit cell volumes for the higher-pressure phase (γ) up to 12 GPa, by assuming an orthorhombic symmetry for this phase.

Figure 3.1 shows both Yoo and Cynn's data and Olinger et al.'s data. Olinger's data is within experimental error of Yoo and Cynn's data. There are three solid lines in the plot, which are; third-order fit of the Birch-Murnaghan EOS to the data from the α-phase, the data from the γ-phase, and the combined data obtained by Yoo and Cynn. In this case, when assuming that the γ-phase is of orthorhombic symmetry, the transition volume change is found to be 1.5–1.6% [11, 12]. Assuming this volume change to be negligible, a single EOS can be fitted as shown in Fig. 3.1. Yoo and Cynn report a bulk modulus of 13.0 GPa with a pressure derivative of 6.3 for the data from both phases.

Olinger et al.'s original work also included the compression of β HMX, the polymorph of HMX stable at ambient pressure and temperature. Using the Hugoniot relationships, they obtained an isothermal bulk modulus of 13.5 GPa with a pressure derivative of 9.4 for the β-phase of HMX, compressed in a pressure medium of 4:1 methanol:ethanol solution. Yoo and Cynn also published an ambient-temperature EOS for β HMX, compressed in Ar, which is such a soft solid in this pressure region that it retains near hydrostatic conditions [12, 13]. They used the third-order Birch-Murnaghan EOS formalism to fit their data. For hydrostatic compression below 27 GPa, they obtained a isothermal bulk modulus of 12.4 GPa with a pressure derivative of 10.4.

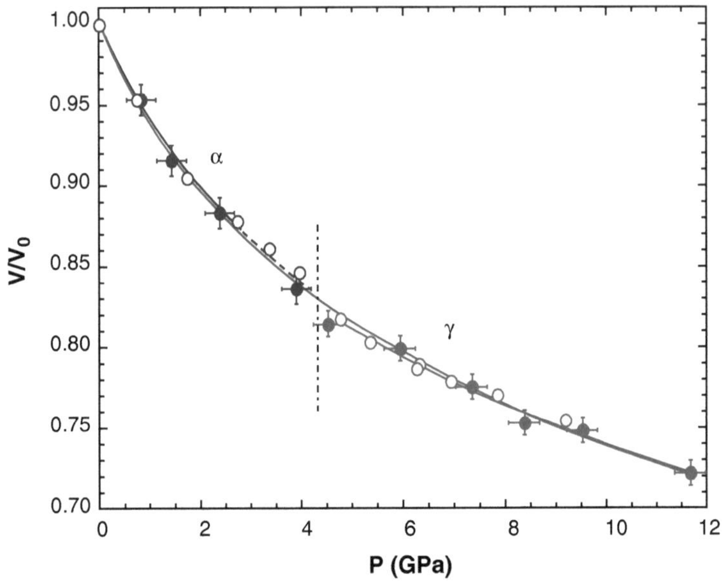

Fig. 3.1 Isotherms of RDX to 12 GPa from Ref. [12]. The solid circles are Yoo and Cynn's data and the open circles are Olinger et al.'s data. The vertical dashed line at about 4 GPa indicates the phase transition from α to γ. The solid lines represent fitted third-order Birch-Murnaghan EOS

These two sets of data from Olinger et al. and Yoo and Cynn were analyzed by Menikoff and Sewell [14]. They fit the data of Olinger et al. with a third-order Birch-Murnaghan (BM) EOS to compare it to Yoo and Cynn's data using the same fitting technique. This fit is shown in Fig. 3.2 as a dotted line. They obtained a bulk modulus of 10.6 ± 1.7 GPa with a pressure derivative of 18.1 ± 13.4. Further, they used Yoo and Cynn'a data to just 12 GPa, in an attempt to compare fits from the same data domain. That fit yielded a bulk modulus of 16.0 ± 2.5 with a pressure derivative of 7.3 ± 1.4. They concluded that using a BM EOS to fit such P–V data requires more data points at lower pressures [14]. We obtained more than 30 — P–V data points up to 5.5 GPa using hexane as a pressure medium [15]. Our bulk modulus, obtained from fitting a third-order BM EOS, is considerably higher, at 21.0 ± 1.0 GPa with a pressure derivative of 7.5 ± 0.9.

Considering these three sets of data on HMX, indications are that it is not necessarily the fitting form that gives such widely disparate results for HMX. It is possibly the HMX itself, in that, the compression of HMX is hugely influenced by non-hydrostatic pressure. Our pressure medium of hexane freezes at 1 GPa [16], thereby introducing non-hydrostatic conditions and shear stresses in the samples from that pressure. Olinger's pressure medium also stiffens sufficiently above 4 GPa to demonstrate stiffer HMX compression above 4 GPa. Whereas both the stiffness and shear strength of the Ar pressure medium used by Yoo and Cynn remains much lower than that of HMX, thereby providing the most hydrostatic compression conditions of these three studies. The bulk moduli obtained from the less hydrostatically

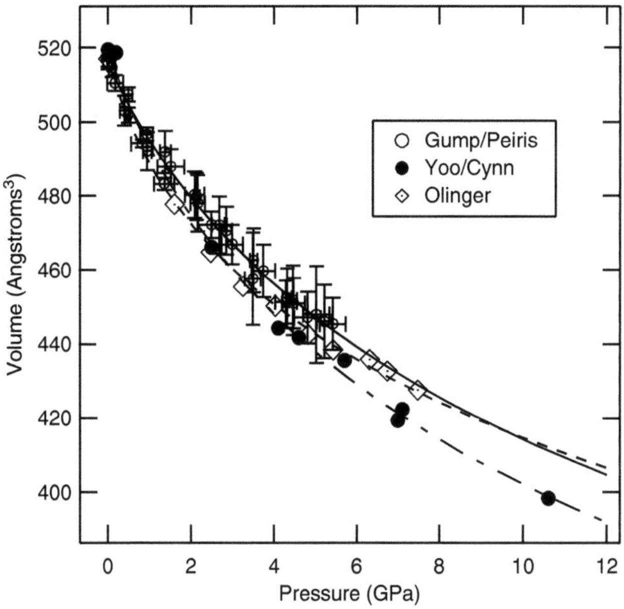

Fig. 3.2 The three sets of data on β HMX described in the legend are from Refs. [11, 13, 15], respectively. The solid curve is a third-order BM EOS fit to our data [15], the dotted curve is a third-order BM EOS fit to Olinger et al.'s data [14], and th e dashed curve is a third-order BM EOS fit to Yoo and Cynn's data to 27 GPa [13]

compressed studies indicate that the non-hydrostatic pressure felt by HMX stiffens its lattice. Yoo and Cynn speculate that this trend, where increased hydrostaticity results is higher bulk moduli, is due to chemical reactions occurring in the absence of hydrostatic conditions [13]. If chemical reactions were occurring in our samples at such pressures as 5 or 6 GPa, they are reversible because our samples decompressed from 5 or 6 GPa do return to the β structure [15]. Clearly, this behavior where non-hydrostatic stress stiffens the β HMX lattice bares further investigation.

The ambient-pressure volumes (V_0) we obtained from samples at room temperature, 100°C, and 140°C were plotted as a function of temperature [15]. Volume thermal expansion was calculated assuming linearity in the V, T relationship. The slope of the V vs. T plot was determined by a linear least-squares fit. Thermal expansion (α) was then calculated by dividing the slope by the ambient-pressure volumes because $\alpha = (1/V)(V/T)_P$. The average volume thermal expansion obtained from our data in this temperature range at ambient pressure is: $0.00027\,K^{-1}$ [15]. An older publication by Hermann et al. [17] reports a volume thermal expansion of $0.00013\,K^{-1}$ and a newer proceedings article by Saw [18] shows the same parameter to be $0.00020\,K^{-1}$. It is possible that the thermal expansion of HMX is very sensitive to the amount of impurities within the lattice (such as RDX inclusions) explaining this wide range of figures obtained for thermal expansion of β HMX. However, neither Refs. [17] or [18] discuss the purity of their samples.

Table 3.1 Isothermal EOS parameters at ambient temperature

Material	Bulk modulus (GPa)	Pressure derivative of bulk modulus	Reference
α RDX in Meth:Eth	13	6.6	[11]
α RDX in Ar	13.9	5.8	[12]
β HMX in Meth:Eth	13.5	9.4	[11]
β HMX in Ar	12.4	10.4	[12, 13]
β HMX in Hexane	21.0 ± 1.0	7.5 ± 0.9	[15]
β HMX in Hexane at 100°C	14.1 ± 0.82	11.6 ± 1.41	[15]
β HMX in Hexane at 140°C	13.5 ± 0.56	9.0 ± 0.85	[15]
TATB in Meth:Eth	16.2 ± 2.0	5.9 ± 1.4	[19]
TATB in Meth:Eth (to 8 GPa)	13.4 ± 0.8	12.8 ± 1.1	[20]
TATB in Meth:Eth (to 14 GPa)	16.9 ± 1.0	8.2 ± 0.7	[20]
PETN in Meth:Eth	8.70	9.9	[19]
PETN in Ar	12.3	8.2	[12]
AP in Meth:Eth	20.3 ± 0.5	4	[25]
AP in FC-75 or NaCl	12.7 ± 0.7	11.0 ± 1.6	[26]
εL-20 in Dow-Corning 200 fluid	13.6 ± 2.0	11.7 ± 3.2	[30]
εL-20 in Dow-Corning 200 fluid at 75°C	11.0 ± 1.3	14.0 ± 2.7	[30]
γ CL-20 (to 0.9 GPa with no pressure medium)	18.9 ± 0.5	4	[30]

Isothermal equations of state at ambient temperature have also been obtained for other energetic materials, as listed in Table 3.1. Another explosive, that is also the most insensitive of those in use today, is triaminotrinitrobenzene (TATB). The isothermal compression of TATB to 7 GPa was also initially reported by Olinger and Cady [19]. They used the methanol:ethanol solution as a pressure medium, and measured pressure using NaF included in the sample and its EOS. Very recently, another group, headed by Dattlebaum, has done x-ray diffraction analysis of the TATB structure under high pressure [20]. They also used a 4:1 ethanol:methanol mixture as a pressure medium and obtained data to 14 GPa. Both data sets are shown in Fig. 3.3.

The variation in Dattlebaum's data in comparison to Olinger and Cady's data above 2 GPa, may be attributed to the analysis of the x-ray patterns used to calculate unit cell volumes. TATB crystallizes in a unit cell with triclinic unit symmetry. Therefore, a minimum of six peaks from a cross section of different *h*, *k*, and *l* lattice plane configurations are necessary to obtain a good unit cell volume. Unfortunately, Olinger and Cady could only see four peaks above 3 GPa pressure, and assumed that the *a*, *b* axis compression is uniform to reduce the parameters necessary to obtain a unit cell volume from a triclinic cell. Dattlebaum's data shows this assumption to be too lenient. Though Dattlebaum too had one problem, in that they could not observe *l* reflections together with all the other *hk* planes from the same samples. Therefore, they used peak positions from different samples (set to the same pressure) to obtain triclinic unit cell volumes. These volumes are experimentally more reliable, because no assumptions about the compression of the unit cell was made.

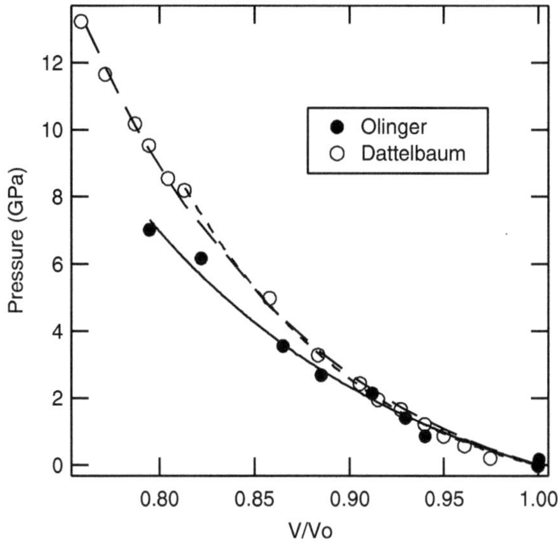

Fig. 3.3 TATB compression with third-order BM EOS fits (with no reciprocal-of-error weighting) to data from Refs. [19] and [20]. The solid line includes all the data in Ref. [19], the short-dashed line includes data to 8 GPa in Ref. [20], and the long-dashed line includes all the data in Ref. [20]

Fitting a third-order BM EOS to data from Ref. [19], results in a bulk modulus of 16.2 ± 2.0 GPa and a pressure derivative of 5.9 ± 1.4. This fit is shown in Fig. 3.3. Dattelbaum's data from Ref. [20] is fitted from ambient pressure both up to 8 GPa and up to 14 GPa. Previous experiments have reported a cusp in the EOS at 8 GPa [20]. However, the data included here does not show an obvious volume change considering the possible error bars for each of the P and V values. To 8 GPa, the third-order BM EOS fit gives a bulk modulus of 13.4 ± 0.8 GPa with a pressure derivative of 12.8 ± 1.1. To 14 GPa, the third-order BM EOS fit gives a bulk modulus of 16.9 ± 1.0 GPa with a pressure derivative of 8.2 ± 0.7. Interestingly, the later two fits with such widely disparate EOS parameters also highlight the vagaries of fitting P–V data to the BM EOS formalism. This problem is less pronounced if each data point is weighted by the reciprocal value of the error in P, as further discussed in the Discussion section of this chapter.

PETN (pentaerythritol tetranitrate) a less utilized explosive, was also initially studied under static pressure by the group of Olinger, et al. [21]. They later published a correction to their pressure measurements which depended on the compression of NaF, as available at the time [19]. Yoo and Cynn's report in 1998 also included the isothermal compression of PETN in Ar at room temperature to 15 GPa [12]. They compared their data to the EOS obtained from shock compression of PETN single crystals, as shown in Fig. 3.4. Their third-order BM EOS fit resulted in a bulk modulus of 12.3 GPa with a pressure derivative of 8.2. Yoo and Cynn used an orthorhombic space group $P2_12_1$ to analyze their diffraction data as published in 1947 [22], while more accurate determinations claim a tetragonal space group

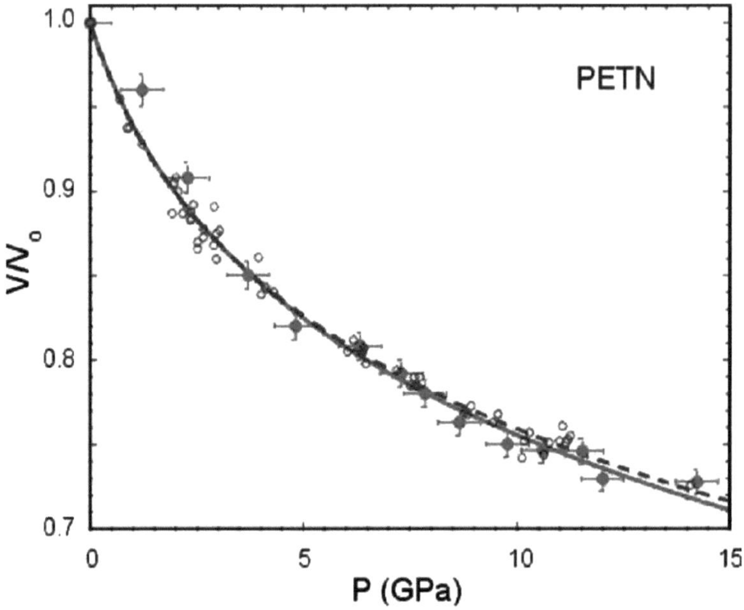

Fig. 3.4 Static compression data (solid circles) together with shock compression data (open circles) of PETN from [12]. The solid line is a third-order BM EOS fit to the static data and the dashed line from the Shock Hugoniot of PETN

$P\bar{4}2_1c$ [23]. More recent experiments confirm the tetragonal symmetry stable to 6 GPa, interestingly, converting to an orthorhombic symmetry (similar to the 1947 structural analysis) above that pressure [24–26].

Ammonium perchlorate (AP) is also widely used in explosive formulations, particularly as an oxidizer. There are many, sometimes contradictory, reports of phase transitions of AP at various pressures. The first high-pressure study on AP was reported by Bridgeman who observed a cusp-like maximum at 3.1 GPa, attributed to a phase transition at that pressure [27]. The change in volume due to this transition was reported to be too small to quantify. Later, ambient-temperature x-ray diffraction measurements with isothermal compression in 4:1 ethanol:methanol mixtures were recorded up to 5.0 GPa [28]. They report the ambient-temperature transition to occur above 3.57 GPa with the new phase seen at 4.70 GPa. Our study of the compression of AP using x-ray diffraction, Raman and IR spectroscopy, illuminated subtle changes different from those reported till then [29]. Compressing AP in different pressure media such as NaCl and FC-75, we saw new diffraction peaks starting from 0.9 GPa and above, with a complete phase change occurring at around 3.0 GPa. Fitting the orthorhombic ambient-pressure unit cell to diffraction patterns up to 3 GPa, results in a third-order BM EOS with a K_0 of 12.7 ± 0.7 GPa and K_0' of 11.0 ± 1.6. The data and fits are shown in Fig. 3.5. Fitting the six data points below 4.7 GPa from Ref. [28] to a second-order BM EOS gives a bulk modulus of 20.3 ± 0.5 GPa with the pressure derivative held constant at 4.0.

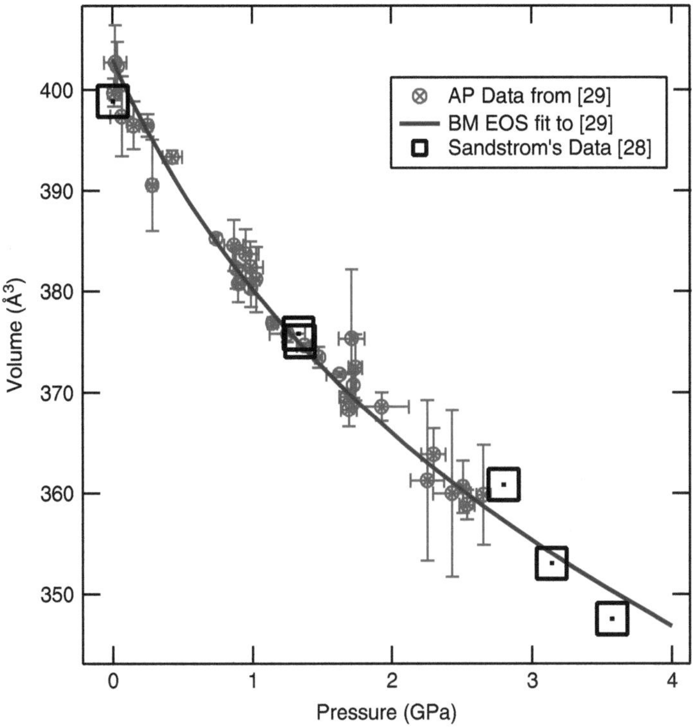

Fig. 3.5 Static compression of AP from [29] with the third-order BM EOS fitted to 3 GPa because of a phase transition observed above that pressure. Also included are data from [28] where the phase transition was observed above 3.57 GPa

Hexanitrohexaazaisowurtzitane (HNIW or CL-20), a relatively recently synthesized material that is not commonly used (mostly for economic reasons), has four known polymorphs that have been found to be stable at ambient conditions. These polymorphs are labeled α, β, γ, and ε, and of these, the ε- and γ-phases have been extensively studied at static high pressure from the early 1990s. However, no isothermal compression data were published until 2006, when we reported the compression of ε CL-20 in Dow-Corning 200 fluid (0.01 $cm^2\,s^{-1}$ kinematic viscosity) to 5.6 GPa, both at ambient temperature and 75°C [30]. The third-order BM EOS fit to the data yields an ambient-temperature isothermal bulk modulus of 13.6 GPa $\pm$ 2.0 with a pressure derivative of 11.7 ± 3.2 for the ε-polymorph. Compressing the γ-polymorph at ambient temperature results in a phase transition at about 0.7 GPa [31]. Since the γ-polymorph only existed for such a small pressure range samples were loaded without the pressure medium, to obtain P–V data of the γ-phase to 0.9 GPa [30]. The third-order BM EOS fit to these non-hydrostatic, ambient-temperature data yields a bulk modulus of 18.5 ± 2.6 GPa with a pressure derivative of 5.4 ± 7.9. The high error in the pressure derivative value is primarily due to the very small pressure range of the data set. Fitting the data with a fixed value of 4 for the pressure derivative of the bulk modulus, (which reduces the BM EOS to second order), results in a bulk modulus of 18.9 ± 0.5 GPa. The data and EOS are shown in Fig. 3.6.

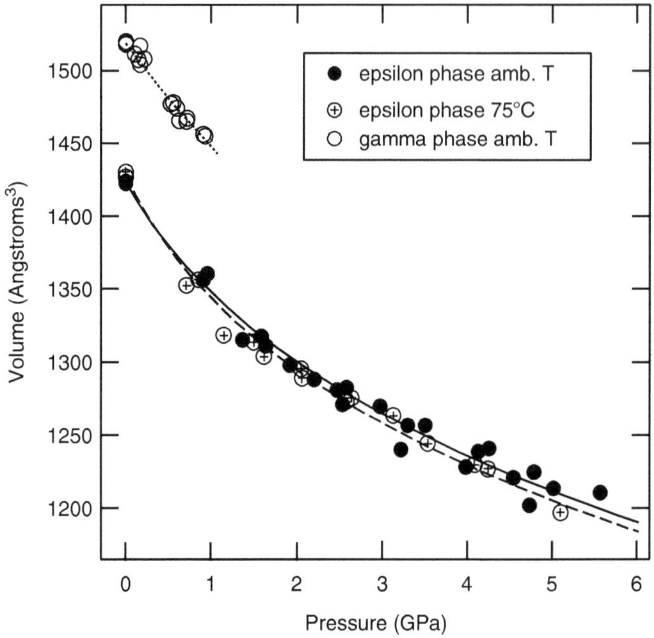

Fig. 3.6 Compression of the ε-polymorph of CL-20 in Dow-Corning Fluid 200, at ambient temperature and 75°C. The γ-phase of CL-20 is only stable to 0.9 GPa, and its compression had to be obtained without a pressure medium

Figure 3.6 also shows the compression of ε CL-20 at 75°C. Thermal expansion of ε CL-20 at ambient pressure, from ambient temperature (of about 27°C) to 75°C is $0.00014 \pm 0.00002\,\mathrm{K}^{-1}$. However, when compressed at 75°C, the expansion of the unit cell becomes negligible, and the compression curves at both temperatures merge at around 1 GPa. A third-order BM EOS fit to the 75°C ε-data yields an ambient-pressure isothermal bulk modulus of 11.0 ± 1.3 GPa with a pressure derivative of 14.0 ± 2.7. The third-order BM EOS for ε CL-20 at room temperature and 75°C are equal to within the error associated with each curve, indicating the merged compression curves and that heating to 75°C does not change the compressibility of ε CL-20.

3.3 High-Pressure Phases

3.3.1 Background

Most solid high explosives have positive Heats of Formation. This results in very fast chemical reactions which usually start with a small insult (such as a shock or laser pulse or other radiation) and persist with supersonic liberation of chemical energy, culminating in the production of stable gaseous reaction products. The positive heats of formation also imply that the ambient-pressure, ambient-temperature solid-state structure is itself meta-stable; resulting in multiple meta-stable crystal structures at ambient pressure and temperature or various polymorphic forms. Therefore, heating or compression of these crystals also results in multiple phase transitions and crystalline phases that are often meta-stable over long but finite timescales.

Further complication is added by the rate-dependent nature of phase transitions. For instance, fast heating may result in a phase transition at a lower temperature than if the sample were heated slowly. Or slow decompression may result in retention of a high-pressure phase, while fast decompression results in a phase transition to the ambient-pressure stable structure. In DAC studies, the presence of shear stresses may also affect compression and the pressure at which phase transitions are observed. Therefore, as concluded in the previous section, it is important to minimize the effect of shear stress dependence on compression and phase-transition pressure by using pressure media that are much softer at all pressures than the explosives being studied.

The most definitive methods for solid structural analysis being neutron and x-ray diffraction, and explosives being organic solids with molecules composed of C, H, O, and N in each lattice site, structural analysis of these materials become non-trivial. In addition, most of the solid-state structures have low-crystal symmetries such as monoclinic or triclinic, requiring high-resolution diffraction methods to resolve multiple peaks occurring in close 2θ or d-space. All these factors, together with the small samples in DACs, and the cone restrictions imposed by DAC configurations culminates in old and well-utilized materials whose high-pressure dependent phases are still under investigation and are not completely clarified.

3.3.2 Method

Typical experiments consist of loading diamond anvil cells with the sample and a pressure medium, similar to the EOS studies. Samples are then heated and compressed to explore the P–T phase space while using x-ray diffraction to detect phase transitions or temperatures of thermal decomposition. Many explosives are not stable in a liquid phase, so continuous heating to higher and higher temperatures usually results in direct decomposition of the material to product gasses and solid residues. In our studies the temperature of the samples was determined from thermocouples in contact with each diamond near the culet tip. Pressure was determined using the technique discussed in Section 3.2.2.

While diffraction is the definitive method for investigation of lattice structure, studies using Raman and Infrared (IR) spectroscopy that provides information about the point group symmetry of the molecule and the local lattice site are also helpful. Both Raman and IR spectroscopy require incident lasers, either single-crystal or powder samples, spectrometers, and detectors, all of which are commercially available. While synchrotron incident radiation for Raman and IR studies can be helpful, excellent results have been obtained from lab setup equipment.

Raman and IR data can only be understood when assigned to vibrational modes calculated from Density Functional Theory (DFT) or other such theories. Commercially available software such as Gaussian (currently 2003 version) have been successfully utilized to calculate vibrational modes.

3.3.3 Data

Almost every explosive undergoes at least one phase transition under compression to about 15 GPa. The bulk moduli or stiffness of explosives being between 8 and 20 GPa (with pressure derivatives of the bulk moduli between 4 and 14) by 15 GPa most of these materials are compressed down to about 70% of their initial volume. Isothermal compression studies of various materials including earth materials that are significantly stiffer than explosives, show that almost all materials undergo phase transitions when compressed down to 60% of their initial volumes. Explosives then, are not an exception to this general observation. However, explosives, in comparison to earth materials, display lower symmetry structures, and multiple phases or polymorphs.

The ambient-pressure structure (α-phase) of RDX, has orthorhombic symmetry (P_{bca}), and a unit cell approximately $11 \times 10 \times 13$Å, containing eight molecules per unit cell. In this structure, the individual RDX molecules possess essentially C_s symmetry, with two of the NO_2 groups being axial (A) and the third equatorial (E) with respect to the triazine ring, in the molecular conformation AAE [32]. Under hydrostatic compression above 3.8 GPa at any temperature between room temperature and 225°C, RDX is reported to undergo a phase change to the high-pressure phase labeled γ [11, 33]. All studies indicate that the γ-phase is of similar orthorhombic symmetry to the ambient-pressure α-phase, and no definite structure for the high-pressure γ-phase had been published until February 2008 [47]. A recent study by Goto et al. of Japan using DFT calculations, FTIR spectra, and powder x-ray diffraction studies, suggests that crystal symmetry of the γ-phase remains P_{bca} with the small volume change caused by the rotation and translation of molecules but with no change in the structure of the molecule itself from the AAE conformation [34]. Another very recent detailed study by Dreger et al. [35] using Raman spectroscopy of single crystals speculate that the point group symmetry of the γ-polymorph seems to be the same as for the α-polymorph, and that the observed increase in the number of modes or the splitting of internal vibrational modes may be due to factor group coupling. Another study to higher pressure, shows new vibrational features above 18 GPa, suggesting yet another phase transition to a new δ-phase [36].

Under compression between 3 and 3.8 GPa at temperatures above 225°C (and below the melting temperature of 250°C at that pressure), the α-phase of RDX is reported to achieve the high-temperature phase β [33] This α–β transition is reported as rapid and reversible, while the γ–β transition achieved by heating at pressure above 3.8 GPa is reported to be sluggish and irreversible. The β-phase was never observed to revert to the γ-phase, and was found to revert to the α-phase only when cooled to room temperature and decompressed almost to ambient pressure.

The crystal structure of the β-phase is hereto unknown, but with significant changes in the molecular and lattice structures [33].

The common ambient P–T-stable phase of HMX, labeled β has a monoclinic $P2_1/c$ structure with two molecules per unit cell and a density of $1.90\,g\,cm^{-3}$. Compression of this β-phase above 12 GPa together with Raman spectroscopy shows many modes splitting at that pressure, but with no accompanying change in unit cell volume [13]. A phase transition was reported above 27 GPa under hydrostatic compression, though recent isentropic compression experiments to 40 GPa in 200–500 ns timescale, shows that this transition, if it occurs, is slow [37].

HMX is also known to crystallize in less-common, perhaps meta-stable phases labeled α with a density of $1.84\,g\,cm^{-3}$ and δ with a density of $1.79\,g\,cm^{-3}$. (The γ-phase was found to be a hydrate.) The effect of high pressure and high temperature on the β- and δ-phases and transitions between them have been studied by many investigators. These studies report that at ambient and pressures to 0.12 GPa, β HMX transforms to δ between 149°C and 190°C, with the lower transition temperatures achieved when RDX impurities are present within the HMX crystals or when the heating rate is very slow [38, 39]. Above 0.2 GPa, the β-phase is stable at high temperature to decomposition temperatures above 278°C [38, 40]. Our compression of the β-phase at various temperatures showed that the δ-phase is obtained upon decompression from above 4 GPa to ambient pressure even at temperatures as low as 140°C [15]. We were not able to hold samples of δ at 140°C for very long to see if that phase really was stable at that temperature and ambient pressure. Instead the samples were cooled to room temperature (at ambient pressure), and after a few hours, the δ-phase obtained from decompression at 140°C would convert to β. This is in keeping with previous observations that when δ is prepared from β or with remnant β in the structure, it slowly converts back to β under ambient conditions [38].

Complicating the phase diagram of HMX is then also the effect of the rate of heating and the rate of compression. In each case, faster insult (heating or compression) results in lower transition pressures or temperatures. Decompression or cooling rates too result in vast hysteresis of phase transition conditions. In addition, as noted in the previous section on equations of state within this chapter, non-hydrostatic pressure also changes the compression of HMX. Such varied observations lead to the ambiguity surrounding the stable pressure and temperature regimes of the various solid phases of HMX. However, given extremely slow heating or compression rates and perfect hydrostatic conditions, the stable phase of HMX may be easily predicted by simply considering the density. Under ambient conditions HMX is at equilibrium in the β structure with a density of $1.90\,g\,cm^{-3}$, and as heating drops the density below $1.79\ g\ cm^{-3}$ the β-phase transitions to the δ-phase. Similarly, upon compression at high temperature as the density again increases above $1.90\,g\,cm^{-3}$, the δ-phase transitions to the β-phase.

The ambient P,T structure of TATB is composed of graphite-like layers of molecules, in triclinic symmetry with two molecules per unit cell. X-ray diffraction studies of commercially available TATB powder show a very high intensity from the layers (*00l* direction) with almost no intensity (<2%) from the other crystallographic planes. Further, finely ground powders show low crystallinity and degraded diffraction patterns. The combined effects makes studying high P–T phases of TATB using x-ray diffraction extremely challenging. Los Alamos National Labs has different methods for preparing crystalline TATB, and is able to make material whose *00l* planes do not overwhelm the diffraction pattern (see Fig. 3.7 for example). Therefore, both x-ray studies of TATB at high pressures reported to date were conducted by LANL [19,20]. Both studies agree that at room temperature the ambient-pressure phase is stable to at least 8 GPa, while the later study hints at a non-uniform volume change around 8 GPa. However, in view of Fig. 3.7, no corresponding diffraction pattern changes are obvious.

Considering the difficulties of defining the structure of TATB, to date, there are no high-pressure studies at high temperature.

The room temperature high-pressure phases of PETN have been investigated in detail. Ambient pressure-structural analysis of PETN published in 1947 claimed a structure with $P2_12_1$ orthorhombic symmetry [22], while more recent determinations established a tetragonal space group of $P\bar{4}2_1c$ [23] with a molecular point group of S_4 [24] with two molecules in the unit cell. Early compression studies by Olinger et al. in 1975, and by Yoo et al. in 1998, did not indicate any phase transitions to 15 GPa [19, 21]. A 1997 detailed study using neutron diffraction and deuterated (d_8-) PETN also showed no change in symmetry to 4.28 GPa, the highest pressure achieved in that study [41]. More recent high-pressure studies report Raman peak splitting starting just above 5 GPa reducing the molecular point group symmetry to C_2 [24]. Since C_2 symmetry is not compatible with the ambient-pressure tetragonal $P\bar{4}2_1$ space group, these changes are interpreted as a

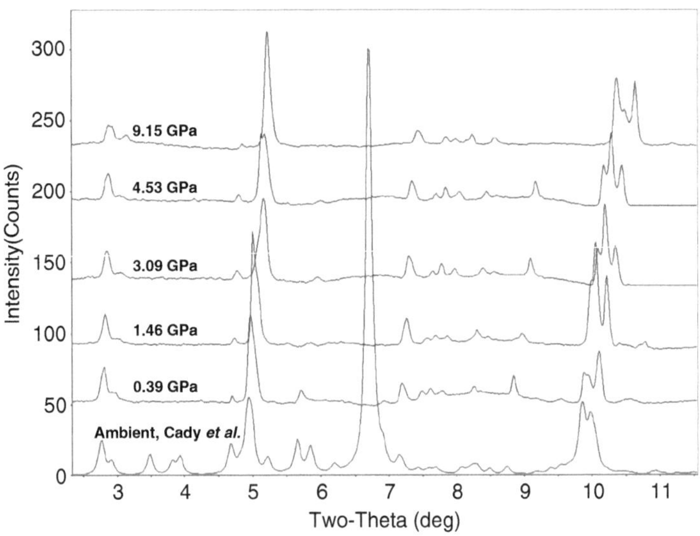

Fig. 3.7 TATB diffraction spectra at various pressures. Note the intense 002 peak between 6.5° and 7° 2θ in the ambient-pressure spectrum from Ref. [19], and the lack thereof in the spectra from Ref. [20]

phase transition to, most likely, the orthorhombic symmetry $P2_12_12$ [24–26]. This later transition is still under investigation, and there are no high-temperature static compression studies of PETN to date.

Ammonium Perchlorate (AP) crystallizes in an orthorhombic structure at ambient P,T. Bridgman observed a cusp-like maximum, thought to be a phase transition, at 3.1 GPa using shear experiments at elevated temperature [27]. Richter and Pistorius clarified the phase-transition pressures and temperatures using heating rates of about $1^\circ C\,s^{-1}$. They published a phase diagram from 1 atm to 4 GPa and ambient to 300°C [42]. However, another study done in DACs reports no solid–solid phase transition to 26 GPa at ambient temperature [43]. Our compression of AP at ambient temperature observed by utilizing x-ray diffraction showed new diffraction peaks appearing from 0.9 GPa with a complete phase change occurring at around 3.0 GPa [29]. Increased pressure probably "freezes" or slows the freely rotating motions of the NH_4 ion, concentrating electron density within the NH_4 units in particular directions or orientations thereby losing local site symmetry resulting in a few additional diffraction peaks as observed at and above 0.9 GPa. When heated at ambient pressure the known orthorhombic-to-cubic phase transition occurs at 513 K. Under static compression the transition temperature is reported to decrease with increasing pressure to 4 GPa, above which pressure this transition was not observed [43]. The solid-to-liquid transition with an ambient pressure onset at about 550 K is reported to have a strong inverse-pressure dependence, in that the onset temperature decreases with increasing pressure. The phase diagram published indicates that at 25 GPa AP melts at 300 K [43]. After this book went to press, the AP structure above 3.0 GPa was reported to be *Pnma*, [48].

HNIW or CL-20, with four known polymorphs α, β, γ, and ε, has been extensively studied at static high pressure [30, 31, 44]. Of these, the α-polymorph (space group P_{bca} with eight molecules in the unit cell) has only been reported as a clathrate, where decomposition products such as H_2O, N_2, CO, or CO_2 are interstitial within the structure. Hence this is also the phase that is observed with vibrational spectroscopy at high temperature just before thermal decomposition occurs [31]. The β-polymorph with a density of $1.99\,g\,cm^{-3}$ (space group $Pb2_1a$ with four molecules in the unit cell) has been observed in a narrow P,T range, while the γ- and ε-polymorphs with densities of 1.92 and $2.04\,g\,cm^{-3}$, respectively (both structures showing space group $P2_1/n$ with four molecules in the unit cell) are reported stable at a range of pressures and temperatures [30, 31]. Of these polymorphs, ε is considered the most stable structure at ambient pressure and temperature [45, 46].

A comprehensive study of the γ-phase using FTIR spectroscopy to 14 GPa and 340°C revealed the phase transitions shown in Fig. 3.8. The γ-polymorph is stable from ambient to 0.4 GPa, and temperature up to 250°C when complete thermal decomposition to gas products occurs. Between 0.40–0.50 GPa and 190°–240°C, γ shows a transition to the α-phase together with partial decomposition products of CO_2 and CO. The results suggest that CO_2 and CO are trapped within the α-phase lattice, because the phase transition or the partial decomposition of γ to $\alpha + CO_2 + CO$ is reversible in the P,T range of 0.40–0.50 GPa and 190–240°C [31].

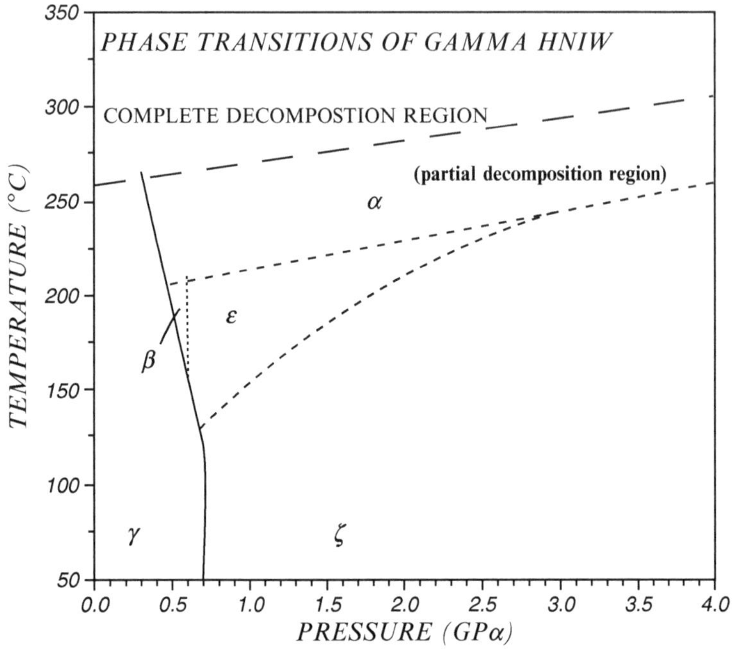

Fig. 3.8 Phase transitions of γ CL-20, using data in Ref. [31] The solid lines indicate reversible phase transitions and the dashed lines indicate irreversible phase transitions. The reversible or irreversible nature of the β–ε transition, shown by a dotted line, could not be established

At pressures between 0.60 and 0.65 GPa and temperatures between 150°C and 170°C the γ-phase shows a reversible and sluggish (requiring several hours) transition to the β-phase. When the β-phase is heated to 200°C at 0.6 GPa, it irreversibly, partially decomposes to the $\alpha + CO_2 + CO$ phase. Furthermore, when β is pressed to pressure little higher than 0.6 GPa, it transforms to the ε-phase. This transition is also reported extremely sluggish [31].

At pressure between 0.65 and 0.7 GPa and between 120°C and 140°C, γ transforms to ε. In this pressure range, when the ε-phase is heated above 200°C it irreversibly, partially decomposes to form CO_2 and CO and the α-phase, which then completely decomposes at 260°C [31].

Above 0.7 GPa and ambient temperature, the γ-phase transforms to the ζ-phase [43]. At 0.7 GPa the two phases coexist, and heating both phases to 110°C results in a transition to the ε-phase, making that P and T a triple point in the phase diagram of CL-20. Above 0.7–2.5 GPa when the ζ-phase is heated between 110°C and 190°C, it transforms (irreversibly) to the ε-phase. When ε is heated further to 220°C in this pressure range, it partially decomposes to the $\alpha + CO_2 + CO$ phase. However, this partial decomposition was not found to be reversible like the γ-phase partial decomposition. Between 0.7 and 2.5 GPa, additional heating above 260°C results in complete decomposition of the α-phase [31].

From 2.5 to 10 GPa the ζ-phase is stable from ambient temperature to above 230°C when it undergoes irreversible transition or partial decomposition to the α-phase, CO and CO_2. In this pressure range, the α-phase completely decomposes to the product gases above 280°C. From 10 GPa up to the maximum pressure of 14 GPa achieved in the study, the ζ-phase fully decomposes between 300°C and 340°C [31].

Since the ε-phase is considered most stable [45,46], we studied the ε-phase using x-ray diffraction experiments to 5 GPa and 175°C [30]. Unlike in the case of in-lab FTIR studies as in [31], our study obtained data at a synchrotron source where conservation of beam time required heating rates as high as $3°C\,min^{-1}$ and quick pressure increases. Therefore, the phase-transition pressure and temperatures achieved, which are often heating/compressing rate dependent, may not be under ideal equilibrium conditions.

At ambient pressure, heating the ε-phase to above 120°C resulted in a phase transition to the γ-phase. This high-temperature transition agrees with the densities of the two phases (ε of 2.04 and γ of $1.92\,g\,cm^{-3}$) in that heating lowers density. Compression of the γ-phase thus obtained even to pressures of 0.4 GPa, at 140°C, converts the γ-phase back to the ε-phase. Decompression of the ε-phase to ambient pressure at 140°C returns the γ-phase, indicating that the $\gamma \leftrightarrow \varepsilon$ transition is reversible at high temperatures between 120°C and 140°C. These observations are in keeping with the observations from the FTIR study, that at pressure between 0.65 and 0.70 GPa and between 120°C and 140°C γ transforms to ε. The 0.4 GPa from Ref. [30] does not include thermal pressure, so the real pressure inside the cell could well be as high as 0.65 GPa. However, cooling the γ-phase from 140°C to ambient temperature did not return the ε-phase. That is, at ambient pressure, when heating is started from the ε-phase transformed to γ above 120°C and cooled, the sample remains in the γ-phase. Even when we retained the ε-phase back to ambient conditions from 140°C and 0.75 GPa, by decompressing very slowly, after a few hours the sample converts back to the γ-phase. This indicates that it is perhaps the γ-phase that is the ambient P–T stable phase, instead of the ε-phase. However, it must be stressed that fast compression and fast heating rates were used, and that samples may not be reaching true equilibrium states.

At ambient pressure, further heating of the γ-phase above 150°C resulted in thermal decomposition between 150°C and 175°C. The decomposition of the γ-phase too is in keeping with Ref. [31], though the temperature of decomposition is a lot lower. The reduced decomposition temperature can be attributed to the faster heating rates employed in the x-ray diffraction study [30].

Compressing the ε-phase at ambient temperature does not result in any phase transitions to the maximum pressure of 5.6 GPa achieved in the study. For comparison with the FTIR study, the γ-phase was also compressed at ambient temperature. The γ–ζ-phase transition started at 0.7 GPa and the sample was completely ζ by at 0.9 GPa, as shown in Fig. 3.9. Similar to [31] we found the $\gamma \leftrightarrow \zeta$ phase transition reversible.

The ζ-phase obtained at ambient temperature and 1.4 GPa was heated and by 180°C, new peaks started appearing. The new peaks at 2θ of 3.97°, 4.62°, 5.78°,

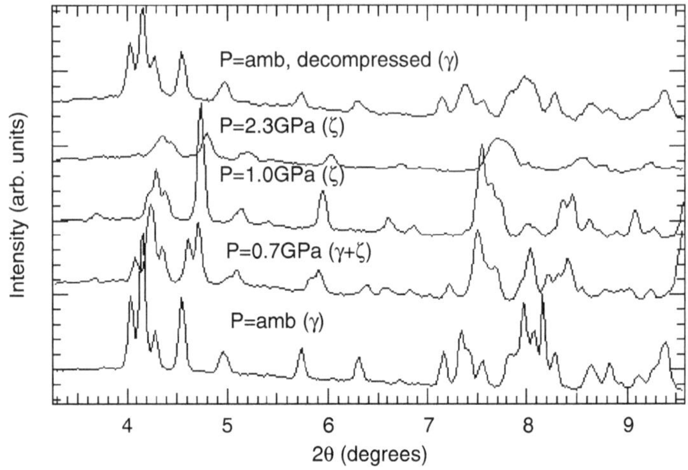

Fig. 3.9 Compression of the γ-phase at ambient temperature illustrating the γ ↔ ζ phase transition beginning around 0.7 GPa

6.54°, and 9.00° is shown in the pattern at 180°C in Fig. 3.10. For comparison, the previous pattern at 160°C of the ζ-phase has also been included in the figure. According to the phase diagram in Fig. 3.8, the phase at 1.4 GPa and 180°C should be the ε-phase. Therefore, Fig. 3.10 also includes the pattern for the ε-phase at ambient pressure and 100°C (highest temperature of this phase that we have a clean pattern at). However, the "new" peaks do not correspond to peaks in the ε-phase pattern. Instead the new peaks do correspond to peaks from the α-polymorph, calculated (shifted) to this pressure and temperature, and shown in Fig. 3.10 as sticks. Because heating adds thermal pressure (which is not evaluated in our method), it is possible that the sample which was heated at 1.4 GPa is at higher pressure by 180°C. Then, consistent with the phase diagram in Fig. 3.8 this new phase could be the partially decomposed $\alpha + CO_2 + CO$ phase. Cooling from 180°C down to ambient did show a slightly yellow-colored sample, also indicating partial decomposition to the $\alpha + CO_2 + CO$ phase. Diffraction obtained from these slightly yellow samples decompressed to ambient pressure show low crystallinity and a mixture of peaks that cannot be assigned to any of the known phases of CL-20. Instead of cooling, heating from 180°C further, to above 220°C, ensures complete thermal decomposition.

3.4 Discussion and Conclusions

This chapter detailed experimentally determined isothermal equations of state and high-pressure, high-temperature phases of energetic materials. The bulk moduli or stiffness of the most commonly used explosives listed in Table 3.1 are between 8 and 20 GPa (with pressure derivatives of the bulk moduli between 4 and 14) similar

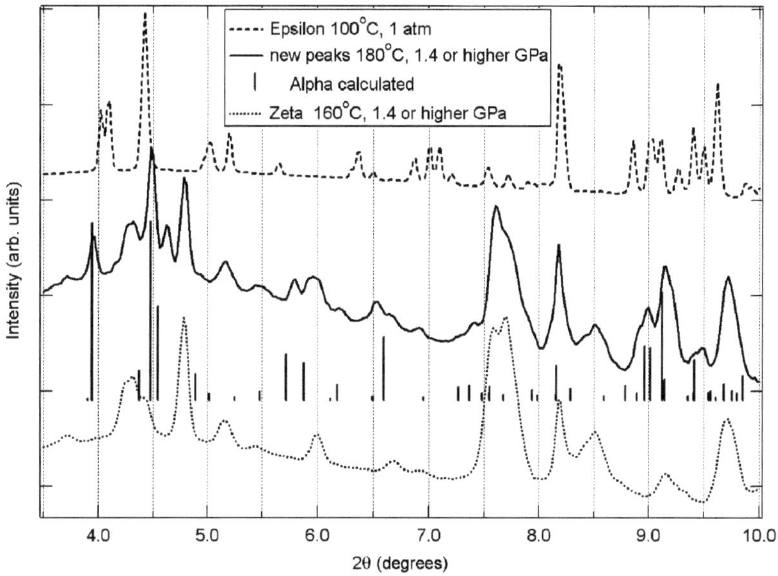

Fig. 3.10 XRD patterns obtained when heating ζ from ambient temperature and 1.4 GPa. The pattern at 160°C is the ζ-phase. The pattern at 180°C shows some new peaks indicating the start of a phase transition

to other molecular solids. In comparison, isothermal EOS of earth materials that are inorganic minerals indicate they are much stiffer than explosives. In addition, explosives, in comparison to earth materials, crystallize in structures with lower symmetry and exhibit multiple phases or polymorphs.

Isothermal EOS are determined by measuring the volume, usually of the unit cell, at various pressures to which the high-pressure device is compressed. When the Birch-Murnaghan EOS is used to obtain thermodynamic moduli, the range and the scatter in the data can result in various different values for the parameters. Such an example is shown in Table 3.1 for TATB data from Ref. [20], where data points to 8 GPa give a bulk modulus of 13.4 ± 0.8 GPa with a pressure derivative of 12.8 ± 1.1, while fitting the data points to 14 GPa gives a bulk modulus of 16.9 ± 1.0 with a pressure derivative of 8.2 ± 0.7. These values are disparate even within the ranges of the their error. Considering that the bulk modulus obtained is the value at $P = 0$, and that any curvature in the data points close to $P = 0$ are not affected by the data points between 8 and 14 GPa, the values for the isothermal $P = 0$ bulk modulus should be identical no matter what the pressure range of the data used in the Birch-Murnaghan EOS fit. Such discrepancies with fitting forms are described in Ref. [14] using HMX data as an example. An important finding noted in Ref. [14] is the need for many data points at lower pressures to constrain the uncertainty in moduli obtained from EOS fitting forms.

Consider the Birch-Murnaghan EOS fitting form:

$$P = \frac{3}{2}K_0\left[\left(\frac{V}{V_0}\right)^{-\frac{7}{3}} - \left(\frac{V}{V_0}\right)^{-\frac{5}{3}}\right]\left\{1+\frac{3}{4}\left[K_0'-4\right]\left[\left(\frac{V}{V_0}\right)^{-\frac{2}{3}}-1\right]\right\}$$

Experimentally, pressure is the independent variable that is easily controlled in DAC experiments while volumes are obtained from assigning *hkl* indices to diffraction patterns assuming certain lattice symmetries. However, in the Birch-Murnaghan EOS and most other EOS formalisms, pressure is the variable whose value deviations are minimized during P–V data fitting. Therefore, to minimize discrepancies, attempts should be made to include the effects of deviations and errors in pressure and volume in to the EOS fitting procedure. One simple solution it to weight the fit with the reciprocal of the error in each pressure measurement. Since the uncertainty in the measurements of lower pressures is low, usually, the reciprocal-error weighted fit favors the low-pressure points. Then this weighting also helps achieve statistically more dependable moduli, similar to obtaining lots of low-pressure data points noted in [14].

Another important consideration highlighted in this chapter is that of hydrostatic pressure conditions during compression. Energetic crystals, being of low lattice symmetry and showing asymmetric axis compression, are easily influenced by non-hydrostatic stress in the DAC, as shown for HMX. Therefore, in an attempt to find the best isotherm for HMX, the data obtained under the most-hydrostatic conditions from Refs. [11, 13, 15] were combined. That is, P–V data from [11] is limited to 4 GPa, and from [15] is limited 1 GPa. Combining the data thus also increases the number of low-pressure data points, while ensuring that the data set is limited to the most hydrostatic compression conditions.

To also understand the importance of weighted fits, attempts were made to fit the combined HMX data with weighting. Unfortunately, no deviations in pressure are published in Ref. [11] or [13]. Therefore, pressure errors were assigned considering that pressure measurements in DACs using most common methods of measuring pressure are only accurate to 0.05 GPa. Usually deviations are about 5% of the value of the measured pressure, and minimally 0.05 GPa. These data are in Table 3.2. By weighting the Birch-Murnaghan EOS fit of these data to 26.0 GPa with the reciprocal of the error in pressure, moduli of $K_0 = 16.6 \pm 0.4$ and $K_0'' = 7.2 \pm 0.3$ were obtained. These values are remarkably close to those obtained for Yoo and Cynn's data to just 12 GPa in Ref. [14]. When the fit is performed without weighting, the values obtained to 26.0 GPa are comparable to those published by Yoo and Cynn in Ref. [13]. These values are tabulated in Table 3.3 with the fitted curves displayed in Fig. 3.11.

This analysis demonstrates the value of weighting when using BM EOS fitting procedures to obtain isotherms. Reference [14] discusses the variation in isotherms obtained by Yoo and Cynn's data to 26 GPa in comparison to when the same data is limited to 12 GPa. However, when a weighted fit is used, as in the combined data set and fits herein, the isotherm obtained to 26 GPa is similar to that obtained even when

Table 3.2 Hydrostatic P–V data for HMX combined from [11, 13] and [15]

Pressure (GPa)	Volume ratio	Assigned pressure deviation (GPa)	Reciprocal error in P
0.00	1.000	0.05	20
0.10	0.998	0.05	20
0.17	0.999	0.05	20
0.20	0.998	0.05	20
0.39	0.975	0.05	20
0.49	0.984	0.05	20
0.83	0.958	0.05	20
0.95	0.963	0.05	20
0.96	0.954	0.05	20
1.61	0.924	0.08	12
2.47	0.899	0.12	8.1
2.50	0.897	0.13	8.0
3.24	0.881	0.16	6.1
4.10	0.855	0.21	4.9
4.60	0.850	0.23	4.3
5.70	0.837	0.29	3.5
7.00	0.807	0.35	2.9
7.10	0.813	0.36	2.8
10.6	0.767	0.53	1.9
14.1	0.742	0.71	1.4
15.8	0.733	0.79	1.2
22.9	0.685	1.15	0.9
24.9	0.682	1.25	0.8
26.0	0.675	1.30	0.7

Table 3.3 EOS Parameters for HMX from data combined from Refs. [11, 13] and [15]

Data description	Bulk modulus (GPa)	Pressure derivative
Combined un-weighted to 26 GPa	14.8 ± 0.7	8.5 ± 0.5
Combined weighted to 26 GPa	16.6 ± 0.4	7.2 ± 0.3
Yoo/Cynn data to 12 GPa [14]	16.0 ± 2.5	7.3 ± 1.4
Yoo/Cynn from [13]	12.4	10.4

data was limited to 12 GPa as in [14]. Thus thermodynamic parameters obtained by considering errors and obtaining reciprocal-error weighted fits are statistically most dependable.

This chapter also details the phases and phase diagrams of energetic crystals at high pressure and temperature. At room temperature, each of the materials RDX, PETN, and AP, undergo phase transition at or below 5 GPa, while the ambient-pressure phases of both HMX and TATB are stable to higher pressure. The phase diagram of CL-20 is complicated by the numerous phases of CL-20 that are either stable or meta-stable at room temperature. As discussed, the phase diagrams of these materials are complex and are also compression-rate and heating-rate dependent.

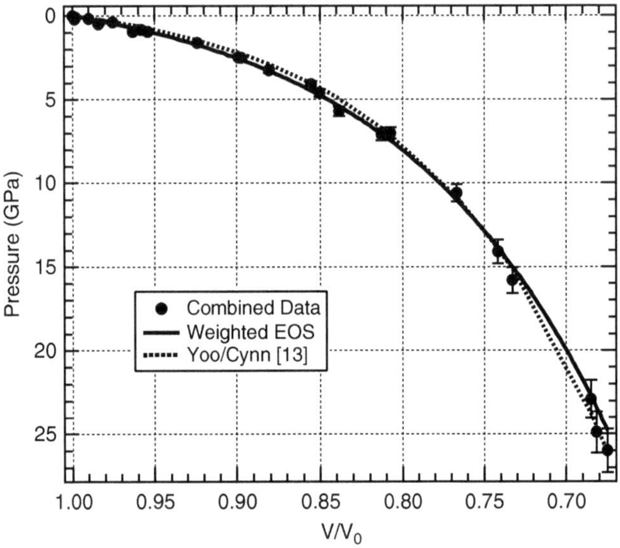

Fig. 3.11 P–V data obtained under most-hydrostatic conditions from Refs. [11,13], and [15] combined, together with the isotherm obtained from fitting the third-order BM EOS weighted by the reciprocal of error in pressure and the un-weighted isotherm obtained in [13]

Perhaps because of this, the high-pressure and high-temperature phases and phase diagrams of most of these materials (including TNT) have still not been investigated even though the materials have been utilized for decades.

Acknowledgments The first author gratefully acknowledges support from the Defense Threat Reduction Agency's Advanced Energetics Program during the compilation of this manuscript.

References

1. F. D. Murnaghan (1937) "Finite Deformations of an elastic solid", *American Journal of Mathematics* **59**, 235–260; F. Birch (1947) "Finite Elastic Strain of Cubic Crystals", *Phys. Rev.* **71**, 809–824.
2. F. Birch (1978) "Finite strain isotherm and velocities for single-crystal and polycrystalline NaCl at high pressures and 300 K", *J. Geophys. Res.* **83**, 1257–1268.
3. P. Vinet, J. Ferrante, J. H. Rose, J. R. Smith (1987) "Compressibility of Solids", *J. Geophys. Res.* **92**, 9319–9325.
4. Raymond Jeanloz (1988) "Universal equation of state", *Phys. Rev. B* **38**, 805–807.
5. S. M. Caulder, M. L. Buess, A. N. Garroway, P. J. Miller (2004) "NQR Line Broadening Due to Crystal Lattice Imperfections and its Relationship to Shock Sensitivity", *Proceedings of the APS Topical Group on Shock Compression of Condensed Matter Conference*, 2003, Editors M. D. Furnish, Y. M. Gupta, J. W. Forbes, American Institute of Physics, Melville, NY, pp. 929–933.
6. D. S. Watt, R.M. Doherty (2004) "Reduced Sensitivity RDX - Where are we?", 35th Annual Conference of ICT, Karlsruhe, Germany.

7. G. J. Peirmarini, S. Block, J. D. Barnett, R. A. Forman (1975) "Calibration of the pressure dependence of the R1 ruby fluorescence line to 195 kbar" *J. of Appl. Phys.* **46** 2774–2780.
8. H. K. Mao, J. Xu, P. M. Bell (1986) "Calibration of the ruby pressure scale to 800 kbar under quasi-hydrostatic conditions", *J. Geophys. Res.* **91**, 4673–4676.
9. G. J. Piermarini, S. Block, J. D. Barnett (1973) "Hydrostatic limits in liquids and solids to 100 kbar", *J. Appl. Phys.* **44,** 5377–5382.
10. W. A. Bassett, A. H. Shen, M. Bucknum, and I.-M. Chou (1993) "A new diamond anvil cell for hydrothermal studies to 2.5 GPa and from $-190°$ to 1200°C", *Rev. of Sci. Instrum.* **64**, 2340–2345.
11. B. Olinger, B. Roof, H. Cady (1978) *Actes du Symposium International sur le Comportement des Milieux Denses sous Hautes Pressions Dynamiques*, Paris, France, pp. 3–8.
12. Choong-Shik Yoo, Hyunchae Cynn W. Michael Howard, Neil Holmes (2000) "Equations of State of Unreacted High Explosives at High Pressures", *Proceedings of the Eleventh International Detonation Symposium*, 1998, Snomass, CO, Office of Naval Research, pp. 951–957.
13. C.-S. Yoo, H. Cynn (1999) "Equation of state, phase transition, decomposition of β-HMX (octahydro-1,3,5,7-tetranitro-1,3,5,7-tetrazocine) at high pressures", *J. Chem. Phys.* **111**, 10229–10235.
14. R. M.nikoff, Thomas D. Sewell (2001) "Fitting Forms for Isothermal Data", *High Pressure Research* **21**, 121–137.
15. Jared C. Gump, Suhithi M. Peiris (2005) "Isothermal equations of state of beta octahydro-1,3,5,7-tetranitro-1,3,5,7-tetrazocine at high temperatures", *J. Appl. Phys.* **97**, 53513–53520.
16. P. W. Bridgeman (1926) "The effect of pressure on forty-three pure liquids", *Proc. Am. Acad. Arts Sci.* **61**, 57–99; P. W. Bridgeman (1942) "Freezing parameters and compression of twenty-one substances to 50,000 kgcm^2", *Proc. Am. Acad. Arts Sci.* **74**, 399–424.
17. M. Hermann, W. Engel, N. Eisenreich (1992) "Thermal expansion, transitions, sensitivities and burning rates of HMX", *Propel. Explosiv. Pyrotech.* **17**, 190–195.
18. Cheng K. Saw (2005) "Kinetics of HMX and Phase Transitions: Effects of Particle Size at Elevated Temperature", *Proceedings of the Twelfth International Detonation Symposium*, 2002, San Diego, CA, Published by Office of Naval Research, pp. 70–76.
19. B. Olinger, H. H. Cady (1976) "The Hydrostatic Compression of Explosives and Detonation Products to 10 GPa (100 Kbars) and their Calculated Shock Compression: Results for PETN, TATB, CO_2, and H_2O", *Proceedings of the Sixth International Detonation Symposium*, 1976, Naval Surface Weapons Center, White Oak, MD, pp. 700–709.
20. Lewis L. Stevens, Nenad Velisavljevic, Daniel E. Hooks, Dana M. Dattelbaum (2008) "Hydrostatic compression curve for triamino-trinitrobenzene (TATB) determined to 13.0 GPa with powder X-ray diffraction", *Propel. Explosiv. Pyrotech.* (submitted).
21. B. Olinger, P. M. Halleck, H. H. Cady (1975) "The isothermal linear and volume compression of pentaerythritol tetranitrate (PETN) to 10 GPa (100 kbar) and the calculated shock compression", *J. Chem. Phy.* **62**, 4480–4483.
22. A. D. Booth, S. J. Llewellyn (1947) "The crystal structure of pentaerythritol tetranitrate" *J. Chem. Soc.* **1947**, 837–846.
23. H. H. Cady, A. C. Larson (1975) "Pentaerythritol tetranitrate II - its crystal-structure and transformation to petn I - an algorithm for refinement of crystal-structures with poor data", *Acta Crystallogr.* **B31**, 1864–1869.
24. Y. A. Gruzdkov, Z. A. Dreger, Y. M. Gupta (2004) "Experimental and Theoretical Study of Pentaerythritol Tetranitrate Conformers", *J. Phys. Chem. A* **108**, 6216–6221.
25. M. Pravica, et al. (2006) "Studies of phase transitions in PETN at high pressure" *J. Phyc. Chem. Solids* **67**, 2159–2163.
26. O. Tschauner, B. Kiefer, Y. Lee, M. Pravica, M. Nicol, E. Kim (2007) "Structural transition of PETN-I to a ferroelastic orthorhombic phase PETN-III at elevated pressures", *J. Chem. Phys.* **127**, 094502.
27. P. W. Bridgeman (1937) "Polymorphic transitions of 35 substance to 50,000 kg cm^2", *Proc. Am. Acad. Arts Sci.* **72**, 45–130.

28. F. W. Sandstrom, P.-A. Persson, B. Olinger (1995) "Isothermal and Shock Compression of High Density Ammonium Nitrate and Ammonium Perchlorate", *Proceedings of the Tenth International Detonation Symposium*, 1993, Boston, MA, pp. 766–774.
29. Suhithi M. Peiris, G. I. Pangilinan, and T. P. Russell (2000) "Structural Properties of Ammonium Perchlorate Compressed to 5.6 GPa", *J. Phys. Chem. A*, **104**, 11188–11193.
30. Jared C. Gump, Suhithi M. Peiris (2007) "Phase Stability of Epsilon HNIW (CL-20) at High-Pressure and Temperature", *Proceedings of the Thirteenth International Detonation Symposium*, 2006, Norfolk, VA, pp. 1045–1050.
31. T. P. Russell, P. J. Miller, G. J. Piermarini, S. Block (1992) "High-Pressure Phase Transition in γ-Hexanitrohexaazaisowurtzitane", *J. Phys. Chem.* **96**, 5509–5512.
32. Richard J. Karpowicz, Thomas B. Brill (1984) "Comparison of the molecular-structure of hexahydro-1,3,5-trinitro-S-triazine in the vapor, solution, and solid-phases", *J. Phys. Chem.* **88**, 348–352.
33. P. J. Miller, S. Block, G. J. Piermarini (1991) "Effects of pressure on the thermal decomposition kinetics, chemical reactivity and phase behaviour of RDX", *Combustion Flame* **83**, 174–184.
34. Naoyuki Goto et al. (2007) "High Pressure Phase of RDX", *Proceedings of the Thirteenth International Detonation Symposium*, 2006, Norfolk, VA, pp. 1051–1057.
35. Zbigniew A. Dreger, Yogendra M. Gupta (2007) "High Pressure Raman Spectroscopy of Single Crystals of Hexahydro-1,3,5-trinitro-1,3,5-triazine (RDX)", *J. Phys. Chem. B* **111**, 3893–3903.
36. J. A. Ciezak, T. A. Jenkins, Z. Liu, R. J. Hemley (2007) "High-Pressure Vibrational Spectroscopy of Energetic Materials: Hexahydro-1,3,5-trinitro-1,3,5-triazine", *J. Phys. Chem. A* **111**, 59–63.
37. D. E. Hare, J. W. Forbes, D. B. Reisman, J. J. Dick (2004) "Isentropic compression loading of octahydro-1,3,5,7-tetranitro-1,3,5,7-tetrazocine (HMX) and the pressure-induced phase transition at 27 GPa", *Appl. Phys. Lett.* **85**, 949–951.
38. H. H. Cady (1961) "Studies of the polymorphs of HMX", *Report LAMS-2652*, Los Alamos National Laboratory, 1962, Los Alamos, NM, pp. 1–50.
39. A. G. Landers, T. B. Brill (1980) "Pressure-temperature dependence of the β–δ polymorph interconversion in octahydro-1,3,5,7-tetranitro-1,3,5,7-tetrazocine", *J. Phys. Chem* **84**, 3573–3577.
40. Gasper J. Piermarini, Stanley Block, Philip J. Miller (1987) "Effects of pressure and temperature on the thermal decomposition rate and reaction mechanism of β– octahydro-1,3,5,7-tetranitro-1,3,5,7-tetrazocine", *J. Phys. Chem.* **91**, 3872–3878.
41. J. J. Dick, R. B. von Dreele (1998) *Proceedings of the APS Topical Group on Shock Compression of Condensed Matter Conference*, 1997, Editors S. C. Schmidt, D. P. Dandekar, J. W. Forbes, American Institute of Physics, Woodbury, NY, pp. 827–830.
42. P. W. Richter, C. W. F. T. Pistorius (1971) "Phase relations of NH_4ClO_4 and NH_4BF_4 to High Pressures", *J. Solid State Chem.* **3**, 343–439.
43. M. Frances Foltz, Jon L. Maienschein (1995) "Ammonium perchlorate phase transitions to 26 GPa and 700 K in a diamond anvil cell", *Mater. Lett.* **24**, 407–414.
44. T. P. Russell, P. J. Miller, G. J. Piermarini, S. Block (1993) "Pressure/Temperature Phase Diagram of Hexanitrohexaazaisowurtzitane", *J. Phys. Chem.* **97**, 1993–1997.
45. R. Y. Yee, M. P. Nadler, A. T. Neilson (1990) *Proceedings of the October 1990 JANNAF Propulsion Meeting*, CPIA.
46. M. F. Foltz, C. L. Coon, F. Garcia, A. L. Nichols (1994) "The Thermal Stability of the Polymorphs of Hexanitrohexaazaisowurtzitane, Part I", *Propel. Explosiv. Pyrotech.* **19**, 19–25; "The Thermal Stability of the Polymorphs of Hexanitrohexaazaisowurtzitane, Part II", *Propel. Explosiv. Pyrotech.* **19**, 133–144.
47. A. J. Davidson et al. (2008) "Explosives under pressure - the crystal structure of γ-RDX as determined by high-pressure X-ray and neutron diffraction", *Cryst Eng Comm*, **10**, 162–165.
48. A. J. Davidson et al. (2007) "High-Pressure Structural Studies of Energetic Ammonium Compounds" Proceedings of the 38th ICT conference, 2007, Karlsruhe, Germany pp 41:1–12.

Chapter 4
Equations of State of Binders and Related Polymers

Dana M. Dattelbaum and Lewis L. Stevens

4.1 Introduction

The union of high-explosive molecules (HEs) with polymeric binders to form plastic-bonded (PBXs) explosives was an important advancement in high-explosives science, offering improved safety and reliability, while maintaining performance [1]. The development of PBXs would not have been possible without several timely improvements in explosives technology, including an ability to produce decreased sensitivity materials, improved chemical stabilities, and greater manufacturing reproducibility. From a practical standpoint, the development of PBXs also brought improved machinability, improved engineering properties, and enhanced chemical resistance and long-term chemical stability. Eventually, insensitive high-explosive (IHEs) molecules were incorporated into PBXs. The development of IHE-based PBXs allowed for even greater flexibility in the choice of polymeric binders, permitting the use of higher-density binders such as fluorinated polymers. Examples of modern IHE-containing PBXs used by the Department of Energy are PBX 9502 (Figs. 4.1 and 4.2) (95% TATB/5% Kel-F 800) and LX-17 (92.5% TATB/7.5% Kel-F 800), which use the copolymer poly(chlorotrifluoroethylene-*co*-vinylidene fluoride) or Kel-F 800, as the polymeric binder. The structures of TATB and Kel-F 800 are shown below.

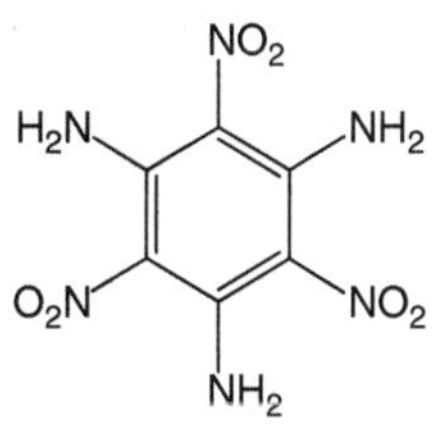

TATB

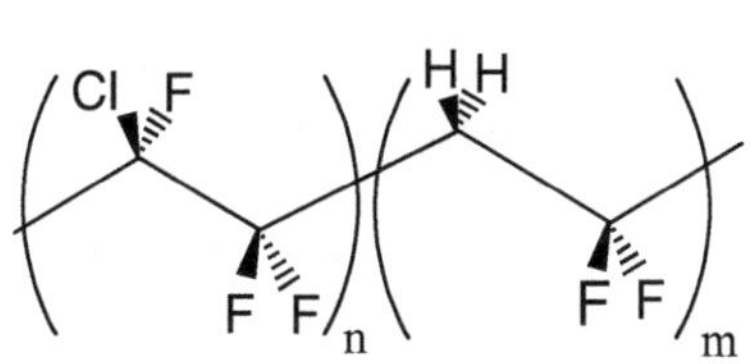

Kel-F 800

S.M. Peiris, G.J. Piermarini (eds.), *Static Compression of Energetic Materials,*
Shock Wave and High Pressure Phenomena,

Fig. 4.1 Optical micrograph of PBX 9501, a conventional high explosive consisting of a bi-modal distribution of HMX crystals in a binder matrix of 2.5% Estane 5703 and 2.5% BDNPF/BDNPA eutectic plasticizer (Micrograph courtesy of P. D. Peterson, Los Alamos National Laboratory)

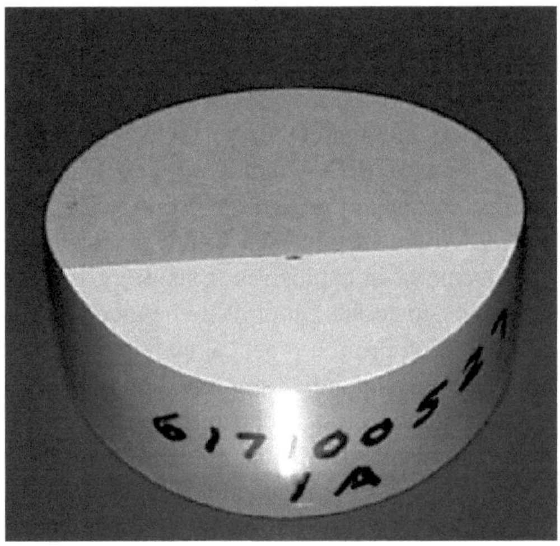

Fig. 4.2 Photograph of a PBX 9502 part machined for plate impact experiments. PBX 9502 consists of 95% TATB, with 5% Kel-F 800 polymer binder (Courtesy R. Alcon, Los Alamos National Laboratory)

The investigation of the properties of plastic-bonded explosives and related propellant formulations will continue to drive research and development at defense laboratories, particularly as simulations demand new models with improved fidelity, and political and applications drivers dictate the development of new energetic material formulations with improved safety. Recent requirements for insensitive and combat safe munitions are currently motivating the development of new PBXs, propellants, and related formulations. For example, the United States Department of Defense recently established a policy that all new munitions must be able to withstand fire, accidents, or attack (such as by gun fire). In addition to conventional weapons applications, plastic-bonded explosives will continue to be used in applications where their flexibility can be exploited including reactive armor applications, cutting explosives, and booster charges.

4.1.1 Plastic-Bonded Explosive Formulations

Polymers offer a benefit of reducing an explosive's sensitivity to impact or friction by providing some mechanical "give" to the charge. To maintain the performance of high explosives, PBX's typically contain high-explosive particles mixed with a polymeric binder at loading percentages exceeding 85%. These high volumetric fills are fairly unique in the materials field and can pose challenges from both materials characterization and modeling standpoints.

The long-chain structures of polymers coupled with their ability to exist in many forms of network and composite structures impart their unique and desirable physical properties. The impact of polymer chemistry and synthetic polymer science since their initial development in the 1940s is profound [2, 23]. In fact, it was recently estimated that over 50% of chemists and chemical engineers are involved in polymer research and development [2]. Polymer comes from "poly" meaning many or multiple, and "mer" describing individual chemical units. Chemical reactions of small chemical species called monomers with one another leads to the formation of oligomers (short chain species), and eventually polymers. What separates polymers from large organic molecules or oligomers is their molecular weight. Low-molecular weight polymers are generally in the 10,000–20,000 amu range, with high-molecular weight polymer capable of molecular weights exceeding 1 million amu.

Polymers can be categorized into a few major classes including thermoplastics and thermosets, and their subclasses. Thermoplastics are materials that soften when they are heated, and many of the polymeric binders encountered in today's PBXs fall into this class. Elastomers, or rubbery polymers, are also commonly found in binder applications. Elastomers are within their rubbery regime, e.g. above the glass transition temperature (T_g), at room temperature and usually for much of the operating temperature regime of PBXs. Additives to improve mechanical or electrical properties of polymers are often encountered in PBX formulations, including stabilizers, antioxidants, and perhaps the most common, plasticizers, or small molecule additives designed to soften the neat polymer network. Generally, thermosets are not widely used as binders in high explosives, as they form insoluble cross-linked networks.

A (non-exhaustive) summary of some plastic-bonded explosive formulations is listed in Table 4.1. Several polymers have emerged as commonly used PBX binders today. They are the fluoropolymers Kel-F 800, PTFE, the Viton (DuPont) and THV (Dyneon) families of fluorinated copolymers, and the elastomers Estane 5703 and hydroxy-terminated polybutadiene (HTPB). Other relevant materials used in recent formulations include the fluoroelastomer Kel-F 3700, Exon 461, the polydimethylsiloxane (PDMS)-based Sylgard formulations (Dow Corning), and HyTemp^{TM} polyacrylic elastomer. Exon 461 is a copolymer of chlorotrifluoroethylene/tetrafluoroethylene/vinylidene fluoride, while Kel-F 3700 is a chlorotrifluorothethylene/vinylidene fluoride copolymer at a 31:69 monomer ratio. Also used in earlier formulations was Kel-F 827, a low-molecular weight version of Kel-F 800. These polymers emerged as commonly used binder materials because of a

Table 4.1 Examples of plastic-bonded explosive formulations [33]. HMX is cyclotetramethylene tetranitramine, RDX is 1,3,5,-trinitro 1,3,5-triazacyclohexane, TATB is 1,3,5-triamino-2,4,6-trinitrobenzene, NC is nitrocellulose, NQ is nitroguanidine, AP is ammonium perchlorate, DATB is 1,3-diamino-2,4,6-trinitrobenzene, Estane is Estane® 5703 a polyester-urethane copolymer.

Common name	High explosive	Binder
X-0007	86 HMX	14 Estane
X-0009	93.4 HMX	6.6 Estane
X-0069	90.2 HMX	9.8 Kel-F 3700
X-0204	83.2 HMX	16.8 PTFE
X-0213	94.6 HMX	2.0 Estane/2.0 BDNPF/1.4 wax
X-0235	94 HMX	2 DNPA/2 NP/2 Estane
X-0143	85.6 HMX/9.2 DATB	5.4 Estane
X-0183	65.7 HMX/26.4 NQ	7.79 Kel-F
X-0118	29.7 HMX/64.9 NQ	5.4 Estane
X-0030	95 DATB	5 Estane
X-0219-50-14-10	50 HMX/40 TATB	10 Kel-F 800
X-0228	90 NQ	10 Estane
X-0224	74 RDX/20 Al	5.4 Elvax, 0.6 wax
X-0250-40-19	40.4 RDX	40.4 cyanuric acid, 19.4 Sylgard
X-0219	90 TATB	10 Kel-F 800
	90 TATB	5 Kel-F 800, 5 Kel-F 820
	85 TATB	15 Kel-F 800
	85 TATB	7.5 Kel-F 800, 7.5 Kel-F 827
AF 902	95 NQ	5 Viton A
AFX-521	95 PYX	5 Kel-F 800
EDC 37	91 HMX/1 NC	8 K10 polyurethane rubber
EDC 35	95 TATB	5 Kel-F 800
EDC 32	85 HMX	15 Viton
EDC 29	95 HMX	5 polyurethane
LX-03	20 DATB, 70 HMX	10 Viton A
LX-04	85 HMX	15 Viton A
LX-07	90 HMX	10 Viton A
LX-09	93 HMX	4.6 BDNPA, 2.4 FEFO
LX-10-0	95 HMX	10 Viton A
LX-10-1	94.5 HMX	5.5 Viton A
LX-14-0	95.5 HMX	4.5 Estane
LX-15	95 HNIS	5 Kel-F 800
LX 16	96 PETN	4 FPC 461
LX-15	95 HNIS	5 Kel-F 800
LX-16	96 PETN	4 FPC 461
LX-17-0, -1	92.5 TATB	7.5 Kel-F 800
PBX 9501	95 HMX	2.5 Estane/2.5 BDNPF/BDNPA
PBX 9502 (X-0290)	95 TATB	5 Kel-F 800
PBX 9007	90 RDX	9.1 polystyrene 0.5 DOP, 0.4 rosin
PBX 9010	90 RDX	10 Kel-F 3700
PBX 9011	90 HMX	10 Estane
PBX 9404	94 HMX	3 NC, 3 tris-betachloroethylphosphase, diphenylamine
PBX 9407	94 RDX	6 Exon
PBX 9405	93.7 RDX	3.15 NC, 3.15 chloroethylphosphate
PBX 9503	80 TATB, 15 HMX	5 Kel-F 800

(Continued)

Table 4.1 (Continued)

Common name	High explosive	Binder
PBX-122, 124, 125	AP, NTO, RDX	HTPB in various concentrations
PBXN-106	RDX	Polyurethane
PBXW-7	36 RDX/60 TATB	5 Viton A
BX1	60 TATB/35 (95 RDX:5 HMX)	5 Kel-F 800
BX2	60 TATB/35 (95 RDX:5 HMX)	5 PTFE
BX3	60 TATB/35 (90 RDX:10 HMX)	5 Kel-F 800
BX4	60 TATB/35 (90 RDX:10 HMX)	5 PTFE
CX-84-A	84 RDX	16 HTPB
CX-85	84.25 HMX	15.75 HTPB

balance of their chemical and physical (mechanical) properties with processibility when combined with explosive crystals of interest, as well as cost. The repeat structures of Estane 5703, PTFE, HTPB, and Viton A are given below.

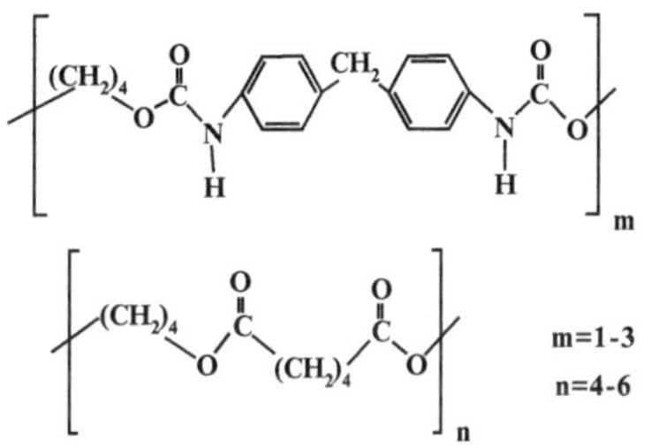

Estane® 5703

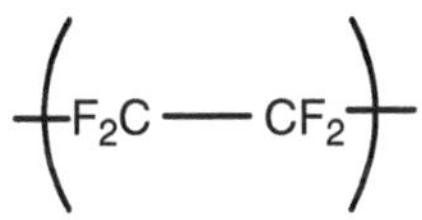

PTFE

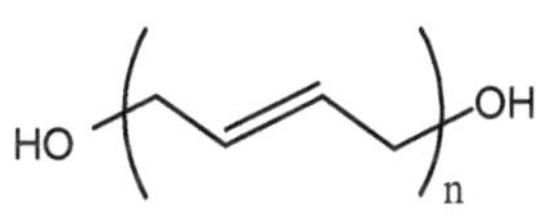

HTPB

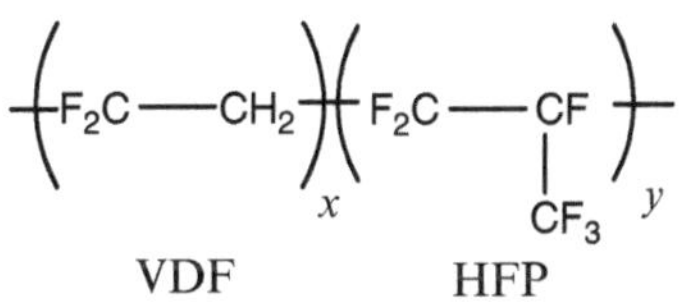

Viton A

Fluorinated polymers have become popular as high-explosive binders because they are not only highly dense, which provides a close match to the density of the high-explosive crystals, but they also exhibit chemical inertness, and robust stability over large temperature ranges. Polytetrafluoroethylene (PTFE), arguably the most common fluoropolymer today, has been used as a binder material in simple formulations. However, PTFE has several disadvantages from a formulations perspective, including a high melt viscosity, high melt temperature, and lack of solubility in common solvents. Copolymerization of tetrafluoroethylene and the related rifluoroethylene with vinylidene fluoride and/or horopropylene dramatically improves the polymer's physical properties for processing, while maintaining high densities and chemical inertness. Examples include the Viton and THV class of resins, and the copolymer of chlorotrifluoroethylene and vinylidene fluoride, Kel-F 800, which is widely used in insensitive high explosives by the United States, and Europe [4,5].

Elastomers, such as Estane 5703, PDMS, and hydroxy-terminated polybutadiene (HTPB), continue to be widely encountered, particularly when combined with some of the more sensitive energetic materials. HTPB is a rubbery polymer used in numerous PBX formulations that allows for high volumetric loading with its low pre-cure viscosity. It is typically cured with an isocyanate, forming polyurethane cross-links. A disadvantage of this polymer is its susceptibility to decomposition under acidic conditions [6, 7].

Thermoplastic elastomers have also gained attention in recent years. In the United States, these materials have gained considerable market share due to their high temperature molding abilities. There are two types of thermoplastic elastomers, those made from ionomers and block copolymers. An example of a block copolymer that has been tested as an explosive binder is the family of styrene–ethylene–butylene terpolymers (Kraton G-6500, G-1650). Kraton G-6500 has been found to provide stability and reduced sensitivity in HMX-based explosives [8–12]. Recently, an evaluation of three binders- Kel-F 3700, Kraton G-1650, and a polyurethane (Impranil DLH)- used in coating HMX was performed. Each polymer was mixed with HMX using a solvent slurry process, in which a lacquer of the polymer dissolved in an organic solvent is mixed with an aqueous slurry of the explosive. It was found that both Kel-F 3700 and Kraton G-1650 posed electrostatic hazards when mixed with HMX, whereas the polyurethane produced the PBX in high yield. In addition, Kel-F 3700 did not dissolve sufficiently to bind the HMX particles.

There have been a few systematic studies published in the open literature linking the choice of polymeric binder with physical or performance characteristics of the bulk PBXs. A comprehensive report describing formulations of TATB with various binders and their performance characteristics was published by Dobratz [13]. Field and coworkers investigated the bulk mechanical properties and initiation sensitivities of PBXs as a function of binder [14]. They found that some polymer binders actually increase the sensitivity of explosives due to the way that they fail, forming shear bands or crack tips, which, in combination their low thermal conductivities, allow local temperatures to rise, causing initiation to occur. Examples of nonideal binder materials that perform in this manner include polycarbonate, polysulphone, and polyester. Comparison of three binders – polyurethane, polyethylene, and Viton – showed that polyethylene was the most brittle of the three materials, followed by Viton and polyurethane, with addition of greater amounts of polymer to the formulation found to increase the strain to failure. The formulation of 95 wt % HMX, 5 wt % polyurethane was found to have the best combination of strength and strain to failure, which is consistent with the ongoing use of the conventional explosives PBX 9501, and the related EDC 37. A study by Arnold has also shown that exchanging a silicon binder for HTPB can have a dramatic influence on sensitivity by the gap test [15]. Additional studies examining the role of binders in influencing the sensitivity of PBX formulations are described in references [16–18].

There have also been efforts to incorporate reactive binders into PBX formulations. Nitrocellulose is an example of one of the original reactive binders, which was often used in formulations with nitroglycerine. A newer class of materials is the substituted polymers containing reactive side-groups. Examples of this class of materials are polymers that have been functionalized with azido- or azidomethyl moieties. Glycidylazide polymer (GAP, poly-(azidomethyl)ethylene oxide) was first synthesized in 1972 [19] and is probably the most widely recognized energetic binder. GAP has been extensively studied as an individual polymer and in energetic formulations [20–29]. The reactivity of GAP is derived from an exothermic reaction that releases the azido groups, and consequently nitrogen gas, through chain scission [30]. Azidomethyl-substituted polyesters have also recently been examined [31]. The oxetane-based binders 3-nitratomethyl-3-methyl oxetane (NMMO, and its polymerized forms poly-NIMMO or NMMO), poly-GLYN (polymerized glycidyl nitrate), and 3,3-bisazidomethyl oxetane (BAMO) are additional examples of energetic binders in widespread use today. Common features of many of these energetic binders are low T_g's, compatibility with common explosive molecules, and miscibility with plasticizers or other inert elastomers.

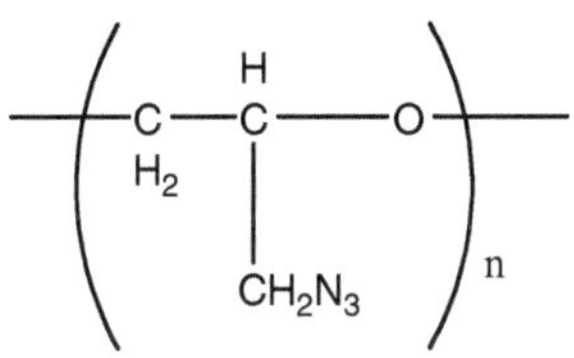

GAP repeat unit

While the advances in explosive science precipitated by the development of PBXs are unquestionable, the use of certain combinations of binders and explosives can also have their disadvantages. Examples include chemical incompatability, or disparate solubilities such that the binder cannot adequately coat the explosive particles. A relevant example is differing thermal-mechanical properties of the two materials. In the TATB-containing materials PBX 9502 and LX-17, the anisotropy of the TATB particles coupled with their growth behavior under thermal cycling causes a phenomenon known as "ratchet growth" in which a 1.5–2.0 vol % growth can occur [32]. It has been found that by using a binder with a higher glass transition temperature (T_g) than Kel-F 800 ($T_g = 28°\text{C}–30°\text{C}$), such the epoxy comprised of bisphenol A and epichlorohydrin, ratchet growth can be inhibited by fivefold [32]. It was postulated that this was due to inhibition of movement between TATB particles. Other binders shown to reduce the bulk CTE of PBXs containing TATB included polystyrene/polyphenylene oxide, and Phenoxy PRDA 8080.

There have been many studies of the mechanical, shock, and initiation properties of plastic-bonded explosives. For the purpose of this chapter, we will focus on experimental methods and recent results pertaining to the equation-of-state properties of binders and related polymers. In most cases, individual polymer materials must be processed to obtain solid samples for experimental studies, as they are often supplied in granular or pellet forms, and so, studies of binders are limited compared with their PBX formulations. In preparation of binders for these studies, special care must be taken with material choice, processing method, and control of thermal conditions, as polymer properties can vary widely with differences in temperature, thermal history, molecular weight, and percentage of crystallinity or phase separation.

4.2 Equations of State

4.2.1 Introduction

The term "equation of state" can have many different meanings in the high-pressure science and shock-wave communities. An "EOS" can be everything from the parameters of a linear fit to shock velocity–particle velocity data from shock-wave compression experiments, to the application of an empirical isothermal fitting form to static high pressure, pressure–volume, or density data derived from x-ray diffraction experiments, to a full description of a materials energy, or so-called thermodynamically consistent EOS, in which the first and second derivatives of the energy with respect to volume or other parameters yield measurable thermodynamic properties (such as heat capacity, pressure, etc.).

While the focus of this book is on the properties of energetic materials at static high pressures, there have been significantly less experimental efforts in interrogating polymers, including high-explosive binders, at high pressures. This is due, in large part, to the inapplicability of many common EOS experimental methods to polymers or other amorphous materials. There is a growing need to understand more about the behavior of polymers and other soft or disordered materials under high static and dynamic pressures. As our EOS and constitutive capabilities improve, so does our need for additional insight into constituent behavior increase. Polymers are used not just as high-explosive binders, but are also found in many critical-function applications such as gaskets, membranes, cushions, and structural supports in engineering applications in which they are subjected to stressing environments either in use or in production and processing.

The purpose of this chapter is to describe current experimental methods for the determination of relevant EOS and thermodynamic properties of polymeric materials, and to give several relevant examples of experimental data, particularly for high-explosive binder materials. In recent years, there has been a resurgence of experimental and theoretical interest in the EOS properties of polymers. It is our hope to capture some recent results and make them available to the reader. The chapter starts with a description of theoretical equation-of-state forms that have been applied to polymers, and continues with a description of both static and dynamic equation-of-state experimental methods.

4.2.2 Equations of State

Given the premise of this chapter, a brief review of equations of state (EOS) provides a general background useful for understanding the analysis available to the high-pressure experimental data discussed herein. While the conceptualization of an EOS is straightforward, its broad application can be limited; particularly in regards to the transferability of an EOS to fundamentally different materials (molecular, polymeric, ionic, etc.). Therefore, EOS and their application become essential for insight into not only macroscopic P–V–T behavior but also the basic forces in condensed media [34–36].

Developing this association is vital to polymer research. While condensed-phase forces range from ionic to dispersive, for polymers the relevant forces are van der Waals forces. This stems from the pronounced affect these long-range, cumulative, dispersive forces have over a polymer's macromolecular structure [37–39]. Subtle variations within the macromolecular backbone, whether configuration or conformation, can give rise to very different physical behavior. Understanding these subtleties highlights the necessity of EOS analyses for polymers. However, EOS analyses can be difficult for polymeric materials given the crystalline, amorphous, phase-separating, or rigid amorphous phases encountered.

EOS can be roughly segregated by their basis, i.e. empirical (and semiempirical) or theoretical. Certainly no attempt will be made here to inclusively describe each type. Our discussion of the empirical EOS will include the Tait [40], Murnaghan [41, 42], Birch-Murnaghan [43], Vinet [44, 45], and Sun [46] equations. This list provides a logical evolution of EOS development ending with Sun's EOS which was applied specifically to polymers. Choosing to begin with the Tait equation is appropriate for several reasons: (1) it has a simple structure, (2) it introduces compressibility/strain-based equations of state and most importantly, (3) it successfully reproduces compression data for variety of systems, including polymers [47–50]. It is also illustrative to discuss the original definition of Tait's EOS, which has a slightly misconstrued interpretation [51]. In the literature, this has led some authors to further discriminate the Tait EOS as either "original" or "usual" [36].

Empirical EOS are beneficial in their capacity to reproduce P–V data for a variety of materials. However, this approach is limited when trying to develop a microscopic, physical understanding of *why* a material behaves a certain way. As previously mentioned, this understanding is necessary, especially for the development of new materials. In part, this has led to the development of several theoretical EOS [52]. These equations are derived through statistical thermodynamics with the assumption of a specific interaction potential, e.g. Lennard-Jones, square-well, or hard-sphere. The widely applied Mie-Grüneisen EOS is derived for solids, comprised of a system of oscillators and how the characteristic frequency changes with volume [52, 53]. In connection to polymer liquids, three main theoretical models are also discussed: the cell, lattice-fluid, and hole theories. The Prigogine [54–56], Flory-Orwoll-Vrij [57], Sanchez-Lacombe [58, 59], and Simha-Somcynsky [60] equations will be used to compare and contrast the theoretical models as applied to polymeric materials.

4.2.3 Thermodynamic Foundation

Before describing different types of EOS forms, it is beneficial to first review the thermodynamic concepts that provide their foundation. This basis begins with thermodynamic potentials, e.g. the Helmholtz free energy, F, internal energy, U, and the natural variables: entropy (S), pressure (P), volume (V), or temperature (T).

Consider U for a closed system, i.e. a system without chemical potential. This not only simplifies the equations but is representative of a typical hydrostatic experiment. Assuming U is a function of V and S, then one can write dU for reversible processes as,

$$\mathrm{d}U = \left(\frac{\partial U}{\partial S}\right)_V \mathrm{d}S + \left(\frac{\partial U}{\partial V}\right)_S \mathrm{d}V \tag{4.1}$$

Using the first and second laws of thermodynamics, Eq. 4.1 reveals that,

$$\left(\frac{\partial U}{\partial S}\right)_V = T \text{ and } \left(\frac{\partial U}{\partial V}\right)_S = -P \tag{4.2}$$

Similar expressions can be found using other potentials, but a common theme remains. Various thermodynamic parameters, S, P, T, and V, can be determined by differentiation of a potential with respect to its natural variables. Given this capacity, equations of the type seen in Eq. 4.1 in conjunction with those like Eq. 4.2 to provide a thermodynamic basis for equations of state [61,62].

Using the partial derivatives in Eq. 4.2 along with Euler's reciprocity relation, the Maxwell relations can be derived. Maxwell relations are advantageous because partial derivatives that are either with respect to or applied to entropy can be equated to partial derivatives in P, V, or T. In the case of U, using the following Maxwell relation for a closed system,

$$\left(\frac{\partial S}{\partial V}\right)_T = \left(\frac{\partial P}{\partial T}\right)_V \tag{4.3}$$

Eq. 4.1 can be recast as,

$$\left(\frac{\partial U}{\partial V}\right)_T = T\left(\frac{\partial P}{\partial T}\right)_V - P \tag{4.4}$$

Equation 4.4 is a thermodynamic equation of state where the right-hand side is composed of state variables or their derivatives. Similar forms can be constructed using different thermodynamic potentials. The partial derivatives are often expressed as thermodynamic parameters such as bulk moduli (inverse of the compressibility), thermal expansion coefficients, or combinations thereof. As will be shown in the following section, the isothermal compressibility is the cornerstone for several empirical P–V equations of state. A useful relationship is the cyclic rule:

$$\left(\frac{\partial x}{\partial y}\right)_z \left(\frac{\partial y}{\partial z}\right)_x \left(\frac{\partial z}{\partial x}\right)_y = -1 \tag{4.5}$$

The specific energy is given by $\mathrm{d}e = -P\mathrm{d}V + T\mathrm{d}S$. The Helmholtz free energy $F(V,T)$ is related to the specific energy by:

$$\begin{aligned} F &= e - TS \\ \mathrm{d}F &= -P\mathrm{d}V - S\mathrm{d}T \end{aligned} \tag{4.6}$$

The free energy can be partitioned between noninteracting degrees of freedom:

$$F(V,T) = F_{\mathrm{c}}(V) + F_{\mathrm{ph}}(V,T) + F_{\mathrm{el}}(V,T) \tag{4.7}$$

In which F_{c} is the energy of the cold curve describing atomic repulsion in the lattice or static molecular structure, F_{ph} is the phonon energy or energy associated with lattice vibrations, and F_{el} is the electronic contributions. The phonon energy can be derived from the quasi-harmonic approximation and the measurement of acoustic and optical modes. The strength and utility of thermodynamically consistent equations of state is that thermodynamic parameters are obtained directly from derivatives of the free energy. For example, the pressure is the sum of partial derivatives of the free energy with respect to volume:

$$P(V,T) = -\left(\frac{\partial F}{\partial V}\right)_T \tag{4.8}$$

Similarly, the entropy can be obtained with the derivative of free energy with respect to temperature:

$$S(V,T) = -T\left(\frac{\partial F}{\partial T}\right)_V \tag{4.9}$$

The Hayes analytical form is a tractable EOS based on the free energy $F(V,T)$, in which a reference isotherm, typically the room temperature isotherm, and thermal component are partitioned according to:

$$F(V,T) = F_0(V) + F_{\mathrm{T}_0}(V) + F_{\mathrm{th}}(V,T) \tag{4.10}$$

The electronic contribution is neglected. In practice, a room temperature isotherm, and ambient pressure and temperature knowledge of the heat capacity and Grüneisen parameter are needed to construct a Hayes EOS.

The development of the Debye EOS begins with a partitioning of the free energy analogous to that described above for the Hayes EOS. In his treatment of the phonon contribution, Debye extended the Einstein single-oscillator model, to include a distribution of frequencies (acoustic phonons), i.e., a phonon density of states, $g(\omega)$. Debye assumed that for a collection of N atoms there are $3N$ vibrational degrees of freedom and as such $g(\omega)$ must be normalized to that constraint. Furthermore, Debye postulated that the available phonon states followed a harmonic progression and asymptotically approached a maximum frequency, ω_D. Through normalization of $g(\omega)$ constrained with the ω_D upper bound, it can be shown that ω_D is found to be,

$$\omega_D^3 = \frac{6\pi^2 N c^3}{V} \tag{4.11}$$

where N is number of atoms, V the volume, and c the sound speed of the material. Sound waves (acoustic phonons) can have three polarizations: two transverse and one longitudinal. For less symmetric materials, these different waves can travel at different speeds and hence c in Eq. 4.11 can be regarded as an average sound speed. The Debye frequency in Eq. 4.11 is commonly represented as the Debye temperature,

$$\Theta_D = \frac{h\omega_D}{2\pi k_B} \tag{4.12}$$

where h and k_B are the Planck and Boltzmann constants, respectively. Through this treatment by Debye coupled with the harmonic approximation the phonon contribution to the thermodynamic variables can be readily derived.

Equations of state provide an analytical relationship between thermodynamic states variables, often P, V, T, and U. Furthermore, utilizing differential calculus and the identities therein, a set of thermodynamic parameters can be developed from these initial state variables. While a rigorous treatment of this development is beyond the scope of this chapter, the discussion of a few specific thermodynamic parameters will prove useful.

The bulk modulus, K, describes the incompressibility of a material under uniform pressure. That is, those materials that have a high bulk modulus are less compressible than materials that have a low bulk modulus. K can be formally defined as,

$$K = -V\left(\frac{\partial P}{\partial V}\right)_{T,S} \tag{4.13}$$

In the above equation, typically the partial derivative is evaluated under either an isentropic or isothermal constraint which respectively yield the isentropic bulk modulus (K_S) or the isothermal bulk modulus (K_T). The interrelation of K_S and K_T can be expressed as

$$K_S = \frac{C_P}{C_V} K_T \tag{4.14}$$

where C_P and C_V are respectively the isobaric and isochoric heat capacities, which are defined as

$$C_P = \left(\frac{\partial H}{\partial T}\right)_P \text{ and } C_V = \left(\frac{\partial U}{\partial T}\right)_V \tag{4.15}$$

where H and U are respectively the enthalpy and internal energy of the system. Since the C_P/C_V ratio is greater than one, K_S will be slightly larger than K_T.

While here emphasis is given the bulk modulus since it often appears in the parameterization of equations of state, another basic thermodynamic parameter is the volumetric thermal expansion coefficient, α, which is defined as

$$\alpha = \frac{1}{V}\left(\frac{\partial V}{\partial T}\right)_P \tag{4.16}$$

Inter-relations between thermodynamic parameters are readily constructed from a small set of independent parameters. To illustrate this, α can be recast as

$$\alpha = \frac{C_P}{K_T V} \tag{4.17}$$

A critical parameter to many equations of state is the Gruneisen parameter, Γ. This thermodynamic quantity often thought of as the "thermal pressure," and can be calculated using:

$$\Gamma = V\left(\frac{\partial P}{\partial e}\right)_V \tag{4.18}$$

Or more simply,

$$\Gamma = \frac{\beta V K_T}{C_V} \tag{4.19}$$

Γ/V and C_V are usually assumed to be constant with pressure in many equation of state approaches, due in large part to a lack of experimental data on the pressure dependence of these values. The basic thermodynamics discussed here serves as a rudimentary introduction into the interrelation of state variables and parameters as they pertain to an equation of state.

4.2.4 Empirical and Semi-empirical EOS

Despite their empirical nature, the equations of state described in this section are encountered in widespread use in the high-pressure community. They are often the starting point for quantification of the bulk modulus and its pressure derivative for importation into more complex EOS forms or direct input into hydrodynamic codes, such as CTH [63].

4.2.4.1 Tait Equation

In 1888, Tait detailed his analysis of sea water density using an EOS of the form,

$$\frac{V_0 P}{V_0 - V} = \frac{B+P}{C} \tag{4.20}$$

where B and C are empirical parameters fit to the experimental compression data with P_0 being implicitly zero. Equation 4.20 is actually the inverse of that reported by Tait, but for all intensive purposes this is the original Tait equation. Inspection of Eq. 4.20 shows that the left-hand side (inverse of the "average compressibility" described by Tait) has a linear dependence in pressure. Hayward details the equivalence of Eq. 4.20 with a linear-secant bulk modulus EOS [51]. Given this dependence, it is evident that B/C is the isothermal bulk modulus, K_0 and $1/C$ is proportional to the pressure derivative of the bulk modulus, K_0'. While bulk moduli determined with the Tait EOS have displayed a range of values, the $1/C$ parameter has shown a much more restricted behavior with values ranging from approximately 0.090 (hydrocarbon liquids) to 0.150 (liquid water) [49]. This narrow range is expected given the marginal pressure dependence of K_0' at lower pressures.

Since the original Tait equation is identical to a linear-secant bulk modulus EOS, it may be limited in application at higher pressures. In this regime, K_0' is no longer constant and the assumption of a linear-dependent bulk modulus fails. Therefore, the Tait equation is generally applicable to liquids (in keeping with its original use), compressible solids and polymeric materials up to approximately 2 kbar of pressure [64–66].

The deviation from the original Tait equation appears to have began with Tamann's reference Tait's equation in 1907 [51, 67]. Tamann removed V_o and replaced the finite difference (assuming P_o is zero in Eq. 4.20) with a derivative. The "cited" Tait equation was of the form,

$$-\frac{\mathrm{d}P}{\mathrm{d}V} = \frac{B+P}{C} \tag{4.21}$$

where again the inverse of the original citation is taken for clarity. Subsequent integration of Eq. 4.21 yields,

$$\frac{V}{V_0} = 1 - C\ln\left(1 + \frac{P}{B}\right) \tag{4.22}$$

This representation of the Tait equation is in the form generally referenced. Hayward argues that the replacement of the finite difference with a derivative is detrimental to the original Tait EOS by virtue of making it unnecessarily more complicated. Inconsistencies aside, either description of the Tait equation reproduce P–V isotherms quite well for a wide range of materials [68].

4.2.4.2 Murnaghan Equation

The modification to the Tait equation seen in Eq. 4.21 shows a tangential slope in P–V-space is fundamental to the development of an EOS. This is the basis for empirical equations based on strain; however, typically the isothermal bulk modulus, which is defined as,

$$K_T = -V\left(\frac{\partial P}{\partial V}\right)_T \tag{4.23}$$

is used instead. Integration of Eq. 4.23 would provide simple P–V equation,

$$P = -K_0 \ln \frac{V}{V_0} \tag{4.24}$$

The utility of this EOS is limited as the approximation of a constant bulk modulus would be restricted to a very narrow pressure range.

The next logical extension of the crude approximation in Eq. 4.23 would be to expand to the bulk modulus in pressure such that,

$$K_T = K_0 + K_0' P + K_0'' P^2 + \cdots \tag{4.25}$$

To simplify Eq. 4.25, terms quadratic or higher in pressure are removed and it is assumed that $K_0{}'$ is constant. Although not outlined in Tait's original paper, these assumptions lead to a result analogous to Tait's equation, i.e., the linear-secant bulk modulus equation. Combining Eq. 4.23 with Eq. 4.25 truncated after the linear term and integrating the result leads to,

$$P = \frac{K_0}{K_0'} \left[\left(\frac{V_0}{V} \right)^{K_0'} - 1 \right] \tag{4.26}$$

This result is the EOS derived by Murnaghan in 1944 stemming from his treatment of finite deformation of elastic solids.

Given the linear dependence on the bulk modulus, a comparison to the Tait equation (modified form shown in Eq. 4.22) seems appropriate. From the Tait equation, K_0 is found to be B/C. Furthermore, the pressure derivative of the bulk modulus evaluated at zero pressure can be shown to be,

$$K_0' = \frac{1}{C} - 1 \tag{4.27}$$

Using these definitions, the Murnaghan equation can be recast as,

$$P = \frac{B}{1-C} \left[\left(\frac{V_0}{V} \right)^{\frac{1}{C}-1} - 1 \right] \tag{4.28}$$

For comparison, consider a mock material whose Tait parameters are $C = 0.150$ and $B = 0.3\,\mathrm{GPa}$. These translates to a K_0 of 2.0 GPa and a $K_0{}'$ of 5.66 GPa. The $P - V$ representation of this mock material is shown in Fig. 4.3. The values determined by both equations are comparable up to 2 GPa. Above that pressure, the $\partial \mathrm{P}/\partial \mathrm{V}$ slope for the Murnaghan equation sharpens considerably relative to the that shown by the Tait equation, even though K_0 and K_0' were assumed identical. This deviation illustrates that when analyzing P–V isotherms considerable attention should be given to the limitation of different EOS forms.

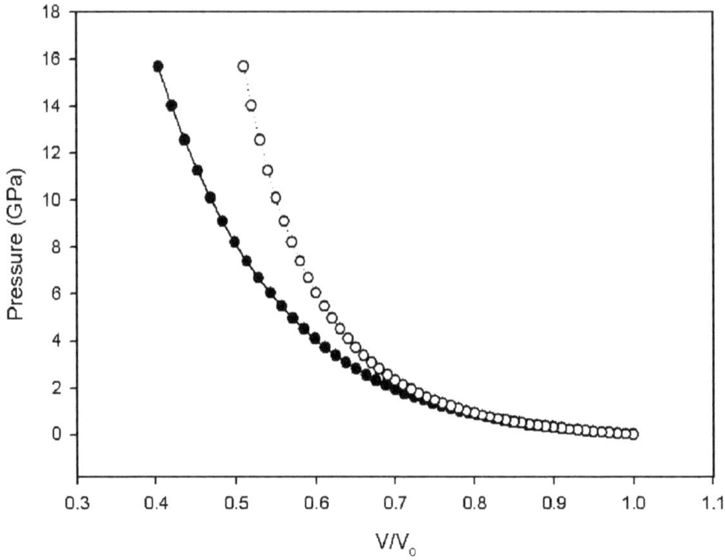

Fig. 4.3 Plotted Tait (•) and Murnaghan (○) equation for a mock material with $K_0 = 2.0\,\text{GPa}$ and $K_0' = 5.66\,\text{GPa}$

4.2.4.3 Sun Equation

Since different phases of a material can have very different intermolecular interactions, EOS can have limited transferability between phases. Given the complex phases often found in polymer materials, it would be improbable for one EOS to accurately describe the P–V–T behavior of polymers in the melt, glassy, and crystalline states. For example, when applied to polymers the Tait equation is predominantly used for the liquid state [66, 68]. As a solution to this problem, Sun *et al.* offered the semiempirical EOS shown here,

$$P = \frac{K_0}{(n-m)}\left[\left(\frac{V_0}{V}\right)^{n+1} - \left(\frac{V_0}{V}\right)^{m+1}\right] \tag{4.29}$$

where n and m are empirical parameters. Equation 4.29 was derived assuming that the internal pressure (usually defined at absolute zero) could be expressed as the same slope at any general temperature. Then adopting a Mie potential for chain–chain interaction energy, it was shown that a generalized EOS could be determined for which the Tait and Murnaghan equations were both approximate solutions. It will also be seen in the following sections that Eq. 4.29 bears a strong resemblance to the Birch-Murnaghan EOS.

The application of Eq. 4.29 was tested by fitting the compression data for several polymers, each in a different physical state (three glassy, three melt, and one liquid) over the pressure range of up to 2 kbar. Sun *et al.* found that the empirical parameters were essentially universal with $n = 6.14$ and $m = 1.16$. While the roots of this

EOS express similarities to the Tait and Murnaghan equations, it is comparatively new and has not been applied to the diverse number of materials its predecessors have. Further application will determine whether or not the n and m parameters are actually "universal."

4.2.4.4 Birch-Murnaghan Equation

Besides expanding the bulk modulus in pressure, the relationship between a material's free energy and its volume can also be used to develop an EOS. For an isothermal, closed system, the change in Helmholtz free energy with volume relates to pressure through,

$$P = -\left(\frac{\partial F}{\partial V}\right)_{T,n} \tag{4.30}$$

Equation 4.30 is the basis for a thermodynamic equation analogous to Eq. 4.6. However, in the treatment by Birch, he expanded the free energy in terms of Eulerian strain. This expansion coupled with Eq. 4.30 allowed for the derivation of the Birch-Murnaghan EOS. Expanding the Helmholtz free energy to a quadratic dependence in strain yielded the second-order Birch-Murnaghan EOS which takes the form,

$$P = \frac{3K_0}{2}\left[\left(\frac{V_0}{V}\right)^{7/3} - \left(\frac{V_0}{V}\right)^{5/3}\right] \tag{4.31}$$

With further expansion to include a cubic term, the resultant third-order Birch-Murnaghan EOS is,

$$P = \frac{3K_0}{2}\left[\left(\frac{V_0}{V}\right)^{7/3} - \left(\frac{V_0}{V}\right)^{5/3}\right]\left[1 + \frac{3}{4}(K_0' - 4)\left(\left(\frac{V_0}{V}\right)^{2/3} - 1\right)\right] \tag{4.32}$$

As would be expected, further expansion terms to the free energy expression could extend the applicable pressure region for the EOS. However, this extension comes with necessarily further parameterization, i.e., K_0'. However, from inspection of Eqs. 4.31 and 4.32, it can be seen that the third-order Birch-Murnaghan EOS reduces to second-order in the limit of $K_o{}'$ approaching $4'$.

With our focus being on polymers, the similar forms of the second-order Birch-Murnaghan and Sun equations suggests an application of the former to polymer-based systems as well. Using the same mock material as previously described, the comparison between these equations is illustrated in Fig. 4.4. Similar to other EOS analyses, significant deviation begins at approximately 1.5 GPa. However, as V/V_o approaches 0.5, the divergence becomes exorbitantly large. This behavior is a direct consequence of the large difference in the V_o/V exponents for the Sun equation ($\Delta = 4.98$) compared to that for the second-order Birch-Murnaghan equation ($\Delta = 0.66$). To determine whether or not the divergence of the Sun EOS is representative of polymer behavior requires further studies beyond 2 kbar.

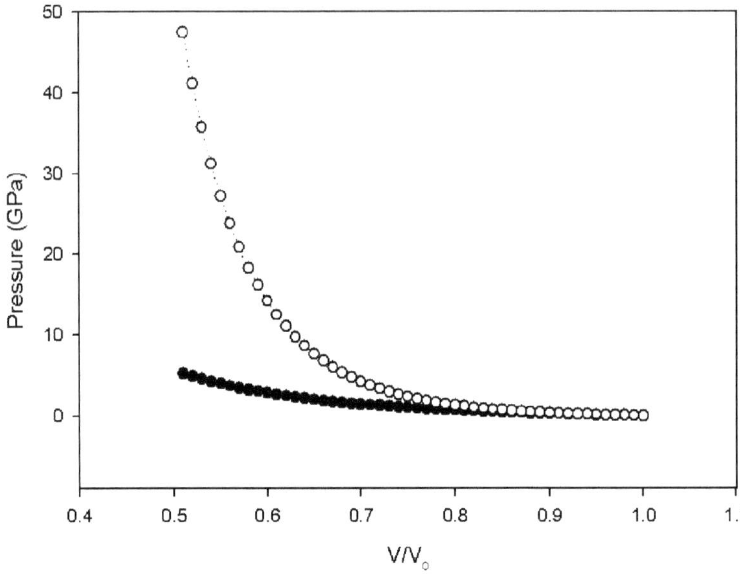

Fig. 4.4 Plotted second-order Birch-Murnaghan (•) and Sun (○) equation for a mock material with $K_0 = 2\,\text{GPa}$ and $K_0' = 5.66\,\text{GPa}$

4.2.4.5 Vinet or "Universal" EOS

Vinet's equation was originally developed for application to all solids ranging from ionic to van der Waals. Analogous to the derivation of the Birch-Murnaghan equations, the free energy relation in Eq. 4.30 is the basis. However, the free energy is not expanded in strain but rather assumed to take the form,

$$F \cong (1+a^*)e^{-a^*} \tag{4.33}$$

where a^* is a scaled distance defined as,

$$a^* = \frac{(a-a_0)}{l} \tag{4.34}$$

where a defines the separation between atoms, a_o is the equilibrium separation and l is a scaling length. Using these definitions in conjunction with Eq. 4.30 yields the Vinet equation,

$$P = 3K_0\left(\frac{V}{V_0}\right)^{-2/3}\left(1-\left(\frac{V}{V_0}\right)^{1/3}\right)\exp\left[\frac{3}{2}(K_0'-1)\left(1-\left(\frac{V}{V_0}\right)^{1/3}\right)\right] \tag{4.35}$$

The universal nature of Eq. 4.35 has been demonstrated on a variety of solids and over a wide range of pressures [45, 69, 70]. However, given the definitions in Eqs. 4.33 and 4.34, it is not surprising that the application has been predominantly to simple monatomic or diatomic solids. Therefore, without modification of the simplified binding-energy expressions, the large number of internal degrees of freedom available to polymeric materials may limit the application of the Vinet equation [70].

4.2.5 Theoretical Equations of State

4.2.5.1 General Development

Unlike purely empirical equations, theoretical EOS can provide *P–V* analysis coupled with a detailed understanding of the fundamental, microscopic forces in materials. This capacity is afforded through statistical thermodynamics which provides the bridge between the microscopic interaction potential and macroscopic properties [71, 72]. Similar to the discussion of empirical EOS, it is worthwhile to briefly introduce the statistics that provide the basis for theoretical methods.

Statistical mechanics is based upon the equal *a priori* probability postulate, i.e., at equilibrium the most probable state is the thermodynamic state. To describe this thermodynamic state, the system must be represented by an ensemble. For isolated, closed or open systems, the respective ensemble is microcanonical, canonical or grand canonical. For detailing a *P–V* isotherm, the canonical ensemble is appropriate. To describe the thermal equilibrium for a closed system, the concept of temperature is introduced using a Boltzmann distribution. Using this distribution, the following partition function may be constructed,

$$Z = \sum_i e^{-E_i/kT} \tag{4.36}$$

In Eq. 4.36, E_i represents the energy of the *i*th microstate. Each possible microstate is comprised of several degrees of freedom including configurational, electronic, nuclear, vibrational, rotational, or translational. If these degrees of freedom are considered independent, then the separate contributions are simply additive in E_i. Using Eq. 4.36, the total partition function can be expressed as a product of individual partition functions each representing a given contribution such that,

$$Z = \prod_j Z_j \tag{4.37}$$

where j spans over all available degrees of freedom and Z_j is an individual partition function describing the distribution in translation, vibration, etc.

With the partition function defined, the connection to macroscopic thermodynamic states is straightforward. For example, the Helmholtz free energy can be calculated from,

$$F = -kT \ln Z \tag{4.38}$$

Applying Eq. 4.30 to the above definition, the pressure is found to be,

$$P = kT\left(\frac{\partial \ln Z}{\partial V}\right) \tag{4.39}$$

Similar expressions to Eqs. 4.38 and 4.39 can be derived for the remaining thermodynamic parameters.

Constructing a partition function by modeling the potential and interactions within the microstates lays the basis to the development and discrimination of various theoretical EOS. With emphasis on polymers, our discussion begins with the Mie-Grüneisen EOS and its application to crystalline polymers. The periodicity of crystalline polymers provides a simplified theoretical treatment often focused on a singular unit cell [52]. The translational symmetry of the unit cell can then be used to develop the equations for the entire solid.

4.2.5.2 Solid Polymers: Mie-Grüneisen Equation

Assuming a system N atoms connected by a harmonic potential, the canonical partition function is given by

$$Z = \exp\left(\frac{-U_0}{kT}\right)\prod_j \frac{\exp\left(\frac{-h\nu_j}{2kT}\right)}{1-\exp\left(\frac{-h\nu_j}{kT}\right)} \tag{4.40}$$

where U_0 is the initial internal energy and ν_j is the jth vibrational frequency. Applying Eq. 4.39 to the above partition function, one finds the pressure can be expressed as,

$$P = -\frac{\partial U_0}{\partial V} + h\sum_j -\frac{\partial \nu_j}{\partial V}\left[\frac{1}{2} + \frac{\exp\left(\frac{-h\nu_j}{kT}\right)}{1-\exp\left(\frac{-h\nu_j}{kT}\right)}\right] \tag{4.41}$$

The first term in Eq. 4.41 indicates the internal pressure and the second term describes how the change in vibrational frequency with volume contributes to the pressure. In this form, Eq. 4.41 is analogous to the thermodynamic EOS (Eq. 4.6) developed in the introduction to this chapter. The second term in Eq. 4.41 defines the increase in thermal pressure comparable to the $\partial P/\partial T$ in Eq. 4.6. The multiplicative term inside the summation is more traditionally cast as,

$$\gamma_j = -\frac{\partial \ln \nu_j}{\partial \ln V} \tag{4.42}$$

where γ_j is the modal Grüneisen parameter for the jth vibrational mode. Under the assumption that all vibrational modes display the same frequency dependence on volume, Eq. 4.41 can be rewritten as,

$$P = -\frac{\partial U_0}{\partial V} + \frac{h\gamma}{V}\sum_j \nu_j \left[\frac{1}{2} + \frac{\exp\left(\frac{-h\nu_j}{kT}\right)}{1-\exp\left(\frac{-h\nu_j}{kT}\right)}\right] \tag{4.43}$$

Furthermore, the definition for internal energy in a canonical ensemble is,

$$E = kT^2 \frac{\partial \ln Q}{\partial T} \tag{4.44}$$

Applying Eq. 4.44 to the partition function in Eq. 4.40 and substituting the result into Eq. 4.43, one arrives at

$$P = -\frac{\partial U_0}{\partial V} + \frac{\gamma E}{V} \tag{4.45}$$

which is the typical Mie-Grüneisen EOS.

While the previous development assumes a monatomic lattice, the theoretical basis can certainly be extended to crystalline polymers. This extension is often centered on the distinction between internal and external vibrations. For a monatomic material, there are no internal vibrational degrees of freedom and thus no distinction is required. However, for polyatomic, molecular, or polymeric materials, the internal and external vibrations arise respectively from the intramolecular and lattice potentials. The former often display little volume dependence in comparison to the external vibration, i.e., the acoustic and optical phonons [73, 74]. This disparate behavior is not surprising given substantial dependence intermolecular potentials have on volume. Given this dependence, the external vibrations and their respective Grüneisen parameters are given emphasis in EOS treatments of polymeric materials.

4.2.5.3 Liquid Polymers

The microscale irregularity inherent to liquids complicates their theoretical modeling [52]. This complexity is highlighted by the number of theoretical models used to describe liquid behavior. Our discussion focuses on three models for liquid polymers: the cell, lattice-fluid and hole theories. For all models, a volume network is imposed onto the material. In the cell model, this volume is flexible; however, the mass within each volume is fixed. In contrast, the lattice-fluid model maintains a fixed volume element, while vacancies are allowed to be introduced. Finally, as an extension to the cell model, the hole theory accommodates vacancies; however, also maintains a compressible volume element. A cartoon representation of the fundamental differences between these three theoretical models is shown in Fig. 4.5. Below, in more detail, representative equations of state are discussed for each of them.

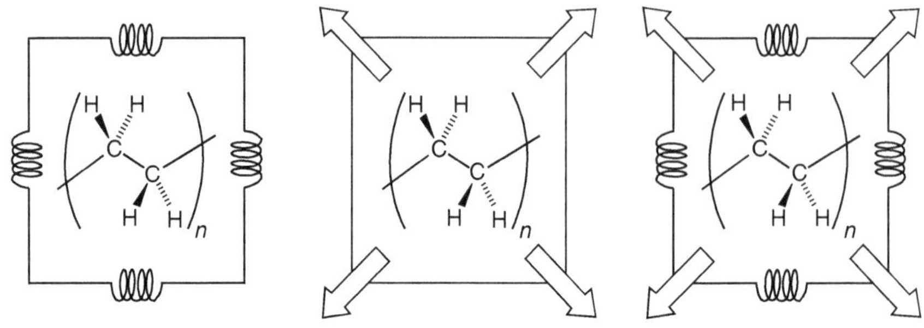

Fig. 4.5 Representation of basic principles in the cell **(a)**, lattice-fluid **(b)**, and hole **(c)** models for polymer liquids. Springs denote compressibility of the unit volume and the arrows denote the possibility of vacancies [60, 75, 76, 78]

Prigogine Equation

Prigogine *et al.* extended the liquid-based cell model for application to macromolecules. The basic premise is that a polymer liquid consisting of r-mer, chain molecules that may be treated as r point centers connected by a harmonic potential. Furthermore, the spatial relation of the r-mers develops a quasi-crystalline lattice where the mean potential energy for a point center displaced from its equilibrium position is assumed to follow a square-well approximation to the Lennard-Jones potential. Other approximations to this mean-field potential were also developed by Prigogine *et al.* [75, 76] but this discussion will be restricted to that described above. Using these potentials, it is assumed that the intramolecular restoring forces are much greater than those between macromolecules. This allows for an effective separation of the internal and external degrees of freedom and thus a separation of the partition function into a form analogous to Eq. 4.37.

Applying Eq. 4.39 to their partition function, Prigogine *et al.* determined the following equation of state,

$$\frac{P_R V_R}{T_R} = \frac{V_R^{1/3}}{V_R^{1/3} - 2^{-1/6}} + \frac{2}{T_R}\left(\frac{A}{V_R^4} - \frac{B}{V_R^2}\right) \tag{4.46}$$

where P_R, V_R, and T_R respectively denote the reduced pressure, volume and temperature.

The reduced variables are defined as,

$$P_R = \frac{P}{P^*}, \quad V_R = \frac{V}{V^*}, \quad T_R = \frac{T}{T^*} \tag{4.47}$$

where P^*, V^* and T^* are characteristic parameters specific to a material. A and B are geometry-dependent constants from the Lennard-Jones potential. Adopting a hexagonal close packing cell, i.e., 12 nearest neighbors, the Lennard-Jones constants become $A = 1.011$ and $B = 1.2045$ [54].

In the development of their EOS, Prigogine *et al.* assumed the distance between monomers (i.e., point centers) within a r-mer was approximately the same to that between adjacent r-mers. Therefore, at low temperatures, a nearly perfect crystalline lattice is formed; however, for non-negligible thermal expansion, the "periodicity" will become increasingly broken with temperature. Coupling this to the necessarily small displacements required by a harmonic approximation to the mean-field potential, Eq. 4.46 may be limited in describing the behavior of branched or cross-linked polymers at higher temperatures.

Flory-Orwoll-Vrij Equation

The treatment by Flory *et al.* departed from the cell model to address the inadequacy of describing a polymer liquid as a quasi-crystalline system. Their development did maintain the separation of internal and external degrees of freedom and therefore derived a partition function of similar form to that found by Prigogine *et al.* However, rather than maintaining an equal intra- and inter-monomer distance, Flory *et al.* segmented the modeled chain molecule into x subdivisions. The only stricture on x was that it be linear in n, the number of repeat units; otherwise, its definition was left open [57]. Furthermore, the modeled polymer contained terminal groups, rather than r identical monomers, to distinguish intermolecular forces exerted at the ends of chains from those in the middle. Flory *et al.* described their assumed intermolecular potential as having, (1) hard-sphere repulsive potential and (2) soft attraction of "unspecified character." Given these considerations, the Flory-Orwoll-Vrij (FOV) equation was found to be

$$\frac{P_R V_R}{T_R} = \frac{V_R^{1/3}}{V_R^{1/3} - 1} - \frac{1}{V_R T_R} \tag{4.48}$$

where the reduced pressure, volume, and temperature are again denoted with a subscript. The fundamental definition of the reduced variables are the same as Eq. 4.47.

In comparing the EOS developed by Prigogine *et al.* and Flory *et al.*, the FOV equation would seem more appropriate in describing liquid phase behavior. To illustrate this comparison let us consider high-density polyethylene (HDPE) at 175°C. In contrast to the empirical EOS comparisons, a specific example is chosen here to illustrate the applicability of specific models. HDPE is a simple hydrocarbon with minimal branching in contrast to its low-density counterpart, LDPE. The lack of branching in HDPE affords a more efficient packing. This behavior may be more aptly described by the spatially restricted cell model of Prigogine *et al.* However, the isotherm used for comparison is at 175°C which is well above the ~130°C melt temperature for HDPE. This endows the material with some liquid character which the FOV equation should more appropriately account for.

The 175°C isotherm for HDPE is shown in Fig. 4.6. The empirical data for HDPE, as well as, the characteristic parameters (P^*, V^*, and T^*) were taken from the recent polymer EOS compendium by Rodgers and references therein [68, 77].

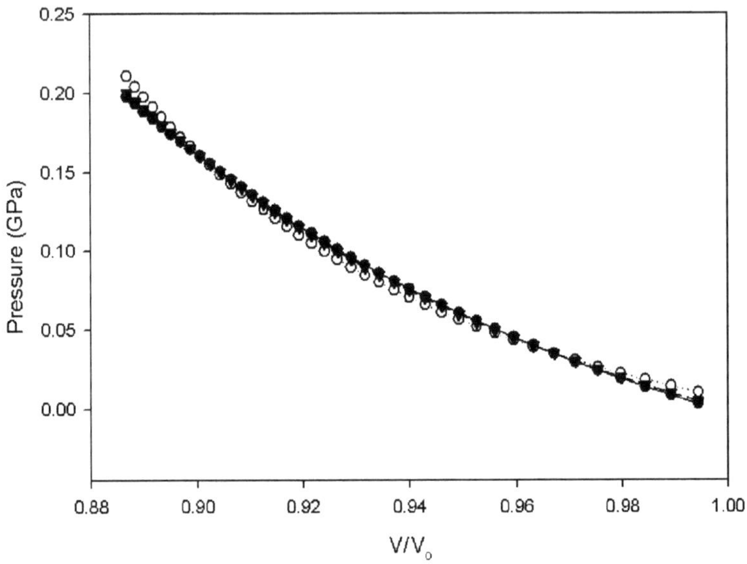

Fig. 4.6 The 175°C isotherm for HDPE (▼) compared with the FOV equation (○) and the cell model equation from Prigogine *et al.* (•)

Inspection of Fig. 4.6 reveals that while the FOV equation represents the *P–V* isotherm well, the cell-model equation by Prigogine *et al.* practically overlays the empirical data. These results may reflect inherent crystallinity of HDPE even at elevated temperatures. Perhaps at still higher temperatures the results would reverse with the FOV equation providing more precise results; however, current experimental data for HDPE is limited to approximately 200°C.

Sanchez-Lacombe Equation

Thus far vacancies in polymeric materials have not been addressed. Clearly, the packing of polymers in random orientations could generate voids in the material, which would consequently affect its *P–V* behavior. To address these voids, there is the lattice-fluid model, in which, the compression of the voids dictates the *P–V*behavior of the material [78]. The lattice itself remains constant and thus changes in volume are brought about by changing vacancy concentrations.

As representative of the lattice-fluid model, the Sanchez-Lacombe equation, which was originally based on an Ising fluid, was developed to establish a general theory for mixed fluids. In connection to polymer liquids, the mixed fluid can be considered a *r*-mer liquid with vacancies dispersed throughout. In their partition function, the only non-zero interaction energy arose from non-bonded, monomer–monomer interactions. Thus, hole–hole, hole–monomer, or intramolecular interaction energies were disregarded. Evaluation of the partition function via the method of steepest descent, Sanchez and Lacombe found in the limit of high molecular weight, i.e., r approaching infinity, the equation of state is given by

$$\frac{P_R V_R}{T_R} = -V_R \left[\ln\left(\frac{V_R - 1}{V_R}\right) + \frac{1}{V_R} \right] - \frac{1}{V_R T_R} \tag{4.49}$$

Equation. 4.49 has been recast from the original form to facilitate comparison with the previously discussed EOS. Although several assumptions are shared in the cell and lattice-fluid models, the potential energy of Eq. 4.49 is not as highly dependent on volume as in the cell model. This more subtle dependency is reflective of liquid behavior. While that of the cell model is more representative of solids than liquids, a criticism already discussed.

Simha-Somcynsky Equation

The final model discussed here for polymer liquids is the hole theory. As the name implies, vacancies are inherent to the model; however, in contrast to a lattice fluid, the cells of the lattice are compressible. This theory is detailed by Simha and Somcynsky in their treatment on the statistical mechanics of spherical and chain-molecule fluids. Using a square-well approximation for their cell potential, Simha and Somcynsky derive an equation of state of the form,

$$\frac{P_R V_R}{T_R} = \frac{1}{1 - 2^{-1/6} y^{2/3} V_R^{-1/3}} + \frac{2y}{T_R}\left(\frac{A}{(yV_R)^4} - \frac{B}{(yV_R)^2}\right) \tag{4.50}$$

where A and B are the Lennard-Jones parameters shown in Eq. 4.46 and y is the fraction of occupied sites. A separate expression for y, which can be found through minimization of the Helmholtz free energy, is given by

$$\frac{s}{3c}\left[\left(1 - \frac{1}{s}\right) + \frac{1}{y}\ln(1-y)\right] = \frac{\frac{y^{2/3}}{2^{1/6} V_R^{1/3}} - \frac{1}{3}}{1 - \frac{y^{2/3}}{2^{1/6} V_R^{1/3}}} + \frac{y}{6T_R}\left(\frac{2B}{(yV_R)^2} - \frac{3A}{(yV_R)^4}\right) \tag{4.51}$$

where s is the number of monomers and $3c$ is the parameterization of the external degrees of freedom. In order to calculate the P–V–T states, Eqs. 4.50 and 4.51 should be solved simultaneously [60].

The rigorous solution to the Simha-Somcynsky equation is unnecessary for the general scope of this discussion. Rather, consider the limiting case of when $y = 1$. Physically, this limit describes a lattice with all sites completely filled, i.e. a cell model. As expected, when $y = 1$ is substituted into the Eq. 4.51 the resultant is Eq. 4.46. The two equations become identical not only because of the full occupancy but also since both treatments used the same interaction potential. Therefore, the y-parameter in the Simha-Somcynsky EOS can simply be seen as a perturbation on the cell model developed by Prigogine *et al.*

4.2.6 Conclusions

Given the complexity of predicting *P–V–T* states over a wide variable range coupled to a limited transferability, there is simply no completely universal EOS. Therefore, a myriad of EOS have been developed from either an empirical, semiempirical, or theoretical basis. Of the empirical EOS, the Tait equation (in either original or modified form) has successfully reproduced the *P–V* isotherms for a variety of materials at pressures up to approximately 2 kbar. Using a mock material with given thermodynamic parameters, all of the empirical EOS fit predicted similar behavior up to ∼1.5 GPa. At still higher pressures, significant deviations were observed in comparing the various empirical EOS. These deviations stemmed from the general assumptions that distinguish the various EOS. Thus, when analyzing an isotherm that spans a wide range of pressures, careful consideration must be given to which EOS is most appropriate.

Beyond empirical equations, theoretical EOS not only allow for the calculation of *P–V–T* states but also provide insight into the fundamental forces in materials. Assuming an appropriate intermolecular potential, the partition function could be constructed. Through statistical thermodynamics, the partition function connects a material's microscopic structure to its macroscopic, thermodynamic potential (G, F, H, etc.). With the thermodynamic potential defined, extension to an EOS is facile.

To simplify the treatment of theoretical EOS often square-well, harmonic, or central potentials are assumed. Furthermore, in relation to polymer liquids, a network is imposed on the material. The various strictures of this network discriminated the cell, lattice-fluid, and hole theories. While the cell model is often criticized for describing a liquid as quasi-crystalline, all models reproduce *P–V* isotherms fairly well for a variety of polymers. However, similar to the empirical EOS, this general predictive capacity is limited in its validation to approximately 2 kbar.

In either case, whether empirical or theoretical, equations of state provide valuable information about the *P–V–T* behavior and intermolecular forces in materials. Understanding these fundamental forces is critical to a broad spectrum of research directions ranging from structure–function relationships, shock-wave dynamics or even the synthesis of new materials.

4.3 Static Experimental Methods

An obstacle to determining equations of state for polymeric materials is the lack of a simple, static experimental method for acquiring P–V–T data for solid networks and liquids at pressures greater than several kilobars. Rather, it is usually a combination of approaches combined with assumptions about material behavior beyond measured pressure and temperature regimes that is used to piece together formulations for the treatment of these materials.

In particular, the use of polymers as binders in explosive formulations means that the shock and detonation physics communities interested in their response are concerned with what are considered very high pressures and temperatures compared with those that the general polymer community would consider "extreme." Furthermore, coupling EOS information with a detailed understanding of network structure, crystalline content and associated crystalline phase transitions, thermal behavior (i.e. glass transition and melting temperatures), damage mechanisms, and degradation chemistry is *essential* for accurately predicting polymer behavior over a broad range of pressure, temperature, and deformation strain rates. Static high-pressure, isothermal results are useful for improving interpretations of dynamic shock experiments, and estimating the effects of shock heating on material response, including shock-induced melting and determination of the onset of degradation or the reactants-to-products transition. Unfortunately, polymer EOS data in the literature is sparse compared to metals and crystalline organics, and for the most part, exists at limited, low pressure regimes (<200 MPa) [68, 79].

4.3.1 Thermodynamic Properties

Density is typically one of the first quantities measured when investigating the high pressure or shock properties of a new material. For polymers, two methods are used extensively, immersion techniques and gas pycnometry. Both are based on volumetric displacement, but of liquid and gas respectively. Application of immersion techniques to polymers should be performed with care to avoid dissolution of the polymer, and consider both polymer-liquid interfacial effects (such as wettability) and void content. For these reasons, He pycnometry is often preferred over liquid immersion methods. This method can also be used to determine skeletal densities of foam materials. Some polymers can absorb He into their networks, giving erroneous results.

One of the most important questions to be asked when investigating a new polymer material concerns the nature of the polymer's phase behavior and thermal transitions. Differential scanning calorimetry (DSC) is an indispensable tool for probing thermal transitions at ambient and low (1–7 MPa) pressures. In addition, with recent advances in experimental capabilities, DSC is now commonly being used to determine absolute or direct (not referenced) heat capacities [80].

A representative DSC trace for annealed Kel-F 800 is shown in Fig. 4.7. A glass transition is visible as step function near room temperature. For Kel-F 800, the T_g ranges from 24°C–27°C. The T_g predicted from the Fox equation and the glass transition temperatures of the homopolymers polychlorotrifluoroethylene (PCTFE) and polyvinylidene fluoride (PVDF) of 27.7°C is in good agreement with the experimentally measured values [81–83]. Two melt endotherms are observable at higher temperature. The melt endotherm is known to be due to melting of the crystalline domains of the chlorotrifluoroethylene runs in the polymer structure, however the origin of the two transitions remains unknown [84]. Incidentally, it was DSC that first

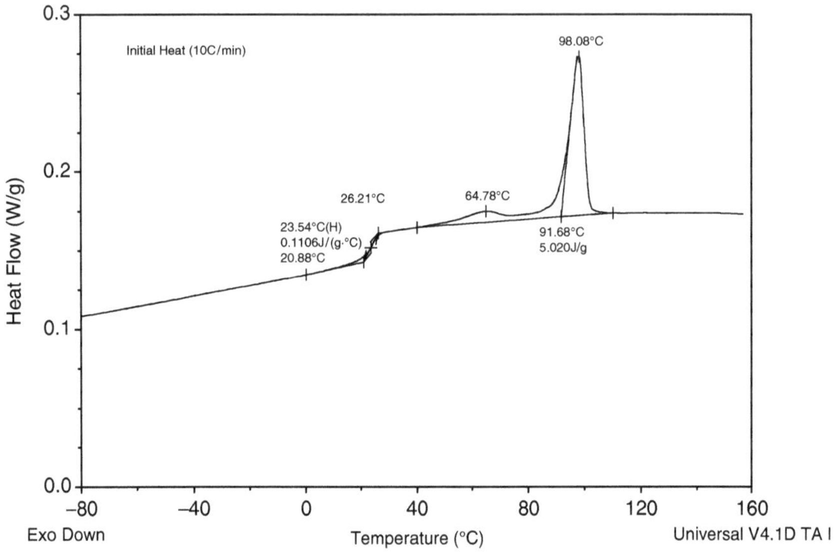

Fig. 4.7 Differential scanning calorimetry trace of highly crystalline Kel-F 800 (courtesy E.B. Orler, Los Alamos National Laboratory).

pointed to incomplete drying of Kel-F 800 during its manufacture. Billets of Kel-F 800 prepared at LANL by compression molding at 90°C and 15,000 psi showed white opaque particles trapped within the bulk sample, Fig. 4.24. DSC of one of these extracted particles showed a melt transition at 0°C, suggesting entrained water. This was later confirmed with coupled thermogravimetric analysis-mass spectrometry [81].

Figure 4.8 shows the measured absolute heat capacity for Kel-F 800 as a function of temperature. From measurement of the heat capacity using DSC, and the bulk modulus and thermal expansion coefficient from dilatometry, one can also calculate the bulk Grüneisen coefficient.

An interesting feature of many polymers is the pressure dependence of the glass and melt transition temperatures. The pressure coefficient, $\frac{dT_g}{dP}$, for the glass transition temperature can be quite large, reaching nearly 400 K/GPa for some materials in the low pressure limit (where the effects of the compression of free volume can be the greatest) [86, 87]. Measurement of the shift of the melt and glass transition temperatures with pressure are typically performed using dilatometry methods (PVT) or dielectric spectroscopy.

4.3.2 Bulk Dilatometry Methods

Bulk dilatometry experimental methods such as piston-cylinder and anvil-based devices use the general operating principle of applying a compressive force through

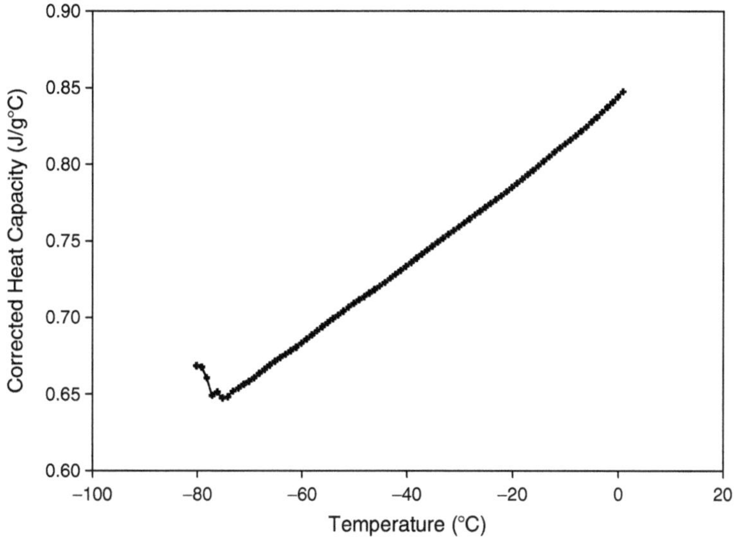

Fig. 4.8 Heat capacity of Kel-F 800 from quasi-isothermal DSC measurements [85]

the use of hardened metal components coupled with the use of confining fluids or powders, such as mercury, nitrogen or argon, to impart a high hydrostatic pressure environment around a sample under study. There is always a tradeoff in high-pressure techniques between the sample size and the obtainable pressure, as the force must be directed over a smaller area to achieve higher pressures. Large volume presses today can accommodate mm to cm-sized samples at 15 GPa [88].

An excellent review of early high-pressure experimental developments is given by Bridgman [89]. Development of high-pressure apparatuses has been driven, in large part, by the desire to reach the pressures deep in the Earth's (and other planet's) interior to interrogate a variety of physical and chemical properties of minerals. A multi-anvil apparatus designed by Kawai and Endo was capable of reaching 25–30 GPa, or the pressure at the top of the Earth's lower mantle [90]. Anvil-based apparatuses are able to reach much higher pressures than most piston-cylinder-based devices, however they are usually applied to solid materials. To reach such high pressures, the anvil materials must be fabricated from high-strength materials such as tungsten carbide, boron carbide, and various types of hardened steels. A review of anvil-based devices is given by Mao and Hemley [88].

In the polymer field, most literature P–V–T data has been obtained using piston-cylinder or "piston-die" methods, due to the means of sample support and confinement offered by these methods. The premise of these devices is the compression of a sample cell assembly enclosed in a cylindrical chamber with either a compressible powder or mercury confining medium. The use of a confining medium imparts hydrostatic pressure around the sample. The prototypical piston-cylinder cell is the Boyd and England design [91]. In these systems, the pressure is determined from the force on the piston, while the volume change is measured directly from measuring the movement of the piston by displacement arms within the assembly. Attainable pressures are typically in the 2–4 GPa range depending on the choice of materials and amount of support. Because of the crush behavior of many of the confining and sample cell materials, data below ~0.5 GPa is usually discarded.

These techniques [50] can be used to obtain pressure–volume curves for the determination of isothermal bulk moduli, and volumetric coefficients of thermal expansion when performed at variable temperatures. Weir and Bridgman began investigating the isothermal compressibility of polymers as early as the late forties and early fifties [92–94]. More recently, with the broad use of polymers in a variety of applications, there have been several recent reports and reviews describing the bulk compressibility and applicable equations of state for a variety of polymers, including poly(dimethylsiloxanes) [46, 68, 95–99, 101–109]. A specialized cell that has been applied extensively to polymers and other soft materials is a piston-die apparatus designed by Zoller, and often referred to as a Gnomix apparatus in its current form [110, 111]. Zoller and Walsh have summarized their efforts cataloging the P–V–T response of numerous polymers, including PTFE, PE, and PDMS materials [112]. We have applied the piston-cylinder approach using a Gnomix PVT apparatus (Datapoint Laboratories) to obtain P–V–T information for several materials of interest, including PTFE, Estane 5703 and Estane-NP mixtures, Kel-F 800, PCTFE, and several elastomers. While piston-cylinder techniques are useful for probing the response of the bulk polymer, they are limited to low pressures. The Gnomix apparatus, for example, is limited to 0.2 GPa. Unfortunately, this is still within the more crushable or non-linear compression pressure regime for most polymers, and the application of the data to higher pressure regimes must be performed with caution.

4.3.2.1 Examples: THV 500

Poly(Tetrafluoroethylene-CoVinylidene Fluoride-Co-Hexafluoropropylene)

The P–V–T data for PTFE obtained from dilatometry is well documented [112]. Copolymerization of PTFE is commonly performed to improve processing and other properties in fluoropolymers. PTFE copolymerized with hexafluoropropylene and vinylidene fluoride can be performed to obtain THV 500, a terpolymer containing 60 wt% tetrafluoroethylene (TFE), 20 wt% hexafluoropropylene (HFP), and 20% vinylidene fluoride (VDF) [113]. The THV 500 molding powder is manufactured by Dyneon, a 3M (Minnesota Mining and Manufacturing) Company.

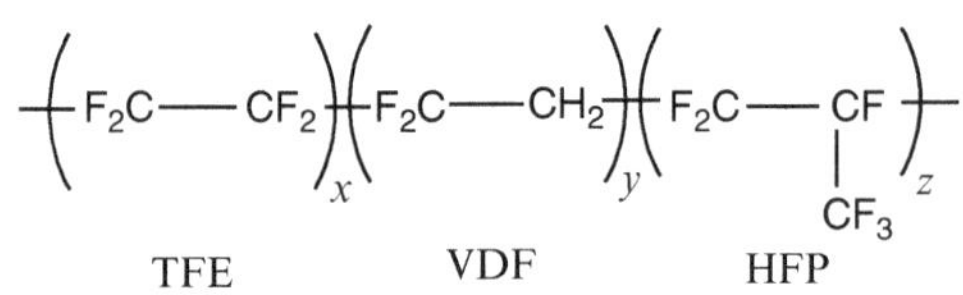

Solid pucks or billets of THV 500 can be prepared by compression molding. The density of the material we have examined is 2.01 gcm^{-3} with values ranging from 2.004 to 2.010 g cm^{-3} measured by He pycnometry, with a T_g at ~28°C–36°C and T_m= 157°C. X-ray diffraction reveals that the crystal structure of THV 500 more closely matches that of phase IV of PTFE in TFE/VDF copolymers than PVDF. This observation and the high TFE content leads us to believe that the TFE domains are crystallizing. For the compression-molded samples studied, the percentage crystallinity is between 23% and 25% within the 60% PTFE domains.

We have applied low pressure dilatometry methods to determine the P–V–T behavior of THV 500. Figure 4.9 shows the isobars for THV 500 at 0, 0.04, 0.08, 0.12, 0.16, and 0.2 GPa, the standard pressures accessed by the Gnomix apparatus [114].

From analysis of the P–V–T data, the volumetric thermal expansion coefficient in the linear regime between 23°C–68°C is 6.89×10^{-4}/°C. As confining pressure is applied, the thermal expansion coefficient decreases to 2.39×10^{-4}/°C by 0.2 GPa. From inspection of the V–T data, there is a clear inflection point associated with the melt transition that shifts to higher temperatures with confining pressure. Pressure decreases chain mobility and is known to shift transitions in polymers such as the glass transition to higher temperatures [115]. For example, in PMMA the glass transition temperature has been reported to shift by 200°C/GPa and asymptotically approach a limiting high-pressure value by ~0.6 GPa [87]. A fit of T_m vs. pressure to a linear fit gives a pressure derivative of 49°C/MPa, which equates to a shift in T_m with pressure from 169°C at ambient pressure to 253°C at only 2 kbar! Further inspection of the data suggests that the glass transition temperature may also be shifting to higher temperatures, though the maximum of the transition is much harder to decipher.

While the room temperature isotherm was not measured, it can be extrapolated by a linear extrapolation from the high temperature data to 23°C. Application of

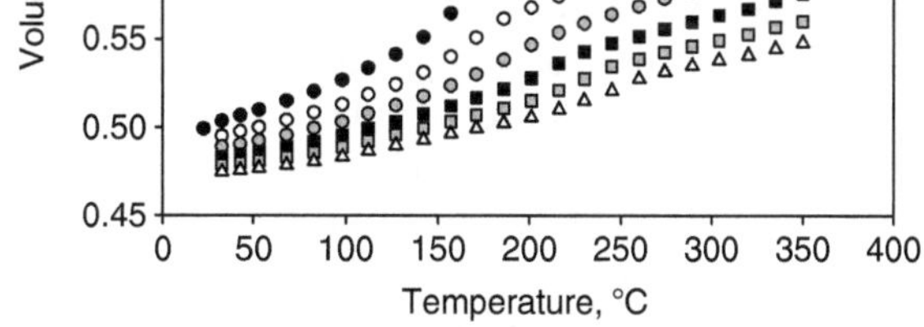

Fig. 4.9 Isobars for THV 500°C to 350°C at 0, 40, 80, 120, 160, and 200 MPa

Table 4.2 Results of application of isothermal fitting forms to the room temperature isotherm of THV 500

EOS form	$K_{T,0}$(kbar)	K'	Comments
Birch-Murnaghan	26.6 ± 0.38	16.2	
Vinet	26.9 ± 0.30	7.3	
Sun	29.3 ± 0.30		$n = 6.14$, $m = 1.16$
Tait	27.2		$C = 0.0744$

isothermal fitting forms to the calculated room temperature isotherm gives a reasonably consistent zero-pressure isothermal bulk modulus and its pressure derivatives, Table 4.2.

Combining the results from dilatometry, with the measured heat capacity also allowed for calculation of the Grüneisen parameter [116].

4.3.2.2 Kel-F 800

Poly(Chlorotrifluoroethylene-Co-Vinylidene Fluoride)

Poly(chlorotrifluoroethylene-*co*-vinylidene fluoride) is known by its trade name FK-800®, formerly Kel-F 800® resin. Kel-F 800 is a copolymer of chlorotrifluoroethylene and vinylidene fluoride in a 75:25 weight ratio. Kel-F 800 was first developed by M. W. Kellog Company in the early 1950s and became commercially available in 1955. In fact, the name "Kel-F" originates from a combination of "Kellogg" and "fluoropolymer". 3M (Minnesota Mining and Manufacturing) continued to market Kel-F 800 under this name until 1994. There has been some confusion about the properties of Kel-F 800 due to its close relatives also bearing the "Kel-F" name, including Kel-F 81 (also known as Kel-F 300), the pure homopolymer of polychlorotrifluoroethylene (PCTFE). PCTFE is more crystalline and has a density of 2.13–2.14 g cm^{-3}, whereas copolymerization of PCTFE with VDF reduces the density to 1.998–2.02 g cm^{-3}, depending on processing method and percentage crystallinity.

The chemical and mechanical properties of Kel-F 800 were first studied extensively by Cady and Caley due to its evaluation as a binder material in PBXs [117]. Kel-F 800 exhibited several advantageous properties that led to its common use in PBXs. Kel-F 800 has a glass transition temperature at room temperature, $T_g = 28°$C, and melting temperatures between 85°C and 110°C. The desirable melt and flow characteristics are derived from the addition of PVDF, which acts to decrease the percentage crystallinity compared to PCTFE. Kel-F 800 is also thermally stable. Cady and Caley found that the onset of weight loss in the copolymer varied from 150°C to 210°C and at low temperatures was attributed to entrapped emulsifier (not the polymer itself) [117]. In fact, there is no significant weight loss below 200°C with decomposition of the polymer beginning at 300°C in air that is not complete until nearly 450°C. The linear copolymer is soluble in several common solvents such as acetone, methylethylketone (MEK), and tetrahydrofuran (THF), and exhibits flow below and approaching its low melting temperature (T_m~110°C).

Another unique feature of Kel-F 800, though perhaps undesirable in its PBX application, is its near-room temperature glass transition temperature. The average T_g was found by Cady and Caley to be ~30°–31°C [117]. In recent studies at LANL, the T_g was observed at ~28°C by DSC. At the T_g, there is a threefold decrease in the storage modulus across the temperature range of the transition.

One of the challenges associated with the determination of mechanical and equation-of-state properties of Kel-F 800 is that it must be processed to produce billets for bulk measurements. Kel-F 800 is supplied from 3M, Inc. as opaque white to yellow colored granules. As with all polymers, processing conditions and thermal history will cause widely varying bulk mechanical and thermal properties. The material reported here was processed by subjecting the granules to cycling between ambient pressure and vacuum at temperatures greater than 115°C for up to 14 days. The resultant material was then annealed at 65°C to promote crystal growth for 40 days, and quenched at 0°C, resulting in a percentage crystallinity of 12% by integration of the melt endotherm and referencing of the value to the heat of fusion of 100% crystalline PCTFE.

P–V–T data to 0.2 GPa at multiple pressures for annealed (~12% crystallinity) Kel-F 800 have also been measured [118]. There is a strong discontinuity in the isobar data near 110°C corresponding to the primary melt transition of the polymer. Surprisingly, the melt transition for Kel-F 800 *does not appear to shift with pressure*, despite the similarities between Kel-F 800, and THV 500, and PCTFE in which the melt transition shifts measurably with pressure. It could be that there is more free volume in the polymer network making the influence of pressure on chain mobility negligible at low pressures. Extension of these experiments to higher pressures is necessary (and in progress) to investigate this behavior further.

Analysis of the room temperature isotherm and zero-pressure isobar allowed for determination of the isothermal bulk modulus and its pressure derivative through the application of isothermal EOS fitting forms, Table 4.3.

The volumetric coefficient of thermal expansion was also determined. Between 24°C and 87°C, $\alpha = 6.73 \times 10^{-4}/°C$. The thermal expansion coefficient is lower below $T_g (\alpha = 5.08 \times 10^{-4}/°C)$, and higher above $T_g (\alpha = 6.81 \times 10^{-4}/°C)$.

Table 4.3 Results of application of isothermal EOS fitting forms to the room temperature isotherm for annealed Kel-F 800

EOS form	$K_{T,0}$ (GPa)	K'
Tait	2.70	
Murnaghan	2.70	15.1
Birch-Murnaghan	2.60	23.53
Vinet	2.66	8.94

4.3.2.3 Estane 5703 and Estane-NP

The phase-separating binder Estane 5703 poly(ester urethane) can be obtained in pellet form from the BF Goodrich Co. Estane 5703 contains approximately 23 wt % "hard" segments which are composed of 4,4′-methylenediphenyl 1,1′-diisocyanate (MDI), and a 1,4-butanediol chain extender. The "soft" segments are comprised of poly(butylene adipate). Estane pellets can be compression molded at 110°C to form a material with $\rho = 1.19\,\mathrm{g\,cm^{-3}}$ and $T_g = -35$°C (by dynamic mechanical analysis) [119]. In PBX formulations, such as PBX 9501, the binder is often a 50:50 mixture of Estane and "nitroplasticer," with a small quantity of stabilizer added. Nitroplasticizer (NP) is the common name for the 50:50 eutectic mixture of bis-(2,2-dinitropropyl)formal (BDNPF) and bis-(2,2-dinitropropyl)acetal (BDNPA), a liquid at room temperature with a density of ~1.385 g $\mathrm{cm^{-3}}$. The structures of BDNPF and BDNPA are given below.

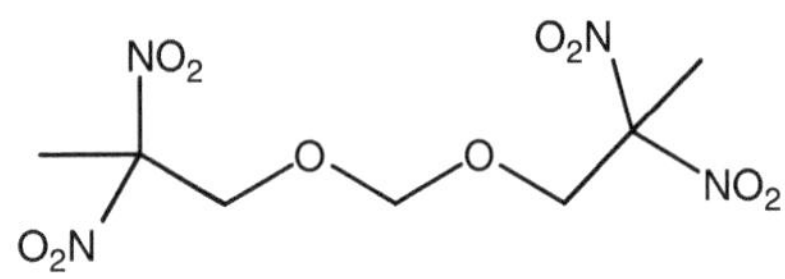

BDNPF

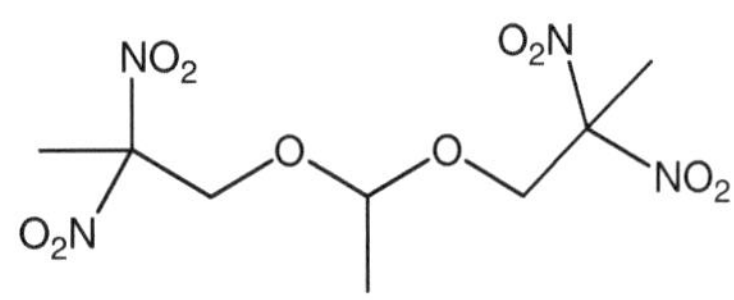

BDNPA

The results of dilatometry measurements for Estane 5703 are shown in Fig. 4.10 to 122°C. The data are reasonably linear at all of the pressures studied across the temperature range from 33°C to 122°C. From a linear fit to the V–T data at zero confining pressure, the thermal expansion coefficient $\alpha = 6.64 \times 10^{-4}$/°C.

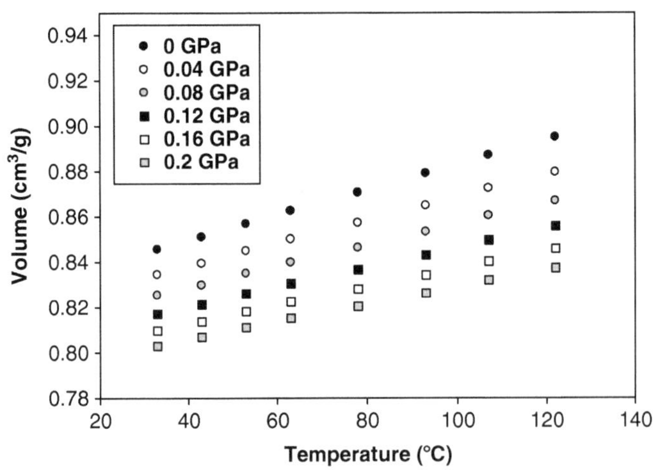

Fig. 4.10 Isobars for Estane 5703 with an initial density of 1.19 g $\mathrm{cm^{-3}}$

Table 4.4 Results of third-order Birch-Murnaghan EOS fits to isotherms for Estane 5703

Temperature (°C)	$K_{T,0}$(GPa)	K'
33	2.86	13.1
43	2.68	14.7
53	2.65	12.8
63	2.55	12.8
78	2.38	13.0
93	2.24	13.3
107	2.18	11.9
122	2.08	11.8

Table 4.4 gives the results of fits to several isotherms to the Birch-Murnaghan EOS form as a function of confining pressure for Estane 5703. As expected the zero-pressure isothermal bulk modulus decreases as the material softens with an increase in temperature.

For a 50:50 mixture of Estane 5703 and nitroplasticizer, P–V–T data were also measured at three temperatures. Interestingly, the isothermal bulk moduli are similar to that for neat Estane 5703. However, the pressure derivative of the modulus, which gives information about the curvature of the compression curve at high pressures, is lower for the plasticized material. The moduli and their pressure derivatives for the Estane-NP material as a function of temperature are as follows: $33°\mathrm{C} K_{\mathrm{T},0} = 2.90\,\mathrm{GPa}$, $K' = 9.47$, $43°\mathrm{C} K_{\mathrm{T},0} = 2.82\,\mathrm{GPa}$, $K' = 9.9$, $53°\mathrm{C} K_{\mathrm{T},0} = 2.7\,\mathrm{GPa}$, $K' = 10.2$.

4.3.3 Diamond Anvil Cell Methods

The successful development of laboratory-scale static high-pressure-generating experimental devices relied on the use of high-strength anvil and supporting materials. In the late 1950s, the hardest natural material on Earth, diamond (from the Greek *adamas*; invincible [120]), was first used in Bridgman anvil assemblies [88]. Extension of the use of diamonds rapidly progressed into today's modern palm-sized diamond anvil cells, advancing the pressure regime safely accessible in laboratory experiment to over 100 GPa [121, 122].

The transparency of diamonds also greatly advanced the static high-pressure field by allowing for numerous probes of material properties under high pressure to be applied. These include visible inspection of sample states, electronic and vibrational spectroscopies, x-ray and neutron diffraction and spectroscopic methods, *in-situ* temperature and pressure measurements, and even shock-wave compression from initial high-pressure states. Here, we describe examples of the application of some of these methods to high-explosive binders and related polymers.

4.3.3.1 X-Ray Diffraction

X-ray diffraction methods paired with high-pressure diamond anvil cells has by far become the paradigmatic method for determining P–V isotherms for metals, minerals, and other crystalline materials. The commissioning of second and third generation synchrotron sources both in the US and abroad have brought with them an explosion of research activities into materials properties at high pressure. Included in this is the recent application of synchrotron-based x-ray spectroscopy methods, such as inelastic x-ray scattering, and extended edge x-ray absorption fine structure. However, most polymers do not possess significant order in the form of crystalline domains requisite for application of diffraction (x-ray or neutron) methods. This includes the large class of elastomers that are often used as high-explosive binders.

X-ray diffraction using both synchrotron and laboratory x-ray sources have been applied to numerous semicrystalline polymers, with PTFE and polyethylene being materials that have been extensively studied and relevant to explosive binders. Previous high-pressure studies on polyethylene have reported mixtures of orthorhombic and metastable monoclinic phases at high pressures [123–125]. Likewise, the phase diagram of PTFE is rich, with at least four known phases, including a high-pressure planar zigzag structure at pressures >6 kbar. The phase behavior of PTFE is described in detail below.

The general approach to developing P–V curves for materials at high pressure is to index the phases to a space group with known lattice parameters, or to apply a structure-fitting program to determine the space group from knowledge of the x-ray wavelength, observed peaks, and Bragg's law:

$$n\lambda = 2d_{hkl}\sin(\theta) \tag{4.52}$$

The lattice parameters, and hence crystal cell volume, are then tracked as the pressure is increased, with careful inspection of the data for possible x-ray degradation or presence of high-pressure phase transitions or chemical reactions.

While the advent of synchrotron sources with increased brightness, more tightly collimated monochromatic beams, and advances in detector technology have significantly improved high-pressure x-ray diffraction measurements, semicrystalline polymers still may not possess enough crystalline domains by volume for extracting useful information. Probing the crystalline domains by using x-rays does give direct insight into the compressibility of those domains, but also neglects the remaining volume percentage of amorphous domains. We have recently applied x-ray diffraction to the polymers Kel-F 800, polychlorotrifluoroethylene, and THV 500 at high pressures. While the determination of space groups remains elusive, known low-pressure crystal structures, and similarities of these materials to PTFE and PVDF allow us to draw some conclusions about the crystallizing domains, and the presence or absence of phase transitions. An example of an x-ray diffraction pattern for Kel-F 800 is shown in Fig. 4.11.

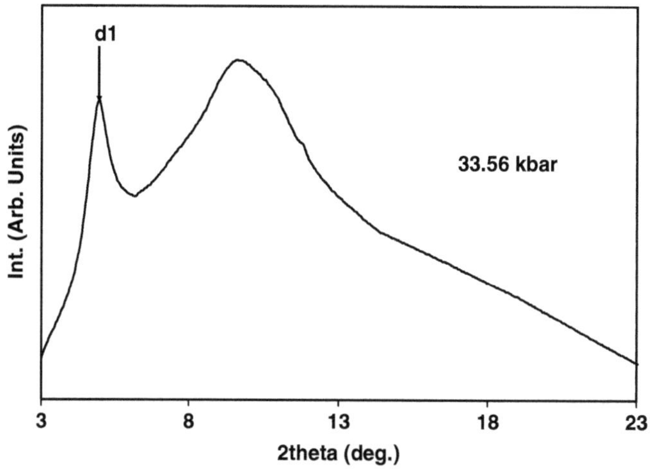

Fig. 4.11 X-ray diffraction pattern of Kel-F 800 taken by angle-dispersive x-ray diffraction with monochromatic x-rays at $\lambda = 0.4234$Å and ~34 kbar. The hydrostatic medium used in the experiments was 4:1 methanol:ethanol solution, with pressure determined by the fluorescence of a small ruby chip co-loaded with the sample

4.3.3.2 Vibrational Spectroscopy

Infrared and Raman spectroscopy are invaluable tools for understanding the behavior of materials at high pressure, and are arguably necessary methods for the discovery of high-pressure phase transitions and changes in symmetry in molecular species, the investigation of chemical reactions and high-pressure bonding changes, and pressure-dependent intermolecular interactions. The availability and high-pressure accessibility of diamond anvil cells, along with their innate optical window properties, have made the coupling of spectroscopy and high-pressure conditions commonplace. While a review of vibrational spectroscopy of polymers, or even polymers at high pressure, is beyond the scope of this chapter, it is worth noting that many of the polymers used as high-explosives binders have been studied with infrared and/or Raman spectroscopy at high pressures or under dynamic deformation. These include Estane 5703 [126–129], Sylgard 184 [50], Kel-F 800 [118, 130], THV 500 [131], and PTFE (see below). Notable features of some of the spectroscopic work reported on these binders include an observation of increased hydrogen bonding in Estane 5703 at elevated pressures, the presence or absence of high-pressure phase transitions associated with the polymer's crystalline domains (PTFE, Kel-F 800, THV 500), and understanding which modes, and by inference which portions of the polymer chain, were more or less sensitive to pressurization through individual mode-Grüneisen analysis (PMMA, polycarbonate, Sylgard 184).

Vibrational Spectroscopy as a Reporter of Material Behavior Under Extreme Conditions: Examples

PTFE is one of the most well-studied polymers at high pressure, and is a good illustrative example of the application of vibrational spectroscopy at elevated pressures to detecting phase transitions in polymers (polyethylene is another good example). PTFE is an interesting material in that its long unbranched chains exist in both crystalline and amorphous domains, with the crystalline domains adapting at least four known crystallographic phases near ambient temperature and pressure [132–135].

The phase diagram of PTFE was first reported by Bunn and Howells in 1954 [132], and subsequently refined by others [133–136]. Two crystalline phase transitions are observed at 19°C and 30°C at atmospheric pressure. Below 19°C, at atmospheric pressure, the linear chains of PTFE adopt a helical conformation with 13 CF_2 groups per 180° turn, known as Phase II. The first-order transition at 19°C between phases II and IV is associated with an unraveling of the helical chain conformation from a well-ordered triclinic structure with 13 atoms/180° turn to a ordered hexagonal phase with 15 atoms/turn. The symmetry of the helical chains in Phase IV is $D(14\pi/15)$. Further rotational disordering and untwisting of the helices occurs above 30°C to give Phase I, a pseudohexagonal structure with dynamic conformational disorder and long-range positional and orientational order, (i.e., the interchain repeat structure is largely retained but the intrachain helical structure adopts a random repeat twist angle). From a group theoretical analysis of phase IV, there are 24 normal modes for PTFE: $4A_1$(Raman-active), $3A_2$ (IR-active), $8E_1$(IR and Raman-active), and $9E_2$ (Raman-active) [137–140]. The phase II–III phase transition has been well-interrogated under high hydrostatic pressure conditions [134, 135, 141–143]. Phase III is the known "high-pressure phase," adapting a planar zigzag conformation, with both orthorhombic and monoclinic phases being reported in the literature [144]. The II–III phase change is "reported" by the disjointed frequency-pressure shifts of two the PTFE vibrational modes. The band associated with the t(CF_2) mode at $\sim$295 cm^{-1} shifts dramatically to lower pressure (by up to 15 cm^{-1}) at the phase transition, in addition to a disjointed shift in the $\rho(CF_2)$ band at 578 cm^{-1} to *ca.* 615–620 cm^{-1} Further pressurization reveals a nearly linear shift in the vibrational bands to higher pressure, with coincident band broadening. Between ambient pressure and 40 kbar, $\frac{d\bar{\nu}}{dP}\left(\frac{cm^{-1}}{kbar}\right) = 0.205$ for $t(CF_2)$, 0.409 for $\Delta(CF_2)$, 0.287 for $\Delta(CF_2)$, and 0.471 for $\nu(CF_2)$, from a linear fit of vibrational energies vs. pressure, Fig. 4.13. Examples of Raman spectra for PTFE below, at, and above the II–III phase transition in the low-frequency region are shown in Fig. 4.12. The kinetics of this transition have been previously studied at cold temperatures by Nicol *et al.* and analyzed using the Avrami equation [145].

Extending our studies to examine fluoropolymer relatives of PTFE revealed different phase behaviors at high pressures. Copolymerization of TFE with monomers such as vinylidene fluoride (VDF) and hexafluoropropylene (HFP) generally

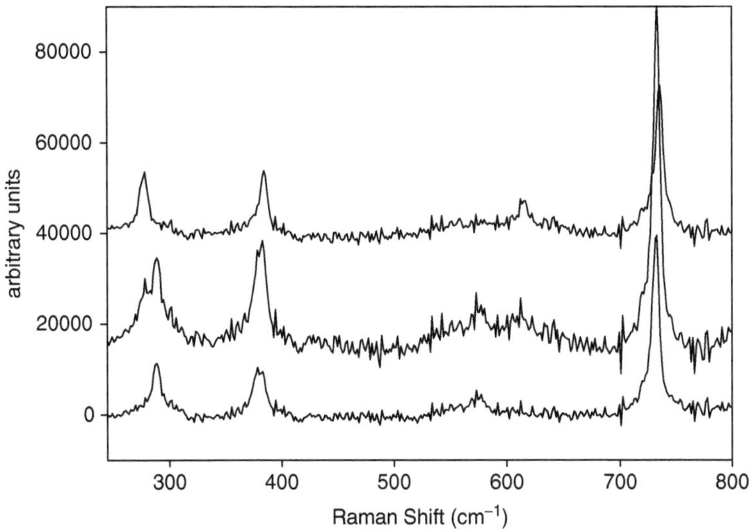

Fig. 4.12 Spectra of PTFE at three elevated pressures near the phase II–III phase transition at 6.5 kbar. The bottom curve is below the transition pressure, between 3–5 kbar. Increasing the pressure to the phase boundary (6.5 kbar), middle curve, reveals the presence of both phases within the focal volume of the laser probe. Note the shoulder that has grown in on the low frequency side of the band at ~295 cm^{-1}, and the presence of two peaks near 580–620 cm^{-1}. Increasing the pressure further completes the transition, top curve

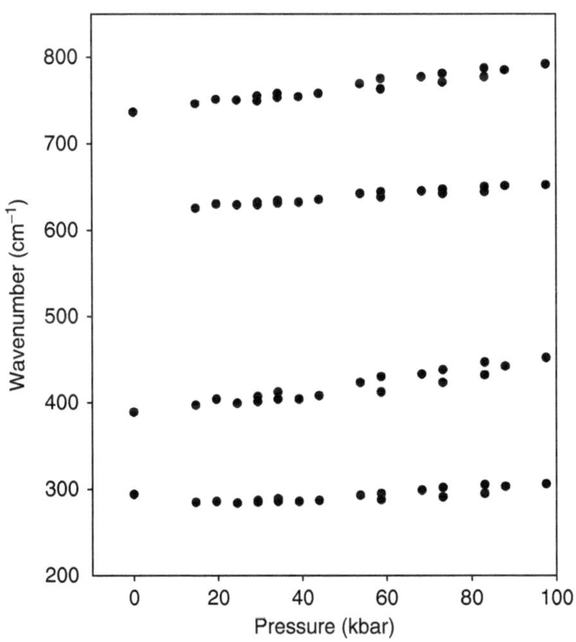

Fig. 4.13 Raman band frequencies (cm^{-1}) vs. pressure for the four low frequency bands of PTFE to 10 GPa

reduces the crystallinity, though more so for the bulky HFP compared with VDF. Copolymerization of TFE with HFP has also been shown to reduce material creep by prevention of the slippage of chains past one other easily [146].

The Raman spectrum of THV 500 (Fig. 4.14) can be assigned based on its components and the band assignments for the homopolymers PTFE and PVDF [137–139]. There is no reported data on a homopolymer of hexafluoropropylene, but one can use assignments from hexafluoropropene as an additional reference [147].

Interestingly, despite containing 60% TFE content, of which ~25% is in a crystalline form, there is no definitive evidence of a phase transition associated with the crystalline domains. While there is a change in the slope of $\frac{d\bar{\nu}}{dP}\left(\frac{cm^{-1}}{kbar}\right)$ near 40–50 kbar, which is consistent with a possible structural change indicated by x-ray diffraction, dramatic vibrational reporting of a phase transition was lacking. Similar results have been observed for the copolymer Kel-F 800, in which the crystallizing CTFE units do not appear to undergo phase changes to 4 GPa [118]. This example reveals some of the complexity in interrogating and interpreting the high-pressure behavior of polymers, particularly when they contain multiple monomers, or phases.

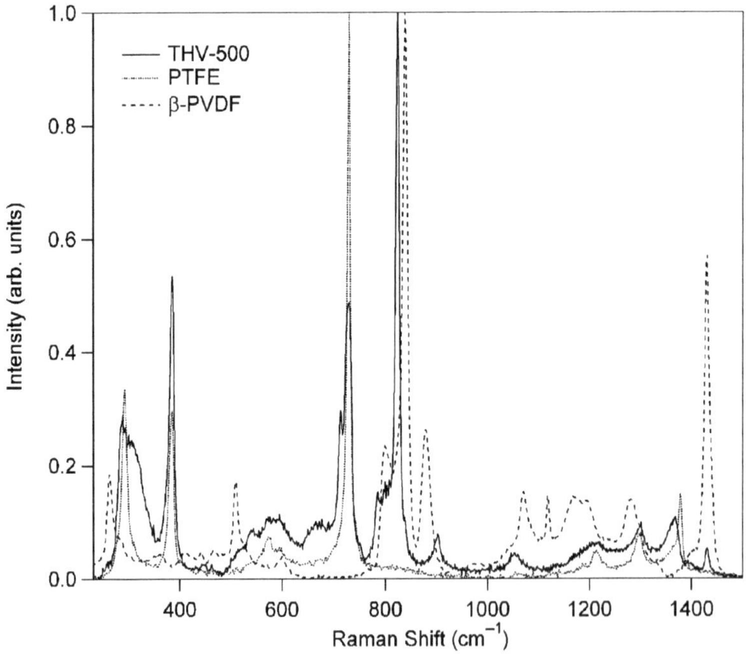

Fig. 4.14 Raman spectra of THV 500, β-PVDF, and PTFE overlaid between 200 and 1,500 cm^{-1} (Reprinted with permission)

Individual Vibrational Mode Grüneisen Analysis

The application of vibrational spectroscopies to polymers at high pressure can be extended beyond the investigation of phase behavior and identification of molecular species to the direct interrogation of thermodynamic properties. The Helmholtz free energy can be determined by summing over all of the normal modes of a molecular species. First and second derivatives of the Helmholtz free energy can then be directly related to relevant thermodynamic parameters, such as pressure. Both the heat capacity and Grüneisen parameter can be directly determined from the measurement of vibrational modes. The Grüneisen parameter is used in multiple equation-of-state forms, including the Mie-Grüneisen approach, and is used in calculating the thermal energy under shock conditions [148]. The normal modes can therefore give direct insight into both material behavior and thermodynamic properties under extreme conditions.

Mode Grüneisen analysis not only provides a quantitative measure of the bulk Grüneisen parameter, but also lends additional insight into the sensitivity of different normal modes to pressure through the use of an appropriate equation of state for the material being studied. Shifts in individual vibrational modes with pressure can be related to changes in volume by Eq. 4.53, in which the subscripts 0 and P refer to the vibrational energies (in wave numbers) and volume at ambient and high pressure, respectively, and γ_i is the individual vibrational mode Grüneisen parameter [149–151].

$$\frac{\bar{v}_0}{\bar{v}_P} = \left(\frac{V_0}{V_P}\right)^{\gamma_i} \tag{4.53}$$

The isothermal individual mode gamma, $\gamma_{i,T}$ can also be defined as

$$\gamma_{i,T} = \frac{B_{0,T}}{v_i}\left(\frac{\partial v_i}{\partial P}\right)_T \tag{4.54}$$

Which can then be combined through weighted summation to compute the bulk Grüneisen parameter: $\gamma_0 = \frac{\sum_i \gamma_i \varepsilon_i}{\sum_i \varepsilon_i}$ in which ε_i are the vibrational energies of the individual oscillators, or $\gamma_0 = \frac{\sum_i \gamma_i C_{Vi}}{\sum_i C_{Vi}}$, in which C_{Vi} are the individual mode heat capacities. The heat capacity, C_V, can also be determined from the heat capacities of individual oscillators, $C_v = \sum C_i$. The heat capacity for polymers has been described by $C_v = \sum_{i=1}^{4N} C_i + N\sum_{i=1}^{14} C_i = 4Nk + N\sum_{i=1}^{14} C_i$. [152]

This analysis has been previously applied to both PMMA and polycarbonate at high pressures [149–151]. We have recently applied this analysis to the Raman modes of Sylgard 184, a cross-linked polydimethylsiloxane polymer, between 200–1,400 cm^{-1} [50]. Figure 4.15 shows Raman data for Sylgard 184, converted to $\ln[\nu_p/\nu_0]$ vs. $\ln[V_0/V_p]$ using a Tait equation-of-state representation of static P–V

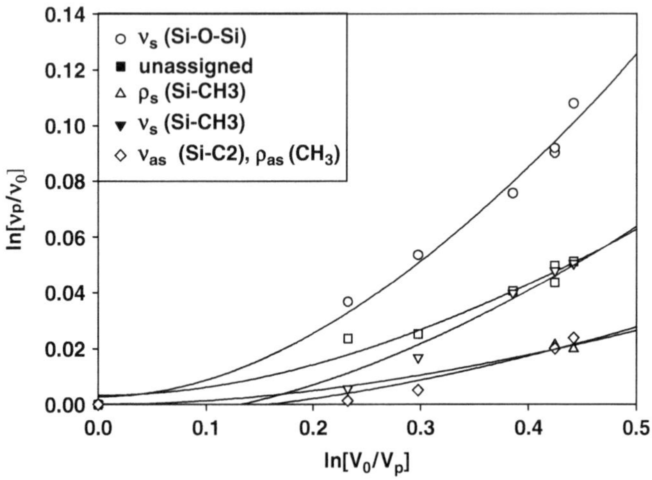

Fig. 4.15 Vibrational frequencies for Sylgard 184 as a function of pressure converted to $\ln[\nu_p/\nu_0]$ vs. $\ln[V_0/V_p]$ using a Tait equation of state fit to static high-pressure P–V data to convert the data to volume. The lines are the fits to a volume-dependent mode Grüneisen form, Eq. 4.53, and show that the symmetric Si–O–Si stretch is most sensitive to pressure (Reprinted with permission from Dana M. Dattelbaum, Journal of Chemical Physics, 122, 144903 (2005). Copyright 2005, American Institute of Physics.)

data [50]. From the behavior of the vibrational modes in this plane, one can see that the dependency is not linear, an indicator that the commonly assumed relation $\gamma\rho$ (the bulk Grüneisen parameter times density) or γ/V = constant with increasing pressure (or $\gamma = \gamma_0(\frac{V}{V_0})^{q=1}$) does not hold for this material [148, 153]. From $\ln(\frac{\bar{\nu}_0}{\bar{\nu}_P}) = \gamma_i \ln\left(\frac{V_0}{V_P}\right)$, a linear relation is expected if γ_i is volume-independent [149]. This observation is not surprising, given the known nonlinear compressibility of polymers, and particularly PDMS, at low pressures. A volume-dependent Grüneisen parameter of the form

$$\gamma_i = (a * (1 - \frac{\Delta V}{V_0})) = (a * (1 - \frac{V_P}{V_0})) \tag{4.55}$$

has been proposed [149], and was applied to the data in the figure. From this analysis, the coefficient a was found to be the largest for the symmetric Si–O–Si backbone stretch, indicating that it is the most sensitive to pressurization. It is interesting that the symmetric Si–O–Si stretch associated with the backbone, *not* vibrations associated with the side-groups, was found to be the most sensitive to volume compression. In addition, all of the modes were found to be increasingly sensitive to the compressive volume change at higher pressures. Once the pressure exceeded the free volume crush-up pressure of ~10–15 kbar, the data *may* be fit to a linear function, though further investigation to higher pressures is still required to validate this approach.

Additional examples are needed to better our understanding of the response of this class of materials to the extreme environments of pressure (and temperature), and further compare different types of polymer network structures, determine the influence of filler particles, and extend our understanding of the behavior of polymers at high pressure to even higher-pressure regimes (beyond 10 GPa).

4.3.3.3 ISLS and Brillouin Spectroscopy

The acoustic properties of materials are excellent reporters of elastic properties and by inference, mechanical response. From the acoustic velocities, information can be extracted about material strength, temporal dynamics and relaxation phenomena related to molecular mobility, and equation-of-state properties. While ultrasonic methods are used at ambient conditions, there are two common methods for measuring sound velocities at static high pressures: impulse stimulated light scattering (ISLS) and Brillouin spectroscopy.

The basic components of an ISLS experiment are illustrated in Fig. 4.16. A laser beam is split and recombined at an arbitrary angle, α, both spatially and temporally.

The interference of the overlapped pulses produces an intensity distribution which in turn induces a transient thermal and acoustic grating. The acoustic wavelength, i.e., the spacing of the induced grating, can be readily calculated from,

$$\lambda_A = \frac{\lambda_E}{2\sin(\alpha)} \tag{4.56}$$

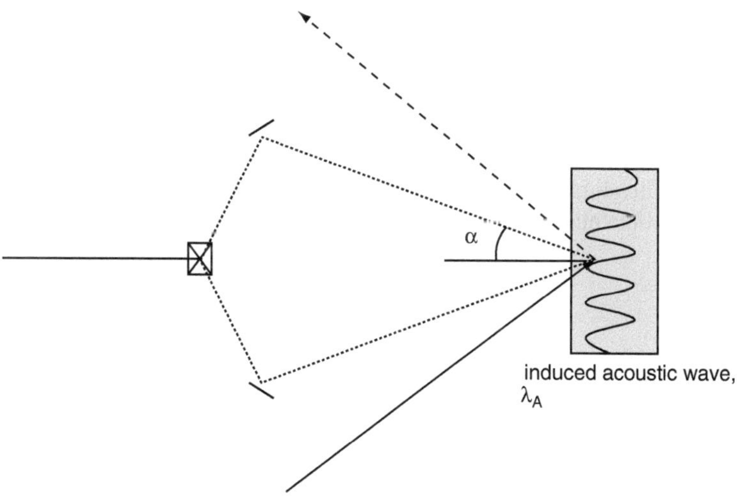

Fig. 4.16 Schematic illustration of the ISLS experiment

where λ_A and λ_E are respectively the wavelengths of the induced acoustic wave and incident laser pulses and α is the angle between the two excitation pulses. A probe laser beam, which typically has a different wavelength than the excitation source, is subsequently diffracted from the induced grating and the intensity of the diffracted probe pulse is monitored as a function of pulse delay. Analysis of the resultant ISLS spectrum yields an acoustic wave frequency and coupled with the wavelength determined from Eq. 4.56 provides a sound speed determination.

A related, though fundamentally very different method for determining sound velocities at high pressure is Brillouin scattering. Brillouin spectroscopy is analogous to its more well-known counterpart Raman spectroscopy, in that it is an inelastic scattering technique. In Brillouin spectroscopy, coherent light is inelastically scattered from acoustic phonons in a material. Liquid samples generally do not support shear, and so only longitudinal modes are observable. In isotropic solids, both transverse and longitudinal modes are commonly observed. The theory of Brillouin scattering is well-documented [154, 155]. As a thermally excited acoustic phonon travels through a material, the molecular displacements generate a periodic fluctuation in the local dielectric. Bragg reflection from this "thermal grating" gives rise to the Doppler-shifted frequencies observed in Brillouin scattering. The acoustic velocities are then calculated from these frequency shifts using the Brillouin shift equation. For scattering from a general material, the Brillouin shift equation is given by,

$$\Delta v = \frac{\upsilon v_i}{c}\sqrt{n_i^2 + n_s^2 - 2n_i n_s \cos\theta} \tag{4.57}$$

where n_i and n_s represent, respectively, the refractive indices of the material along the incident and scattered light directions, θ is the angle between these directions, Δv is the frequency shift relative to the incident frequency v_i, and υ is the acoustic velocity.

In application to polymer systems, often the polymeric material is assumed to be optically and elastically isotropic. Furthermore, in high-pressure Brillouin-scattering experiments, symmetric, forward scattering is typically used [156, 157]. Descriptions of standard and high-pressure Brillouin instrumentation can readily be found in the literature [158–160]. The application of this scattering geometry to an optically isotropic material reduces the Brillouin shift equation to,

$$\Delta v = \frac{2\upsilon}{\lambda_o}\sin\left(\frac{\theta}{2}\right) \tag{4.58}$$

where Δv is the inelastic shift from the Rayleigh (elastically scattered) line. As a consequence of forward, symmetric scattering, the effects of refraction at the interface with the diamond, pressure medium and sample, mutually cancel. Thus, the refractive index dependence in the Brillouin shift equation is removed. For high-pressure Brillouin scattering this is particularly convenient because the acoustic velocities can be essentially calculated directly from the measured frequency shift, using Eq. 4.58. Measuring these velocities and their pressure dependence affords detailed insight into, for example, the mechanical and thermal properties, equations of state analysis, and fundamental intermolecular interactions.

Recently, the acoustic properties of three polymer elastomers, a cross-linked poly(dimethylsiloxane) (Sylgard 184), a cross-linked terpolymer poly(ethylene-vinyl acetate-vinyl alcohol) (VCE), and Estane 5703 were measured from ambient pressure to approximately 12 GPa by using Brillouin scattering in high-pressure diamond anvil cells [119]. To our knowledge, these results were the first Brillouin scattering data on polymers to pressures exceeding 10 GPa. A second study examined the pressure-dependence of the acoustic velocities for three fluoropolymers, Kel-F 800, PCTFE, and PTFE, also for the first time [161]. Representative Brillouin spectra for Estane 5703 at several pressures are shown in Fig. 4.17.

For isotropic materials, elastic properties are described by two elastic constants, C_{11} and C_{12} [8]. The elastic constants relate directly to the longitudinal and transverse acoustic velocities through Eqs. 4.59 and 4.60:

$$\rho\, \upsilon_L^2 = C_{11} \tag{4.59}$$

$$\rho\, \upsilon_T^2 = \frac{C_{11} - C_{12}}{2} \tag{4.60}$$

The elastic constants can be related to the mechanical properties of solids through computation of the bulk, shear, and Young's moduli. For isotropic solids, the bulk modulus can be calculated from the sound velocities by:

$$B = \rho \left(\upsilon_L^2 - \frac{4}{3}\upsilon_T^2 \right) \tag{4.61}$$

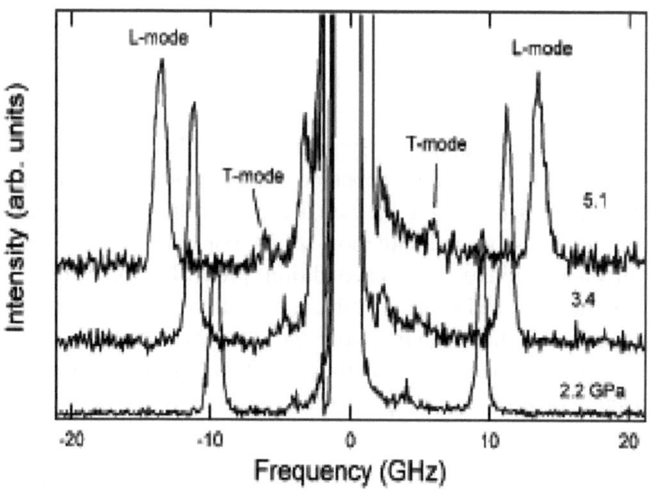

Fig. 4.17 Brillouin spectra of Estane 5703 at high pressures (Reprinted with permission from Lewis L. Stevens, Journal of Chemical Physics, 127, 104906 (2007). Copyright 2007, American Institute of Physics.)

Similarly, the shear (G) and Young's (E) moduli for an isotropic system can be determined by:

$$E = \frac{(C_{11} - C_{12})(C_{11} + 2C_{12})}{C_{11} + C_{12}} \tag{4.62}$$

$$G = \frac{C_{11} - C_{12}}{2} \tag{4.63}$$

Since both shear and longitudinal acoustic waves were observed at elevated pressures, elastic constants, mechanical property, and EOS analyses were performed for the elastomers. The elastic constants were determined directly from the acoustic velocities and Eqs. 4.59 and 4.60 for three elastomers [119]. The C_{11} and C_{12} elastic constants for all three polymers were quite similar in both magnitude and pressure dependence. Application of a linear regression to the C_{11} and C_{12} pressure dependences showed that in the low-pressure limit, C_{11} and C_{12} converge and may be approximately equal, i.e., the transverse velocity would approach zero.

B, G, and E were also calculated using Eqs. 4.61–4.63 for all three elastomers, and of relevance here, for both Sylgard 184 and Estane 5703. They show the following pressure dependencies (with y intercepts or zero-pressure extrapolated values): Sylgard 184$\partial B/\partial P = 5.6(y_0 = 9.42\,\text{GPa})$, $\partial G/\partial P = 1.6(y_0 = 3.18\,\text{GPa})$, $\partial E/\partial P = 4.3(y_0 = 8.54\,\text{GPa})$; Estane 5703 $\partial B/\partial P = 5.8(y_0 = 7.81\,\text{GPa})$, $\partial G/\partial P = 1.9(y_0 = 0.06\,\text{GPa})$, $\partial E/\partial P = 5.3(y_0 = 0.62\,\text{GPa})$. The data also reveal a common misconception about polymers. While the elastomers have low strength at ambient pressure, they dramatically strengthen/stiffen with applied pressure.

Further insight into the elastic properties can be gained through the analysis of the Poisson's ratio. Poisson's ratio relates the transverse strain to the strain of elongation, and can be calculated for isotropic materials from:

$$\sigma = \frac{\upsilon_L^2 - 2\upsilon_T^2}{2\upsilon_L^2 - 2\upsilon_T^2} \tag{4.64}$$

The calculated Poisson's ratios for Sylgard 184 and Estane 5703 were found to decrease with pressure. The values appeared to asymptote a nominal value for elastomers at ambient pressure (0.4–0.5), then decrease with increasing pressure to values more commonly associated with most solids. For reference, $\sigma \sim 0.5$ for both an ideal elastomer and liquids at ambient conditions [162]. The three elastomers appear to behave close to this ideal value at ambient conditions, then change with pressure towards values more in line with those of typical solids, ~0.33.

Measurement of the acoustic properties as a function of pressure also gives direct insight into equation-of-state properties, specifically the pressure-dependence of density by Eq. 4.12.

$$\rho_P - \rho_0 = \int_{P_0}^{P} \frac{\gamma}{\upsilon_B^2} dP \tag{4.65}$$

The bulk sound velocity, υ_B, used in Eq. 4.65, can be calculated from the longitudinal and transverse velocities with:

$$\upsilon_B^2 = \upsilon_L^2 - \frac{4}{3}\upsilon_T^2 \tag{4.66}$$

where υ_L and υ_T designate the longitudinal and transverse acoustic velocities respectively. At pressures where the bulk sound velocity could be calculated for the series of elastomers, Eq. 4.65 was integrated to generate a pressure-density or P–V curve.

Five commonly used isothermal EOS forms applied to the P–V curve yielded a range of zero-pressure isothermal bulk moduli (K_0) for Sylgard 184 and Estane 5703, with average K_0 of 1.1 and 3.2 GPa, respectively. Parallel analyses of dilatometric data below 0.2 GPa yielded extrapolated zero-pressure bulk moduli of 1.0 and 2.9 GPa for Sylgard 184 and Estane 5703. The pressure-dependent disparity in K_0 is likely due to preferential free volume compaction at low to intermediate pressures common to polymeric materials, and the frequency-dependent stiffness or viscoelasticity of the elastomers. The low and differing values of the bulk moduli point toward the different polymer chemical and network structures influencing their compressibilities, e.g., the flexible Si–O–Si backbone of the PDMS-based Sylgard 184, and the organizing feature, e.g., physical cross-linking, crystallization, and phase separation, in Estane 5703. The compaction process results in nonmonotonic changes in $\partial P/\partial V$ for a given isotherm, and reveals limitations in the capacity of an EOS to accurately reproduce the observed $P-V$ states in polymers over a broad a pressure range.

Shear waves (transverse modes) were not observable for Sylgard 184 below 3.5 GPa, and for Estane 5703 below ~1 GPa [119]. Though instrumental factors is one possible explanation to describe this unusual behavior, material features such as possible near-zero Pockels constants, and frequency- and pressure-dependent glass transition temperatures influencing chain mobility and relaxation dynamics may also be factors.

The sound velocities vs. pressure for three fluoropolymers, including Kel-F 800 and Kel-F 81 (PCTFE) show, by comparison, higher acoustic velocities at low pressures than the three elastomers, but are surpassed by those of the elastomers at pressures >5 GPa [161]. Transverse modes appeared at elevated pressures similar to the elastomers, and at just 0.27 GPa for the binder Kel-F 800. The longitudinal and transverse acoustic velocities for Kel-F 800 as a function of pressure are shown in Fig. 4.18. Transverse modes were observed at lower pressures compared with the elastomers studied. In addition, there may be evidence of a "cusp" in the acoustic data for Kel-F 800 around 3–4 GPa.

4.3.3.4 Optical Microscopy

Even with the extension of piston-cylinder methods to higher pressure, and the promising nature of the recent Brillouin results, there still remains a need for improved experimental methods for EOS determination (specifically pressure–volume

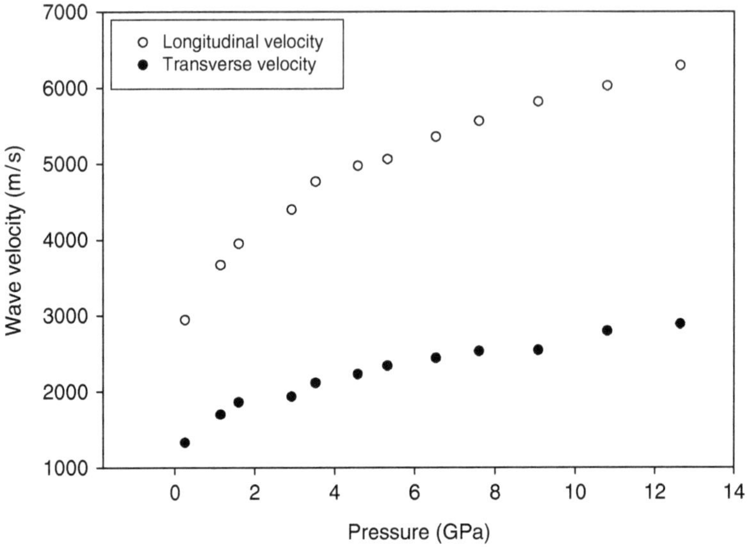

Fig. 4.18 Acoustic velocities for Kel-F 800 to > 10 GPa

relationships) for amorphous materials. One novel approach that we reported recently was the use of optical microscopy in high-pressure diamond anvil cells coupled with image analysis to directly measure changes in cross-sectional areas that could be then related to changes in volume [50]. The approach is similar to the piston-cylinder methods in which unidirectional displacements are related to volumetric strain. Middaugh and Goudey described results for 17 polymers to 10,000 psi to be used in deep-sea, moderate pressure environments using this method [163].

In this direct volume-optical approach, strain in two dimensions (area) is related to that in three dimensions (volume) by assuming that dimensional reduction under pressure occurs isotropically (uniformly) in all dimensions. The cross-sectional area of a polymer sample is determined by capturing an optical image of the sample in a high-pressure DAC. Optical micrographs of the polymer are then taken in a stepwise fashion as a function of increasing pressure, which are subsequently analyzed with image analysis tools. A typical optical image of a polymer sample loaded in a DAC is shown in Fig. 4.19.

The two-dimensional strain can be related to volumetric strain by the following procedure. For a rectangular sample of dimensions h (height), b, and c (dimensions of rectangular sides), the initial volume is given by Eq. 4.67 (the subscript 0 represents an ambient pressure value, and 1 is a high-pressure value).

$$V_0 = h_0 b_0 c_0 \tag{4.67}$$

The change in volume after application of pressure is given in Eq. 4.68.

$$\Delta V = V_1 - V_0 = h_1 b_1 c_1 - h_0 b_0 c_0 \tag{4.68}$$

Fig. 4.19 Optical micrograph of a pre-cut piece of the elastomer Sylgard 184 in a high-pressure diamond anvil cell at elevated pressure. A ^{13}C-labeled diamond chip is located to the right for internal pressure determination. The hydrostatic medium is nitrogen (Reprinted with permission from Dana M. Dattelbaum, Journal of Chemical Physics, 122, 144903 (2005). Copyright 2005, American Institute of Physics.)

Assuming that the thickness of the sample changes in proportion as the length of a given sample dimension, a constant aspect ratio is maintained, Eq. 4.69.

$$h_1 = h_0 \frac{c_1}{c_0} \tag{4.69}$$

The constant aspect ratio reduces the change in volume, Eq. 4.68, to a simple relation of cross-sectional area and sample dimensions by image analysis, where A_0 is the cross-sectional area at ambient pressure, and A_p is the cross-sectional area at pressure.

$$\frac{V}{V_0} = \left(\frac{A_p l_p}{A_0 l_0}\right) = \left(\frac{A_p \sqrt{A_p}}{A_0 \sqrt{A_0}}\right) \tag{4.70}$$

This method was applied to Sylgard 184, a cross-linked silica-filled PDMS network [50]. Comparisons between static and dynamic data were made, and the data yielded several observations. For example, the compression curve obtained by this method is stiffer than might be expected, and overlays with Hugoniot data. Given the low calculated value for the Grüneisen parameter, the offset between the isotherm and Hugoniot in the P–V plane is expected to be small, however one would still expect the isotherm to be softer than the Hugoniot. In addition, Brillouin scattering determination of the isotherm for Sylgard at high frequencies shows a softer curve. The observed "stiffened" response of the polymer could be attributed to several factors including the adsorption of nitrogen into the polymer network, errors in image analysis procedure, and a breakdown of the constant aspect ratio assumption.

This method was also recently applied to another elastomer, the cross-linked terpolymer poly(ethylene-vinyl alcohol-vinyl acetate). Poly(ethylene-vinyl alcohol-vinyl acetate) is prepared by a saponification of poly(ethylene-vinyl acetate) (EVA) using potassium hydroxide until 6% alcohol content is obtained to yield a terpolymer containing 57 wt % ethylene, 37 wt % vinyl acetate, and 6 wt % vinyl alcohol. This material is then cross-linked with a bisphenol adduct of methylene bis-(4-phenyl-isocyanate). The final product has a $T_g \sim -25°C$, and $\rho = 0.99\,\mathrm{g\,cm^{-3}}$. Figure 4.20 shows the compilation of experimental data on VCE, including the P-V curve generated using the optical method. Also shown in the figure are the results from recent Brillouin scattering results, the low pressure isotherm from dilatometry, and Hugoniot loci from plate impact experiments.

Again, the results derived from the optical method are stiffer than what might be expected, and overlay with the Hugoniot loci. There is a noticeable discontinuity in the isotherm determined from Brillouin scattering below 20 kbar which is due, in part, to an elevated pressure on-set of transverse modes, which affect the P–V curve through integration of Eq. 4.65. A Murnaghan EOS fit to the combined Brillouin and dilatometry P-V curve gives $K_{T,0} = 3.5\,\mathrm{GPa}$ and $K\prime = 5.1$. When these values are applied in a Hayes EOS formulation, the predicted Hugoniot matches the experimental results quite well (Fig. 4.20).

The coupling of microscopy methods with high-pressure diamond anvil cells is an intriguing tool for direct determination of material volume in the future, particu-

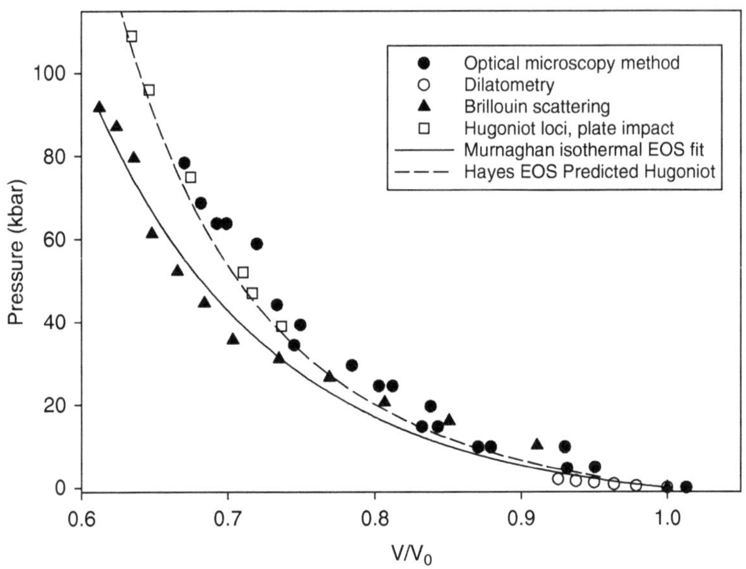

Fig. 4.20 Summary of equation-of-state data for VCE, including results from the static high-pressure methods (optical microscopy method, dilatometry, and Brillouin scattering)

larly as microscopy methods continue to be refined. Recent examples of extensions of this work include the confocal imaging of water in a diamond anvil cell, [164] and the application of x-ray microtomography to interrogate sample volume at high pressures [165]. Direct measurement of volume continues to be a challenge to be addressed for amorphous materials at high pressures.

4.4 Dynamic Experimentation

Shock compression science over the past 30+ years has provided a reasonably good assessment of the shock response of a number of polymeric materials. There was a large experimental thrust focusing on polymers during the 1960s-1970s, when much of the Los Alamos Scientific Laboratory data was collected, then again over the last 10 years or so in both the US and UK. This is due in part to the general applicability of many shock techniques to the bulk response of polymer samples. There are several challenges to deciphering the effects of shock loading on materials. These include temporal limitations of the short timescales of the experiments (nanoseconds to microseconds), and reliable methods for bridging the gap between what is typically measured in a shock experiment, i.e., the macroscopic features of the shock environment (velocities, pressures, etc.), and the microscopic structural changes that occur during the shock pulse. These changes must be gauged either by fast spectroscopic techniques or by postmortem recovery and analysis.

The thermodynamic parameters associated with Hugoniot locus, pressure, specific volume (density), and specific internal energy, can be determined indirectly through the direct measurement of shock and particle velocity and the Rankine-Hugoniot equations derived from the conservation of mass, momentum, and energy across a shock discontinuity:

$$\frac{V}{V} = \frac{1-u_p}{U_s}$$

$$P - P_0 = \rho_0 U_s u_p \tag{4.71}$$

$$E - E_0 = \frac{1}{2}(P - P_0)(V_0 - V)$$

Hugoniot loci determined from direct shock and particle velocity measurements are often presented in the shock velocity–particle velocity plane. Barring phase transitions or other dynamic processes, this relationship is expected to be linear, $U_s = c_0 + su_p$. The value of $u_p = 0$, or the y intercept of the linear fit to the data in the $U_\mathrm{s} - u_\mathrm{p}$ plane, is the bulk sound velocity, $c_\mathrm{b}^2 = \frac{K_\mathrm{s}}{\rho}$. The slope of the data in this plane, $s = (K' + 1)/4$, where K' is the pressure derivative of the isentropic bulk modulus.

4.4.1 High-Explosive Driven Experiments

Early insights into shock physics of polymers was gleaned from a large number of plane wave lens experiments at LANL and elsewhere [166]. Carter and Marsh performed arguably one of the most comprehensive shock-wave investigations of polymers over the course of several years during the 1960s-1970s. While much of this data is published in the LASL Shock Handbook, a second LANL report specifically details their findings on polymers, and in particular their propensity to undergo reactions under shock-wave conditions, was published in 1995 [166]. Carter and Marsh's compendium detailed data for many important polymers including polyethylene, polyvinyl chloride, PTFE, PCTFE, PMMA, polystyrene, epoxy, and polyurethane.

The pressures accessible in these early experiments result from HE products driving a flyer plate, or alternatively, direct drive through boosters, and buffer materials, such as a PMMA. Shock velocity is often obtained from "flash" associated with argon or other flash gaps (transit time across sample), while the particle velocity is obtained by impedance matching method, using standards with known Hugoniots, or more recently velocimetry methods. A diagram of a plane wave lens assembly is shown in Fig. 4.21. Advantages of this method is that lots of data can be acquired with a single shot, but disadvantages include accessibility to firing sites, reproducibility, planarity of the shock break-out from the plane wave lens, and limited insight into the shock process.

Carter and Marsh discovered several unique features associated with the Hugoniot behavior of polymeric materials. The first is that linear Rankine-Hugoniot behavior in the $U_s - u_p$ plane typically does not extrapolate to the bulk sound speed, e.g., there is curvature in this plane at low particle velocities [167]. Nonlinearity in the $U_\mathrm{s} - u_\mathrm{p}$ plane at low u_p was attributed to the disparity in intermolecular vs. intramolecular forces in the polymeric chains. At low pressures, the compressibility is dominated by H–H interactions between chains, compared with intramolecular (covalent) forces at higher pressures. The disparity in forces gives rise to greater compressibilities at low pressures to a "crush-up" pressure at which the intermolecular forces become comparable to the intramolecular forces, and the polymer behaves similarly to most crystalline solids.

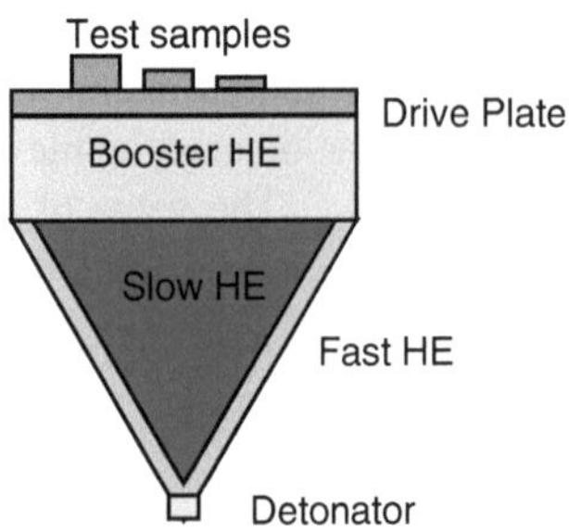

Fig. 4.21 Diagram of experimental setup for plane-wave lens driven experiments

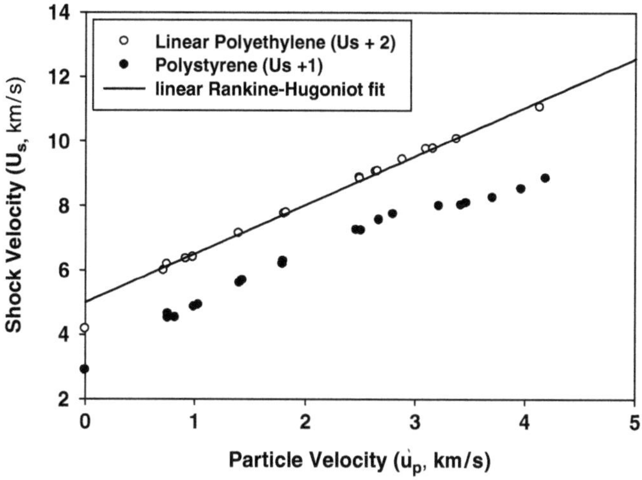

Fig. 4.22 Hugoniot data for linear polyethylene (○) and polystyrene (•) from Carter and Marsh offset from one another in the shock velocity–particle velocity plane. There is a clear break in linearity for the ring-containing polystyrene near $u_p = 3.5\,\mathrm{km\,s^{-1}}$

There is also a nearly universal change in slope in the polymer Hugoniots in the same plane at particle velocities between 2.5 and $3.5\,\mathrm{km\,s^{-1}}$. From the cusp in the P-V plane, Carter and Marsh estimated volume changes for various polymers, and qualitatively associated larger changes in volume at the transition with those containing side chains or pendant moieties, particularly ring structures. As seen in Fig. 4.22, the Hugoniot for polyethylene is fairly linear, while polystyrene has a pronounced cusp at $u_p \sim 3.25\,\mathrm{km\,s^{-1}}$. The cusp has been attributed to interchain chemistry by Carter and Marsh, and to the on-set of shock-induced decomposition processes.

4.4.2 Gun-Driven Experiments

A more commonly used method for obtaining Hugoniot data today is gun-driven plate impact, which uses gas- or powder-driven guns to launch projectiles containing impactors onto stationary targets. The advantage of this method is that the input shock can be well-controlled and well-understood through the use of multiple diagnostics, and even tailorable by use of different types of impactors. There are several active gas gun facilities in the US and abroad. Accessible impact velocities with single- and two-stage light gas guns can range from $\sim 100\,\mathrm{m\,s^{-1}}$ or less, to over $7\,\mathrm{km\,s^{-1}}$. Reaching the higher limits of projectile velocities requires the use of high-performance guns with small bore launch tubes. Common diagnostics on gas gun experiments are shorting or piezoelectric pins, velocimetry probes such as point- and line-VISAR methods [168–171], ORVIS [172], and more recently photon Doppler velocimetry (PDV) [173–175] and embedded gauging methods using materials such as maganin, PDVF, or Al-foil electromagnetic gauges. These methods provide necessary insight into dynamic processes, such as the shock-to-detonation transition in explosives, or phase transitions, viscoelasticity, melting, and spall from the direct measurement of the shock-wave profile as it transits the material.

4.4.3 Hugoniot Data for Binder Materials

4.4.3.1 Estane 5703

Carter and Marsh first measured the Hugoniot for Estane 5703 using plane-wave lens techniques [150]. Experiments were performed on a material with an initial density of $1.186\,\mathrm{g\,cm^{-3}}$. The linear fit to their 17 Hugoniot points between 1.13 and 17.90 GPa is $U_s = 2.32 + 1.70u_p$. There is some evidence of increased scatter, and the on-set of curvature in the $U_s - u_p$ plane at low pressures.

Johnson *et al.* followed the early work on Estane 5703 by examining its dynamic response at low pressures, and adding data on the binder in PBX 9501, Estane-NP [176]. The Estane 5703 studied had an initial density of $1.19\,\mathrm{g\,cm^{-3}}$. The 50:50 Estane:NP mixture has a higher initial density, $1.27\,\mathrm{g\,cm^{-3}}$. Polymer targets were impacted by z-cut quartz and TPX (poly(4-methyl-1-pentane) impactors in gun experiments using a single stage gas gun. The Hugoniot points for Estane at low particle velocities (0.076, 0.124, and $0.291\,\mathrm{km\,s^{-1}}$) reveal the curvature of the data in this plane, approaching the bulk sound velocity ($1.742\,\mathrm{km\,s^{-1}}$) of the polymer at the intercept. The linear Rankine-Hugoniot fit to just the low-pressure data gives $U_s = 2.00 + 2.00u_p$. Just two Hugoniot points are reported for the Estane:NP binder material, at impact velocities of $196\,\mathrm{m\,s^{-1}}$ with TPX, and $179\,\mathrm{m\,s^{-1}}$ with z-quartz [176]. A linear fit to these two data points gives $U_s = 1.74 + 2.57u_p$. There have been no extensions of the Hugoniot for Estane 5703 above a particle velocity of $2.5\,\mathrm{km\,s^{-1}}$, which is just below the region where most polymers undergo a chemical change.

Bourne *et al.* have also studied the shock response of Estane [177]. In their experiments, tungsten, copper, or aluminum impactors mounted in polycarbonate sabots were launched using 50 and 75 mm bore light gas guns. Longitudinal stresses were determined by the response of manganin gauges mounted at the impact and rear surface of the targets. In addition, lateral stress gauges (manganin, MicroMeasurements LM-SS-125CH-048) were embedded into the materials ~4 mm from the impact surface off center-axis.

4.4.3.2 PTFE

PTFE is the most commonly encountered engineering polymer. The shock behavior of PTFE has been extensively studied. In fact, it may be the most examined polymer by shock-wave methods to date. High-pressure experiments using high explosively driven experimental assemblies and gas gun-driven plate impact have been performed by Johnson *et al.* [176], Carter and Marsh [166], Morris and coworkers, [178], and others [179]. By far the most thorough investigation of the shock response of PTFE, including development of equation of state and shock recovery, has been performed by Morris *et al.* [178] PTFE is known to have a complex phase diagram with at least four known crystalline phases near ambient pressure and temperature. Detecting evidence of the phase II–III crystalline phase change in dynamic experiments was pursued by Champion [180] and Robbins [180, 181]. The II–III transition occurs under static high pressure near 0.5–0.6 GPa [182–185]. Recently, this transition was also observed under laser shock conditions using Raman spectroscopy [186].

A selection of Hugoniot data for PTFE is shown in Fig. 4.23. Recently, additional Hugoniot data were reported on well-characterized DuPont-grade PTFE [187].

Generally, there is good agreement in the Hugoniot data measured by multiple laboratories. Scatter in the data is likely due to varying amounts of crystallinity and thus different densities. For example, the material studied by Carter and Marsh had an initial density of $2.152\,\mathrm{g\,cm^{-3}}$, whereas that studied by Morris *et al.* had a density of $2.204\,\mathrm{g\,cm^{-3}}$ indicating a difference in percentage crystallinity of nearly 20%.

At low pressures, Champion reported a cusp attributed to the II–III phase transition associated with a change in volume of 2.2% [180]. His linear fits to the data below and above the cusp are $U_s = 1.258 + 2.434^* u_p$ at pressures <5 kbar and

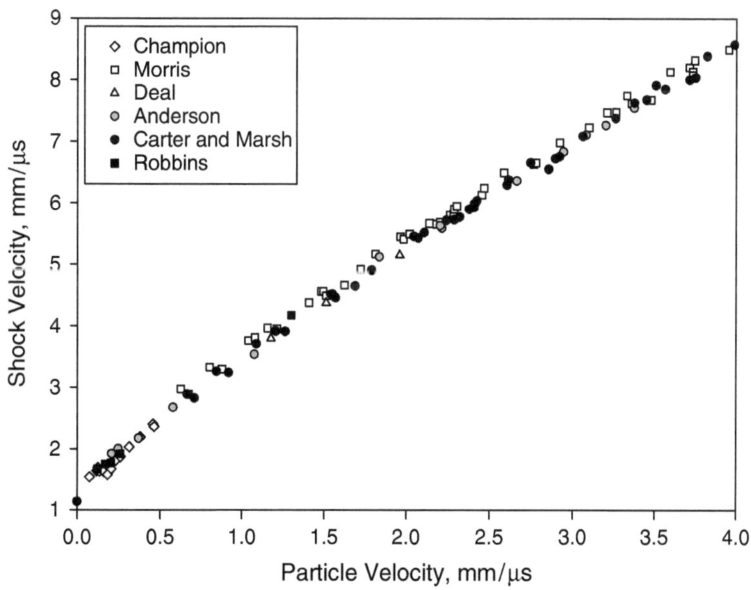

Fig. 4.23 Hugoniot data for PTFE in the shock velocity–particle velocity plane from several sources

$U_s = 1.393 + 2.217^* u_p$ at pressures >5 kbar [180]. Robbins *et al.* also observed a cusp in the Hugoniot consistent Champion's results, though at slightly higher pressure (~7 kbar) [181]. The sample used in Robbins' experiments had a lower density than that used in Champion's work, indicating a lower crystallinity. The purpose of the more recent experiments was to apply the shock-wave methods to a well-characterized PTFE sample, in which the percent crystallinity and other properties of the material were well known. One of the unique aspects of the work performed at LANL was the use of embedded electromagnetic gauges, which give direct insight into shock-wave evolution through a material through the *in-situ* measurement of particle velocity and shock-wave profiles [188]. Inspection of the shock-wave profiles also revealed a dependence of the dynamic response of PTFE to shock input pressure. At low pressures, significant rounding on the front of the wave profile was observed, which lessened with increasing input pressure. Wave profiles from a single high-pressure experiment at 11.7 GPa did not show any appreciable rounding. Additional work is still needed to understand the dynamic of wave evolution in polymers and its relationship to strain-rate dependent mechanical properties as a function of input shock pressure and strain rate.

4.4.3.3 THV 500 and Viton B

The THV and Viton families of polymers are related fluorinated copolymers. THV 500 is the fluorinated terpolymer described earlier. Viton B resins are close relatives and are also terpolymers of vinylidene fluoride, hexafluoropropylene, and tetrafluoroethylene. The Viton A polymers are copolymers of vinylidene fluoride and hexafluoropropylene. DuPont reports that the fluorine content of Viton A is 68% by weight. Viton A has been used as a PBX binder.

The shock compression behavior of THV 500 has been studied by Dattelbaum *et al.* using plate impact methods [189], on the same material as described above. Plate impact experiments were performed by launching Lexan projectile containing PCTFE impactors at instrumented THV 500 targets using single- and two-stage light gas guns at LANL. The experiments used embedded electromagnetic gauges for *in-situ* measurement of shock-wave profiles and extraction of particle and shock velocities. Multiple Hugoniot points were measured and were fit to a linear relationship giving $U_{\mathrm{s}} = 1.62 + 2.02^* u_{\mathrm{p}}$. The intercept, $1.62\,\mathrm{km\,s^{-1}}$, overestimates the bulk sound speed ($c_{\mathrm{b}} = 1.34\,\mathrm{km\,s^{-1}}$) by $\sim 300\,\mathrm{m\,s^{-1}}$. In the case of THV 500, the presence of the hexafluoropropylene in the structure of the polymer is anticipated to increase the free volume in the network due to the side chains, compared with the homopolymer, altering the pressure dependence of inter- and intramolecular forces which are likely responsible for the nonlinearity in the curve at low pressures. An estimate of the pressure derivative of the isentropic bulk modulus can also be obtained from the slope of the linear Rankine-Hugoniot fit, $s = 2.02, \mathrm{K}_{\mathrm{s}}' = 7.1$.

Millet *et al.* studied Viton B by plate impact with aluminum alloy (6082-T6) and copper impactors launched by a single-stage gas gun [190]. Since Viton B contains the TFE repeat unit, the authors also looked for the presence of a low-pressure phase change similar to that observed in PTFE. The density of the Viton B studied was 1.77 g cm^{-3}, much lower than that of THV 500 despite a qualitatively similar repeat structure. The lower density points to lower TFE content, and less crystallinity, though no characterization of the network structure was performed. The Hugoniot of Viton B was linear in the shock velocity–particle velocity plane in the pressure regime studied (Table 4.5). In some of the experiments, the sound velocities at pressure from release wave speeds were also measured, and found to be quite high ($3-4$ km s^{-1} range). Overall, the Hugoniot for Viton B lies above that of both PTFE and PVDF. It was postulated that this could be due to the presence of bulk $-CF_3$ side-groups in the hexafluoropropylene runs that could inhibit dynamic flow during shock loading.

Table 4.5 Summary of the results of linear Rankine-Hugoniot fits to experimental Hugoniot data for binder materials

Polymer name	Initial density	c_0(km s^{-1})	s	Source
Estane 5703	1.186	2.32	1.70	Marsh, LASL Handbook
	1.19	2.00	2.00	Johnson *et al.*
50:50 Estane: NP	1.27	1.74	2.57	Johnson *et al.*
Kel-F 800	1.998–2.02	1.838	1.824	Dattelbaum *et al.*, 3 crystallinities
Kel-F 81	2.14	1.989	1.763	Sheffield & Alcon, LANL gun-driven data
Viton B	1.77	1.88	2.37	Millet, Bourne, Gray
THV 500	2.00–2.01	1.62	2.02	Dattelbaum *et al.*
HTPB	1.46	1.53	2.84	Millett, Bourne, Akhavan "HTPB 1"
	1.43	1.65	2.13	Millett, Bourne, Akhavan "HTPB 2"
PVDF	1.77	2.587	1.575	Carter and Marsh
	1.78	Quadratic fit	$U_s = 2.01 + 3.54u_p - 1.72u_p^2$	Millet and Bourne
				Low pressure
PTFE	2.165	1.35	2.45	Robbins – below cusp
	2.165	1.20	2.78	Robbins – below cusp
	2.175	1.258	2.434	Champion – below cusp
	2.175	1.393	2.217	Champion – above cusp
	2.150	1.14	2.43	Johnson *et al.*
	2.151	1.68	1.79	Carter and Marsh – low pressure
	2.151	2.08	1.62	Carter and Marsh – high pressure
	2.15	1.84	1.71	Bourne
Polychloroprene	1.42	1.40	4.00	Bourne

4.4.3.4 Kel-F 800 and PCTFE (Kel-F 81)

There have been at least five experimental studies detailing the dynamic response of PCTFE (Kel-F 81), including a published report by Carter and Marsh [166], additional Los Alamos experiments by Johnson, Halleck, and Wackerle [191], and most recently, data from Anderson and coworkers [192], Sheffield and Alcon [193], and Weinberg [194]. The first two series of data were obtained via high explosive-drive experiments. The data sets from Sheffield and Alcon, and Weinberg are the result of gas gun-driven plate impact experiments. Anderson's study focused on symmetric impact experiments at low pressures (ca. 0.2 and 2.0 GPa) resulting from plate impact at $130-136\,\mathrm{m\,s^{-1}}$ and $766-783\,\mathrm{m\,s^{-1}}$, respectively [192]. In all of the experiments, the shock velocity was measured directly. The particle velocity is obtained either by impedance matching (in the case of the plane wave lens experiments) or directly from the response of embedded gauges (in the case of the gun driven experiments at LANL). A linear Rankine-Hugoniot fit to the Kel-F 81 Hugoniot data from Johnson, Hallacke and Wackerle, Sheffield and Alcon, and Carter and Marsh below ~8.0 GPa are fairly consistent with one another with $U_s = 1.963 + 1.839u_p$. Above $u_p = 1.0\,\mathrm{mm\,\mu s^{-1}}$, there is a deviation of the data from the linear behavior observed at low pressures.

Recently, shock-wave experiments have been performed on the PBX binder Kel-F 800 with varying percentage crystallinity between 1.3% and 15%, using the embedded gauge technique. The polymer was processed from its granular form (Fig. 4.24) to create solid billets for the gas gun experiments. Particle (u_p) and shock (U_s) velocities were derived from the response of the embedded particle velocity trackers in the gauge package (Fig. 4.25).

The voltage change arising from the electromagnetic gauges is proportional to the particle velocity in the material, surrounding magnetic field, and gauge length. The arrival times of the shock at the embedded gauges as a function of Lagrangian

Fig. 4.24 Photographs of Kel-F 800 binder in its as received granules, and after compression molding at 90°C and 15,000 psi

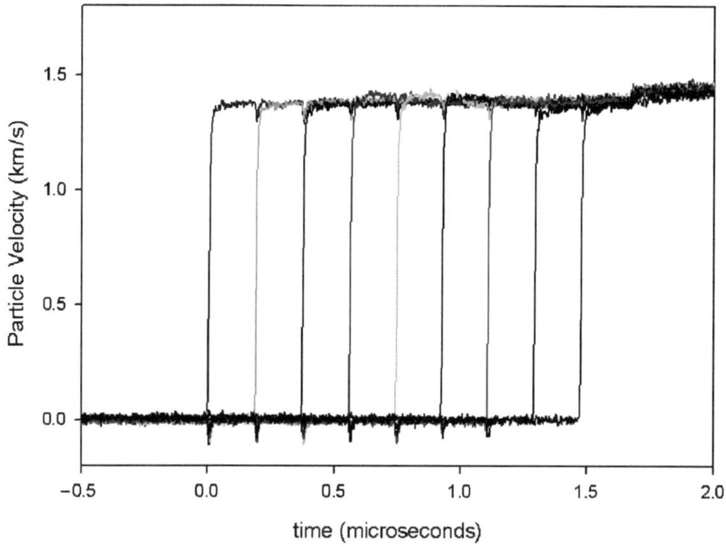

Fig. 4.25 Response of embedded electromagnetic particle velocity gauges for shot 2S-174, Kel-F 81 (PCTFE) impacting Kel-F 800 at 3.2 km s^{-1}

position in the sample were used to derive U_s. Hugoniot loci for all three percentage crystallinities (MP = 1.3%, HX = 12%, CM = 15%) of Kel-F 800 plotted together in the $U_s - u_p$ plane can be fit to the linear Rankine-Hugoniot relationship $U_s = c_0 + su_p$, with $s = 1.824(\pm0.052)$ and $c_0 = 1.838(\pm0.059)$. All of the data are well-represented by this function even with a difference in percentage crystallinity of up to 15%, allowing one to set margins on the shock response and EOS of the polymer at a range of crystallinities that may be observed in the binder in the PBX. Linear Rankine-Hugniot fits to shock data for the homopolymers PVDF and Kel-F 81 are, for comparison, $U_s = 2.587 + 1.575u_p$ and $U_s = 1.963 + 1.839\ u_p$, respectively. Each of the homopolymers is known to exhibit substantially greater crystallization than Kel-F 800. A compilation of shock-wave data for Kel-F 800 and Kel-F 81, and the related fluoropolymers PTFE and THV 500 is shown in Fig. 4.26.

4.4.3.5 HTPB

The Hugoniot of HTPB was first measured by Gupta and Gupta [195]. More recently, the dynamic response of two different HTPB compositions was interrogated by Millett and coworkers. One composition, with a proprietary composition, was obtained from Royal Ordnance, Glascoed, UK. The second material was prepared by Millett *et al.* by the reaction of HTPB (Krahn Chemie GmbH) with isophorone diisocyanate, with dibutyl tin dilaureate as the catalyst. The densities of the two materials were similar: 1.46 g cm^{-3} for the Royal Ordnance formulation, and 1.43 g cm^{-3} for the Royal Military College of Science material. Both compositions were found to

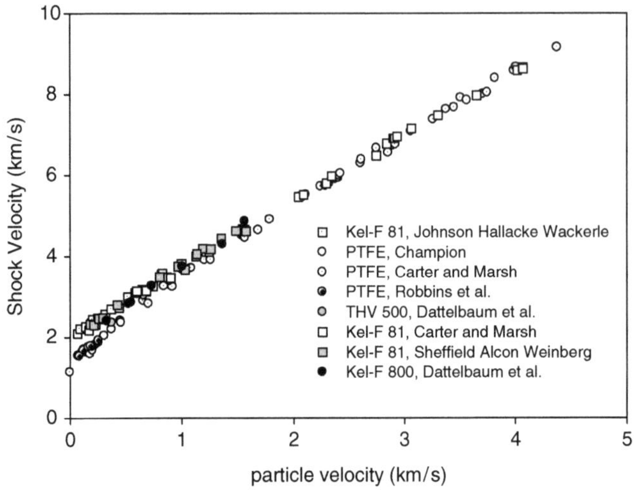

Fig. 4.26 Compilation of Hugoniot data on fluoropolymers Kel-F 800, Kel-F 81, PTFE, and THV 500

have a linear Rankine-Hugoniot response, however, the slopes in the shock velocity–particle velocity plane were measurably different. For the Royal Ordnance formulation, the results of five shots between hydrodynamic pressures of 0.44 and 3.7 GPa gave $U_s = 1.53 + 2.84u_p$. Three experiments were performed on the second material giving $U_s = 1.65 + 2.13u_p$. The difference between the two materials was attributed to additives, including plasticizers, in the Royal Ordance material as it is used in the formulation of PBXs.

As part of the dynamic experiments, samples of HTPB were recovered after being shocked to several pressures. The authors concluded that up to 1.5 GPa, there were no measurable changes in the morphology or chemistry as a result of shock loading.

4.4.3.6 Summary of Hugoniot Data on Binders

Table 4.5 summarizes the Rankine-Hugoniot fits, sometimes referred to Hugoniot-based equations of state, to experimental shock data for the polymeric binder materials summarized in the proceeding sections.

Generally, the Hugoniot behavior of binders fall into a range of values. The bulk sound velocity from the extrapolated y intercept of a linear Rankine-Hugoniot fit to the data in this plane falls between 1.7 and 2.7 km s^{-1} with an outlier for polychloroprene at 4.0 km s^{-1}. Similarly, the slope of the fit, s, is typically between 1.2 and 2.5.

Selected data for the fluoropolymers PCTFE, Kel-F 800, PTFE, and THV 500 are overlaid in the shock velocity–particle velocity plane in Fig. 4.26. The common features described by Carter and Marsh of the dynamic response of polymers are observed here. The materials do not exhibit linear behavior across the range of particle velocities studied, with curvature observed at both low and high pressures. In addition, PTFE gives an example of the influence of phase transitions on the Hugoniot behavior in this plane. The data for PTFE at low pressures lies measurably below the other fluoropolymers in this plane. This is due to the presence of the helical-to-planar zigzag chain conformation change, and a volume collapse of $\sim 2\%$, associated with the low pressure phase II–III transition at 0.5–0.7 GPa.

4.4.4 Isentropic Compression

A complement to shock-wave experiments are isentropic compression experiments, in which the materials are loaded with a "ramp" wave or a more gradual pressure wave. Isentropic compression probes the isentrope, another slice of the EOS surface. This section briefly describes the contribution of isentropic compression experiments (ICE), with emphasis given here to the recent ICE on the Z accelerator at Sandia National Laboratory [196–198].

At low volumetric strain, a Hugoniot curve and an isentrope are essentially identical [199]. To fully describe a Hugoniot curve, several impact experiments are required; however, the path of an isentrope may be developed in a single experiment. For ICE with the Z accelerator, a square short configuration connects the cathode and anode with an evacuated gap running between them. A current pulse is applied and a smoothly increasing pressure load at the surface of the sample is generated by the cross-product of the current density, J, and the resultant magnetic field, B. The time-varying, magnetic pressure is given by [197],

$$P_B(t) = \frac{1}{2}\mu_0 \left[J\right]^2 \tag{4.72}$$

where μ_0 is the permeability of free space. With a typical current pulse reaching a maximum of 20 MA in a rise-time of around 2 µs, the resultant P_B is on the order of 100 GPa. Given the nonlinear response of materials at these pressures, the initial compressive wave would eventually become a shock wave. However, this can be avoided by judiciously choosing sample dimensions, drive pressure, and pulse duration.

By driving two samples simultaneously with the Z accelerator, isentropic EOS measurements can be taken. One sample serves as an internal reference relative to the other. Using a velocity interferometer system for any reflector (VISAR), the velocity histories are measured at the rear surface of both samples. With the sample dimensions and velocity histories known, the P–V isentrope can be constructed [200].

Isentropic compression experiments (ICE) have been performed by a joint Sandia-Los Alamos National Laboratory team on Estane 5703, Estane-NP, a composite of small particle size HMX crystals in Estane binder, PTFE, Kel-F 81, Kel-F 800, and THV 220 (a copolymer of tetrafluoroethylene, vinylidene fluoride, and hexafluoropropylene produced by Dyneon) on the Sandia Z machine [201–205]. In addition, experiments have also been performed on HTPB [202, 206].

4.4.5 Special Topics: Behavior of Polymers under Dynamic Compression

4.4.5.1 Compressibility of Polymer Networks

Curvature in the Hugoniot in the shock velocity–particle velocity plane data at low u_p is common for liquids, polymers, and soft organic materials. It has been attributed to the compression of porosity and free volume that occurs at fairly low pressures [207], as well as to disparities in intra- and intermolecular forces in crystalline materials. In liquids, the curvature is gradual, however it is often more pronounced in solid polymers networks, reaching a "crush-up" pressure at which the free volume is fully compressed out by 1–1.5 GPa. In polymers, free volume can be attributable to interchain spacing and volume-producing features in the network structure (cross-linking, branching, side chain size, etc.). Free volume content is important in many industrial applications, for instance, for gas permeability, and diffusivity of small molecules in polymer networks, and viscoelastic properties [166, 208].

There are several methods for estimating free volume in polymers. A well-known approach is the combined method of Williams-Landel-Ferry (WLF) and Doolittle [209]. This approach is based on an empirical observation that above T_g, the temperature dependence of all mechanical and electronic relaxation processes in amorphous polymers can be described by a single function. There is an empirical "universal constant" in this function that has been related to polymer free volume. The "universal constant" arises from invoking Doolittle's equation, which surmises that the temperature dependence of polymer viscosity is based largely on free volume content [209, 210]. There is an abrupt change in free volume as $T \rightarrow T_g$, and correspondingly a large increase in viscosity [211–214]. The nature of the volume change at T_g was found to be nearly universal for all supercooled liquids [209]. The fractional free volume ($f_g = (\frac{v_f}{(v_0+v_f)})$, where v_0 and v_f are the core and free volume, respectively) at the glass transition temperature T_g has been found to be a nearly universal value of $f_g \sim 0.025$ (or 2.5%). The universality of this value is consistent with the iso-free volume postulate of Fox and Flory, and Simha and Boyer, which assumes a constant coefficient of thermal expansion of the free volume, $\alpha_{FV,\text{glass}}$, below T_g [215, 216]. Above T_g, the fractional free volume in amorphous polymers increases in proportion to the difference in thermal expansion above and below T_g ($\alpha_2 = (\alpha_{>T_g} - \alpha_{<T_g})$ and the temperature as $f = f_g + \alpha_2(T - T_g)$) [209].

An "iso-core" volume model, assumes that the core volume of the polymer stays constant and bulk volume changes with temperature are solely related to the free volume. Recently, Positron Annihilation Lifetime Spectroscopy (PALS) coupled with P–V–T experiments has been used to elucidate the amount of free volume in polymers and its temperature-dependent expansion behavior both below and above T_g. PALS is the only direct means of experimentally determining free volume content [208]. Recent work has discovered that the glassy state for amorphous, moderate molecular weight polymers such as poly(methylmethacrylate) (PMMA), poly(styrene) (PS), and poly(2,6-dimethylphenyleneoxide) (PPO) is, in fact, not an iso-free volume state [217]. There are net increases in both core- and free-volume with temperature below and approaching the glass transition temperature. In fact, the *free volume* expansion, as opposed to "core" volume expansion, dominates bulk thermal expansion below T_g, in contrast with the iso-free volume assumption by Fox *et al.* and Simha *et al.* (α_{FV} can be over an order of magnitude greater than α_{core}) [215, 216]. An abrupt break in free volume content occurs at T_g, followed by much greater thermal expansion of the free volume fraction above T_g. Bulk thermal expansion above T_g is, again, largely dominated by the free volume and is attributed to polymer chain mobility in a "liquid"-like state above T_g. At their respective T_g's the free volume content in PMMA, PS, and PPO ranged from ~6% to 15%, significantly higher than the "universal" 2.5% [217].

A first principles approach derived from van der Waals and molar volumes can also be used to estimate free volume. The van der Waals volume for a molecule or polymer chain can be calculated according to the method of Bondi using recommended values for atomic and group radii [218]. This method assumes spherical shapes and a knowledge of bond distances, angles, and van der Waals radii, r_W. The van der Waals (V_W) volume is related to V^*, the minimum in a Lennard-Jones intermolecular potential, by an experimentally determined empirical linear function $V^* = 1.3–1.4\ V_W$ at 0 K, or more specifically $[V^* = (1.45 * V_W) + 3.88]$ for polymers and $[V^* = (1.6 * V_W)]$ for liquids [219]. The free volume can be estimated from the van der Waals volume, scaling or core volume, and V_f (free volume) = 1 - V_{core}/V_m (V_m is the molar volume or $1/\rho$).

We have applied some of these methods to estimate the free volume of a cross-linked PDMS network [50]. Examination of the volumetric thermal expansion coefficient of linear PDMS and the cross-linked network Sylgard 184 [220], combined with application of the WLF/Doolittle method gave a free volume content, $f_g \sim 10–15\%$ above T_g for linear PDMS depending on molecular weight. A molecular feature of note for PDMS is that the molar volume decreases (and density increases) as the molecular weight increases due to fewer end groups and better chain organization [78]. Application of a modified P–α model described previously to curvature in the shock velocity–particle velocity plane was also performed for Sylgard 184 [207]. The porous EOS analysis predicts a free volume content of approximately 10%, with a complete crush-up pressure of ~1.5 GPa. Previously, this model was applied to Estane 5703, yielding a free volume estimate of only 1.4% [207]. The phase separation and organizing features of the network structure of Estane 5703 could account for the smaller predicted free volume. Similarly, this method yielded an estimate of 3% for Kel-F 800 [221]. Recently, there have been attempts to measure the free volume in binder materials using positron annihilation lifetime spectroscopy (PALS) [222]. For Sylgard 184, the free volume from PALS was ~18%, which is in line with the estimates from the methods described above.

4.4.5.2 Shock-Induced Chemistry in Polymers

Shock waves in both energetic and inert materials can induce damage, fracture, and chemistry by the high pressures and temperatures that are typically generated during the shock. As soft materials, polymers are subject to plastic deformation, shock induced damage, and chemical transformations at lower pressure and temperature thresholds than other classes of materials.

Carter and Marsh found that polymers almost universally undergo a high-pressure transition marked by a change in slope in the U_s, u_p plane between $u_p \sim 2\text{–}3\,\mathrm{km\,s^{-1}}$ [166]. This break in the shock velocity–particle velocity plane has been attributed to dissociation but not rigorously investigated for multiple classes of polymers. There has been some debate about the microscopic molecular origin of this transition, and whether it is a solid–solid phase transition or a dissociation reaction into product species. Morris *et al.* found that the high-pressure phase transition in polyethylene driven by a steady-state Mach compression disk was attributed to the irreversible dissociation into products [223]. Post-shock recovered samples were examined to characterize atomic and molecular species, structure, and morphology. The majority of the bulk product was determined to be amorphous carbon.

Morris *et al.* also performed shock-recovery experiments on PTFE to aid in the interpretation of the Hugoniot, and specifically the presence of cusps or nonlinearities at high pressures [178]. In their experimental apparatus, a PTFE cylinder was confined within a steel tube. A Mach disk was then introduced into the center of the sample, with a large radial distribution of pressures from the center of the cylinder to the confining walls. Post-shot inspection of the PTFE cylinder showed a small hole in the center filled with amorphous carbon. Gas analysis of the products showed that CF_4, C_2F_6, and other perfluorinated gaseous species were formed.

While it is assumed that polymers dissociate into small molecular species like explosives under sufficient shocked pressures and temperatures, preliminary detonation chemistry apparatus experiments (LANL) suggest that only partial fragmentation may occur, such as in polydimethylsiloxane foams in which oligomeric species were observed in the product mixture [224]. There has also been report of persistent radical species observed in shock loaded and recovered samples of PMMA, Kapton, and Vespel [225].

Graham and others have examined a possible effect of partial bond breakage in polymers under shock loading. Several polymers have been shown to exhibit shock-induced conductivity or shock-induced polarization. Shock-induced polarization has been observed in polymers such as polymethylmethacrylate, polystyrene, nylon, epoxy, polyethylene, PTFE, polyvinyl chloride, poly(pyromellitimide) (PPMI, or Vespel SP-1), and others [226–232]. In all cases, there appears to be a percentage compression on-set to the phenomenon. Graham investigated the compression-effective polarization phase space for several polymers and identified three regions: an onset compression, a region of strong polarization generation in which the polarization ranges over three orders of magnitude, and a saturation region [232]. He also found that polymers with complex backbone structures, often incorporating ring structures, exhibited the strongest shock-induced polarization, and suggested that this may be due to an increased inclination to undergo shock-induced bond scission.

Shock-induced conduction measurements date back to Russian work in the late 1940s on PMMA [233]. Conductivity under both dynamic and static high pressures has been investigated for PTFE, PE, polyvinylchloride, polyvinylacetate, polychlorotrifluoroethylene, and polyethyleneterephthalate, with many of these materials remaining insulators or even showing an increase in resitivity at high pressures [234–237]. However, investigations of thin films of PPMI and Kapton have shown an electrical "switching" effect under shock-wave loading [234].

Related to this work is shock-induced piezoelectric effects in polyvinylidene-fluoride (PVDF). PVDF adopts a TGTG (T = trans, G = gauche) configuration in its α-phase, but converts to an all trans, β structure when stretched or sufficiently deformed [238]. The β is spontaneously polarized due to the orientation of the dipoles along the polymer backbone. This deformation-induced phase transformation has been exploited for the use of PVDF a stress gauge in shock-wave experiments [239–242].

Sheffield and Dattelbaum recently examined the shock response of simple molecules that were expected to react under shock loading as models for carbon- and silicon-based polymers [243]. In shock-wave experiments on *tert*-butylacetylene and trimethylsilylacetylene, both materials underwent shock-induced chemical reactions at moderate pressures in which the reaction products were more dense than the liquid reactants. By studying the two materials with a change in structure of only one atom (from C to Si), the authors hoped to find a difference in the compressive energy input required to break the chemical bonds. Interestingly, the energy was nearly identical in both materials, indicating that the first bond to break was likely a C–H bond on the acetylene moiety rather than a C–C or Si–C bonds between the bulky substituent and acetylene group. It was postulated that the shock-induced reactions were likely related to dimerization and trimerization of the acetylene molecules under shock conditions.

Like knowing the first chemical steps in the initiation of explosives, understanding shock-induced chemistry remains a major challenge. The characterization tools for polymers are necessarily different than metals. Unlike metals, one cannot rely on crystallographic information to determine shock-related damage mechanisms. However, one can anticipate and interrogate measurable changes in polymeric materials during and post-shock loading such as melting, changes in crystalline content and density, plastic deformation and related changes in microstructure and texture, phase transitions (and the possibility of metastable phases), and dissociation chemistry.

4.4.5.3 Shock Recovery of Polymers

Shock recovery, or shocking samples with postmortem inspection and characterization, is a common method for probing the effects of shock loading on material properties. Key to the success of shock recovery and shock-induced damage evaluation experiments is the careful design of sample geometries and loading paths to ensure that recovered samples had been subjected to primarily uniaxial shock compression under well-understood conditions (such as peak pressure etc.) [244–249]. Soft recovery experiments use "momentum trapping," in which the test sample is confined by a shock impedance-matched material in order to mitigate radial release waves (and possible spall) at the sample boundaries (edges). To further minimize tension and plastic work from occurring as a result of rarefraction wave interactions, a sample diameter-to-thickness ratio of 7:1 is also recommended for recovery experiments [250]. Several authors have experimented with a variety of sample geometries including confined, star-shaped, square, and composites of the above [245–249]. A measure of the success of the momentum trapping experiment can be determined by inspection of the residual strain in the sample, ε_{res}, which is the ratio of sample thickness before and after the test [251]. Typically, ε_{res} is 0.2% to 0.5% for a successfully trapped experiment on metals [250]. Postmortem analysis of grain structure (metals) and bulk mechanical properties gives insight into damage mechanisms derived from shock passage.

There are numerous reports of shock-recovery of metals and investigation of shock-induced damage and strength changes by postmortem inspection and testing [252–257]. This type of information is much rarer for polymers and other soft materials because of challenges associated with both controlled 1-D shock-wave loading, and the "soft catch." A successful approach for shock recovery of metals introduces a shock wave into a sample typically by explosively launched or gas gun-driven plate impact, through a same-material front surface plate or window. The sample to be studied is contained in an assembly that has several features serving several functions. Momentum trapping rings, impedance-matched to the target, accommodate release waves from the assembly sides, and minimize interface wave interactions between the target and the rings. The sample is released from both the high-pressure state, and mechanically released from the rings into a soft catch material before rarefaction waves from the sides can impinge on the sample. It has been determined that two lateral momentum trapping rings are optimal for this purpose [252]. The sample is further contained by spall plates at the backside of the assembly (i.e., opposite side of sample from impact face), whose purpose is to absorb damage and spall behind the sample. The target assembly also typically has a sabot-stripper on the front end to strip off the heavy sabot that may damage the sample at late times, and a soft catch chamber, consisting of soft cloths, felt, or, as sometimes used in the recovery of explosively driven shock experiments, a large water reservoir.

Bourne and Gray performed Eulerian hydrocode simulations of their soft recovery apparatus used for polymers to validate the experimental design [258]. Their simulations showed nominally 1-D strain during the duration of the shock experiment, and a maximum lateral strain in the sample of ~0.2%. This assembly has been applied to study shock-induced changes in PMMA, HTPE, PTFE, and an HTPB-glass composite.

PMMA was used to test the soft recovery assembly. Impact of an EN3 steel flyer plate into the PMMA at 300 m s^{-1} resulted in a fully recoverable PMMA sample, with measured lateral strains calculated from postmortem dimensions of $<1\%$. The sample itself was also permanently (plastically) compressed in the shock-wave direction.

Shock recovery experiments on HTPB examined the same material subjected to shock loading, [259] a isophorone diisocyanate (IPDI) cross-linked formulation.

Soft-recovered HTPB did not show any changes in its glass or melt transition temperatures. Further, there was no evidence of any chemical changes in HTPB to a shock pressure of 1.5 GPa, including no evidence of any change in molecular weight or cross-link density.

An extension of the work on polymers is the investigation of the response of polymer-glass composites, prepared as stimulants of plastic-bonded explosives. In one experiment, a composite consisting of 30 μm diameter glass beads in an HTPB matrix was impacted by a steel flyer plate traveling at 200 m s^{-1} [259]. Inspection of the composite by scanning electron microscopy (SEM) before and after the experiment showed evidence of cracking and decohesion of the binder from the glass particles following shock loading, with no damage observed in the glass beads themselves.

Brown and coworkers recently described shock recovery of PTFE at pressures below and above the phase II–III crystalline phase transition [260]. PTFE was recovered from a shock recovery fixture similar to that described by Bourne *et al.* with less than 1.3% residual strain. They found that shock loading above the phase transition resulted in an increase in crystallinity from 38% to 53% by differential scanning calorimetry. They were also able to correlate the input shock pressure with changes in microstructure by tapping-mode atomic force microscopy. From post-shock evaluation of the mechanical properties, shocked PTFE below the phase change showed a decreased Young's modulus and yield stress, while shocked PTFE above the phase change showed increases in both the modulus and yield stress.

4.4.5.4 Spall in Polymers

Spall fracture is fracture in a material resulting from the interaction of rarefaction waves placing a material in dynamic tension. Though all fracture is by definition a "dynamic process," spall occurs at high-strain rates and typically after a compression/shock wave has passed through the material. Posttest analysis adds insight into the spall process by looking at the nucleation, growth, and coalescence of microcracks or voids in the material in order for the spall mechanism to be described by appropriate models [261]. Therefore, spalled samples are often "soft-recovered" by being caught in soft rags or other "soft" environments (such as water or felt) for postmortem analysis.

There exists a solid foundation of literature reports with appropriate experimental designs for studying shock recovery in metals [250, 251]. The total flow stress in metals can be attributed to several factors including the strength of dislocation–dislocation interactions, the Peierls distortion barrier, hardening mechanisms, and grain boundaries [250, 251]. Changes in microstructure are dependent on the strain path, its rate, and temperature. Postmortem analysis of metals such as copper, aluminum, zinc, zirconium, and others by scanning electron microscopy (SEM) and/or transmission electron microscopy (TEM) reveals that changes in microstructure induced by high-strain rate deformation can include dislocation, planar slip, stacking faults, and twinning [251]. If the materials are properly shock loaded and trapped, the damage is typically uniformly distributed throughout the material [250]. Metastable phases that exist back to ambient conditions have also been discovered in some cases [250]. Complementary to microstructural analysis, postmortem mechanical testing strengthens the micro- to macro-scale understanding of the shock effects on final bulk material properties. Most metals undergo greater work hardening under shock conditions than under comparable quasi-static deformation to the same strain which results in a greater overall yield strength under reloading. The work hardening has been attributed to a higher concentration of dislocation sites in the shock-loaded metal.

There is comparably little information on shock damage in ceramics or other brittle materials. The primary reason for the lack of data is their low yield strength and consequently high probability of fracture from both radial waves and from stress concentration and fracture under compression waves [261]. Pores, microcracks, and grain boundaries are stress concentrators in the inherent microstructure of ceramic materials, initiating fracture. These materials are inherently inhomogeneous and generally undergo plastic deformation as their dominant damage mechanism. It is also difficult to determine fracture mechanisms in highly fragmented samples because one cannot distinguish between plastic deformation due to the shock process or due to the tensile release in fragmentation.

While there are few reports on shock-induced damage in polymers, there have been reports of spall fracture in poly(methylmethacrylate) (PMMA) and polycarbonate (PC), both glassy, rigid polymers. In PMMA, small-scale oscillations or reverberations were evident in the spall signal and were found to be unique to this polymer [261]. These likely arise from reverberations in the spalled piece and perhaps from the viscoelastic response of this amorphous polymer. As the peak input stress was increased in PMMA, the spall became more viscous, presumably due to heating and plasticization of the polymer [261]. The spall strength of PMMA was determined to be 0.17–0.21 GPa. PC is transparent, making it easy to measure macrocracks in the material. Work by Curran, Shockey, and Seaman showed that spall in PC occurred from the compilation of many short cracks through the specimen [261]. There was very little nucleation, but cracks grew significantly once nucleated and coalesced into a large crack.

Kanel *et al.* [261] have also examined elastomers, and found them to be inherently different than other polymers. In these materials, spall does not occur, rather void formation occurs, which typically proceeds ultimate rupture. This occurs at a low "microvoid nucleation strength" of 15–30 MPa and the material does not undergo complete failure.

Johnson and Dick investigated the spallation behavior of Estane as a follow-on to their work measuring the low-pressure Hugoniot and investigating the viscoelastic properties of the binder [262]. Shock-release wave profiles were measured using VISAR, which were referenced to piezoelectric pins on the impact surface for the determination of the shock velocity. Modeling of the data with a finite-difference wave propagation code containing a simple void-growth model, successfully reproduced giving a void growth threshold of 65 MPa. The void growth kinetics also reproduce a gradual loss of material strength indicated by the VISAR profiles. No attempts at shock recovery were performed to inspect the nature of the failure.

Bourne and Gray also focused on the spall and shear strengths of Estane under increasing shock loading [177]. Estane showed a spall signature at 280 m s^{-1} impact conditions, which would correspond to a low spall strength of 0.015 GPa, however no evidence of damage was observed in the target. From the response of lateral stress gauges located near the impact surface, there is a decrease in the lateral stress as a function of time, which they report as indicative of a decline in shear strength in the material under shock loading.

4.5 Conclusions

Polymers possess intricate behavior. Remaining challenges to understanding their equation of state and dynamic behavior include measurement of material properties at very high pressures (>10–20 GPa), persistently underexplored phase diagrams, interrogation of high-pressure chemistry, and understanding of their highly pressure- and strain-rate dependent properties. The development of new tools and methods for studying the dynamic behaviors and equation-of-state properties of amorphous materials was recently cited as one of the priority research directions for understanding materials under extreme conditions [263]. As we continue to require better models for polymers as a high-pressure community, new discoveries relating to the response of these amorphous (or semi-amorphous) materials under extreme conditions are certain.

Acknowledgments The authors gratefully acknowledge the experimental contributions and working relationships with current and past members of the Shock and Detonation Group at Los Alamos including Stephen Sheffield, David Robbins, Richard Gustavsen, Ray Engelke, David Stahl, Robert Alcon, Michael Weinberg, Nenad Velisavljevic, and Erik Emmons. We are also grateful for continued stimulating discussions with members of the polymer projects teams, particularly E. Bruce Orler, Edward Kober, Eric N. Brown, Philip J. Rae, and Bradford E. Clements. Support for the investigation of polymers behavior under dynamic conditions has been provided by the Department of Energy and NNSA under Campaign 2, and the DOE/DoD Joint Munitions program at Los Alamos National Laboratory. We also acknowledge support from Department of Defense Laboratories: Army Research Laboratory, NSWC-Dahlgren Division, NSWC-Indian Head Division, and the Air Force Research Laboratory at Eglin Air Force Base. Los Alamos National Laboratory is operated by Los Alamos National Security (LANS) LLC for the Department of Energy and NNSA.

References

1. James, E. *The Development of Plastic Bonded Explosives*, Lawrence Livermore National Laboratory, UCRL-12439, 1965.
2. Allcock, H. R.; Lampe, F. W. *Contemporary Polymer Chemistry*; Prentice Hall: Englewood Cliffs, NJ, 1990.

3. Flory, P. J. *Principles of Polymer Chemistry*; Cornell University Press: Ithaca, NY/London, 1953.
4. Dobratz, B. M.; Crawford, P. C. *Lawrence Livermore National Laboratories Explosive Handbook: Properties of Chemical Explosives and Explosive Simulants*; University of California: Livermore, CA, 1985.
5. Eerligh, R.; van Gool, M. A.; Kramer, R. E.; van Ham, N. H. A. 17th International Annual Conference of ICT, 1986, pp. 10.1–10.12.
6. Hercules, *Urethane Technology*; 1 January 1989, p. 33.
7. Hercules, *Chemical Week*, 12 December 1981.
8. Akhavan, J.; Burke, T. C. *Prop. Expl. Pyro. 17*, 271–274, 1992.
9. Johnson, H. D.; Osborn, A. G.; Stallings, T. L.; Anthony, T. R. "MHSMP-77–54 Process Endeavor No. 105," Pantex Plant, 1977.
10. Osborn, A. G.; Stallings, T. L.; Johnson, H. D. "MHSMP-78–57 Process Endeavor No. 105," Pantex Plant, 1978.
11. Benzinger, T. M.; Loughran, E. D.; Davey, R. K. *LA-8436-MS*, Los Alamos National Laboratory, 1980.
12. Stallings, T. L.; Osborn, A. G.; Johnson, H. D. *MHSMP-81-17*, Pantex Plant, 1981.
13. Dobratz, B. "The Insensitive High Explosive Triaminotrinitrobenzene (TATB): Development and Characterization - 1888 to 1994," Los Alamos National Laboratory, 1995, LA-13014-H, UC-741.
14. Field, J. E.; Swallowe, G. M.; Pope, P. H.; Palmer, S. J. *P. Inst. Phy. Conf. Ser.*, Oxford, 70, 381, 1984.
15. Arnold, W. 35th Int. Ann. Conf. ICT, Karlsruhe, Germany, 2005, pp. 188-1–188-14.
16. Eich, T.; Wild, R. 34th Int. Ann. Conf. ICT, Karlsruhe, Germany, 2003, p. 99-1.
17. Mostafa, A. 37th Int. Ann. Conf. ICT, Karlsruhe, Germany, 2006, p. 56-1.
18. Parker, R. 21st Int. Ann. Conf. ICT, Karlsruhe, Germany, 1990, p. 15-1.
19. Vandenburg, E. J. United States, 1972.
20. Vandenberg, E. J.; Woods, F.; Inc., H., Ed. U. S., 1972; Vol. 3, pp. 645, 917.
21. Frankel, M. B.; Flanagan, J. E.; International, R., Ed. U. S., 1981, Vol. 4, pp. 268, 450.
22. Frankel, M. B.; Wilson, E. R.; Woolery, D. O.; Hamermesh, C. L.; McArthur, C. *Gov. Rep. Announce. Index 82*, 2156, 1982.
23. Frankel, M. B.; Wilson, E. R.; Woolery, D. O.; Hamermesh, C. L.; McArthur, C. *Chem. Abst. 97*, 110490, 1982.
24. Frankel, M. B.; Wilson, E. R.; Woolery, D. O.; Hamermesh, C. L.; McArthur, C. *Chem. Abst. 97*, 74986, 1982.
25. Kubota, N. Combustion Mechanism of Azide Polymer (I) and Part II, Japanese Defense Agency, 1988.
26. Lavigne, J.; Lessard, P.; Ahad, E.; Dubois, C. International Symposium of Energetic Materials Technology, Orlando, FL, 1994, pp. 265–271.
27. Leu, A.; Shen, S.; Wu, B. 21st Int. Ann. Conf. ICT, Karlsruhe, Germany, 1990, p. 6–1.
28. Tokui, H.; Saitoh, T.; Hori, K.; Notono, K.; Iwama, A. 21st Int. Ann. Conf. ICT, Karlsruhe, Germany, 1990 p. 7–1.
29. Kubota, N.; Sonobe, T.; Yamamoto, A.; Shimizu, H. *J. Prop. Power 6*, 686, 1990.
30. Kubota, N. *Prop. Explos. Pyro. 13*, 172, 1988.
31. Keicher, T.; Wasmann, F.-W. *Prop. Expl. Pyro. 17*, 182, 1992.
32. Rizzo, H. F.; Humphrey, J. R.; Kolb, J. R. *Prop. and Expl. 6*, 57–62, 1981.
33. Gibbs, T. R.; Popolato, A. *LASL Explosive Property Data*; University of California Press: Berkeley/Los Angeles, CA, 1980.
34. Roy, S. B.; Roy, P. B. *J. Phys.: Condens. Matter 11*, 1999.
35. Godwal, B. K.; Sikka, S. K.; Chidambaram, R.*Phys. Rep. 102*, 1983.
36. MacDonald, J. R. *Rev. Mod. Phys. 41*, 1969.
37. Tonelli, A. E.; Srinivasarao, M. *Polymers from the Inside Out An Introduction to Macromolecules*; Wiley: New York, 2001.
38. Seymour, R. B.; Carraher, C. E. *Structure-Property Relationships in Polymers*; Plenum Press: New York, 1984.

39. Rosa, C. D.; Auriemma, F.; Perretta, C. *Macromolecules 27*, 2004.
40. Tait, P. G. *Phys. Chem. 2*, 1, 1888.
41. Murnaghan, F. D. *Proc. Nat. Acad. Sci. 30*, 1944.
42. Murnaghan, F. D. *Finite Deformation of an Elastic Body*; Wiley: New York, 1951.
43. Birch, F. *Phys. Rev. 71*, 1947.
44. Vinet, P.; Smith, J. R.; Ferrante, J.; Rose, J. H. *J. Phys. C*19,1986.
45. Vinet, P.; Rose, J. H.; Ferrante, J.; Smith, J. R. *J. Phys.: Condens. Matter 1*, 1989.
46. Sun, Z.; Song, M.; Yan, Z. *Polymer 33*, 328, 1992.
47. Dymond, J. H.; Malhotra, R. *Int. J. Thermophys. 9*, 1988.
48. Neece, G. A.; Squire, D. R. *J. Phys. Chem. 72*, 128, 1968.
49. Cutler, W. G.; McMickle, R. H.; Webb, W.; Sciessler, R. W.*J. Chem. Phys. 29*, 1958.
50. Dattelbaum, D. M.; Jensen, J. D.; Schwendt, A. M.; Kober, E. M.; Lewis, M. W.; Menikoff, R. *J. Chem. Phys. 122*, 144903, 2005.
51. Hayward, A. T. J.*Brit. J. Appl. Phys. 18*, 1967.
52. Curro, J. G. *J. Macromol. Sci.- Rev. Macromol. Chem. C*11, 321,1974.
53. Johnson, J. N. *Am. J. Phys. 36*, 1968.
54. Prigogine, I.; Trappeniers, N.; Mathot, V. *Disc. Faraday Soc. 15*, 1953.
55. Prigogine, I. *The Molecular Theory of Solutions*; North-Holland: Amsterdam, 1957.
56. Prigogine, I.; Bellemans, A.; Naar-Colin, C. *J. Chem. Phys. 26*, 1957.
57. Flory, P. J.; Orwoll, R. A.; Vrij, A. *J. Am. Chem. Soc. 86*, 3507, 1964.
58. Sanchez, I. C.; Lacombe, R. H. *Nature 252*, 1974.
59. Sanchez, I. C.; Lacombe, R. H. *J. Phys. Chem. 80*, 2352, 1976.
60. Simha, R.; Somcynsky, T. *Macromolecules 2*, 342, 1969.
61. Lewis, G. N.; Randall, M. *Thermodynamics*; 2nd ed.; McGraw-Hill: New York, 1961.
62. Mortimer, R. G. *Physical Chemistry*; 2nd ed.; Academic Press: San Diego, 2000.
63. Sandia National Laboratory: Albuquerque, NM.
64. MacDonald, J. R. *Rev. Mod. Phys. 38*, 1966.
65. Baonza, V. G.; Cáceres, M.; Núñez, J. *Phys. Rev. B 51*, 1995.
66. Cho, J.; Sanchez, I. C. In *Polymer Handbook*; 4th ed.; Wiley: New York, 1999.
67. Tammann, G. *Über die Beziehungen zwischen den inneren Kräften und Eigenschaften der Lösungen*; Leopold Voss: Hamburg, 1907.
68. Rodgers, P. A. *J. Appl. Poly. Sci. 48*, 1061–1080, 1993.
69. Hemley, R. J.; Mao, H. K.; Finger, L. W.; Jephcoat, A. P.; Hazen, R. M.; Zha, C. S. *Phys. Rev. B 42*, 1990.
70. Cohen, R. E.; Gulseren, O.; Hemley, R. J. *Am. Mineralogist 85*, 338–344, 2000.
71. MacQuarrie, D. A. *Statistical Mechanics*; Harper & Row: New York, 1976.
72. Chandler, D. *Introduction to Modern Statistical Mechanics*; Oxford University Press: New York, 1987.
73. Shen, M.; Hansen, W.; Romo, P. *J. Chem. Phys. 51*, 1969.
74. Wada, Y.; Itani, A.; Nishi, T.; Nagai, S. *J. Polym. Sci. 7*, 1969.
75. Prigogine, I.; Garikian, G. *Physica 16*, 239, 1950.
76. Prigogine, I.; Mathot, V. *J. Chem. Phys. 20*, 1952.
77. Hellwege, K.-H.; Knappe, W.; Lehmann, P. *Kolloid-Z. Z. Polym. 183*, 1962.
78. Dee, G. T.; Walsh, D. J. *Macromolecules 21*, 815, 1988.
79. Zoller, P. *J. of Poly. Sci.: Polym. Phys. Edn. 20*, 1453–1464, 1982.
80. TA Instruments Product Information.
81. Dattelbaum, D. M.; Sheffield, S. A.; Stahl, D. B.; Gustavsen, R. L.; Orler, E. B.; Velisavljevic, N. *in preparation*.
82. Fox, T. G.*Bull. Am. Phys. Soc 1*, 123, 1956.
83. Shen, M. C.; Eisenberg, A. *Rub. Chem. Tech. 43*, 95, 1970.
84. Manzara, A.; Orler, E. B.; Dattelbaum, D. M. *Private communication*, 2006.
85. Orler, E. B. *Unpublished results* Los Alamos National Laboratory, 2007.
86. Roland, C. M.; Hensel-Bielowka, S.; Paluch, M.; Casalini, R. *Rep. Prog. Phys. 68*, 1405–1478, 2005.
87. Skorodumov, V. F.; Godovskii, V. K. *Polym. Sci. U.S.S.R. 29*, 127–133, 1987.

88. Mao, H. K.; Hemley, R. J. In *Reviews in Mineralogy. Ultrahigh-Pressure Minerology: Physics and Chemistry of the Earth's Deep Interior*, Vol. 37; Hemley, R. J., Ed.; Mineralogical Society of America: Washington, DC, 1998.
89. Bridgman, P. M. *The Physics of High Pressure*; G. Bell & Sons: London, 1958.
90. Kawai, N.; Endo, S. *Rev. Sci. Instrum. 41*, 1178, 1970.
91. Boyd, F. R.; England, J. L. *J. Geophys. Res. 65*, 741, 1960.
92. Weir, C. E. *J. Res. Nat. Bur. Stand. 46*, 207, 1951.
93. Weir, C. E. *J. Res. Nat. Bur. Stand. 53*, 245, 1954.
94. Bridgman, P. W. *Am. Acad. Arts Sci. 76*, 71, 1948.
95. Sanchez, I. C.; Cho, J. *Polymer 36*, 2929–2939, 1995.
96. Sachdev, V. K.; Yahsi, U.; Jain, R. K. *J. Poly. Sci. Part B: Polymer Physics 36*, 841–850, 1998.
97. Yan, Z.-T. *Commun. Theor. Phys. 17*, 389–392, 1992.
98. Zoller, P.; Fakhreddine, Y. A. *Thermochimica Acta 238*, 397–415, 1994.
99. Patrickios, C. S.; Lue, L. *J. Chem. Phys. 113*, 5485, 2000.
100. Kontogeorgis, G. M.; Harismiadis, V. I.; Fredenslund, A.; Tassios, D. P. *Fluid Phase Equilibria 96*, 65–92, 1994.
101. Tochigi, K.; Kurita, S.; Matsumoto, T. *Fluid Phase Equilibria 160*, 313–320, 1999.
102. Brostow, W.; Duffy, J. V.; Lee, G. F.; Madejczyk, K. *Macromolecules 24*, 479–493, 1991.
103. Barrett, A. J.; Domb, C. *J. Stat. Phys. 77*, 491–500, 1994.
104. Kang, J. W.; Lee, J. H.; Yoo, K. P.; Lee, C. S. *Fluid Phase Equilibria 194*, 77–86, 2002.
105. Sun, Z. H.; Song, M. *Acta Chimica Sinica 50*, 729–733, 1992.
106. Schmidt, M.; Maurer, F. H. J. *J. Poly. Sci. B: Polym. Phys. 36*, 1061, 1998.
107. Zoller, P. In *The Polymer Handbook*; Brandrup, J., Immergut, E. H., Eds.; Wiley: New York, 1989.
108. Lichtenthaler, R. N.; Liu, D. D.; Prausnitz, J. M. *Macromolecules 11*, 192, 1978.
109. Sachdev, V. K.; Jain, P. C.; Nanda, V. S. *Mat. Res. Soc. Symp. Proc. 22*, 243, 1984.
110. Zoller, P.; Bolli, P.; Pahud, V.; Ackermann, H. *Rev. Sci. Instrum. 47*, 948, 1976.
111. Gnomix PVT Apparatus, P. Zoller, Boulder, CO 80304, USA.
112. Zoller, P.; Walsh, D. J. *Standard Pressure-Volume-Temperature Data for Polymers*; Technomic Publishing: Lancaster, 1995.
113. Ames, R. G. MRS Symp. Proceedings, 2006, 896.
114. Dattelbaum, D. M.; Velisavljevic, N.; Emmons, E.; Stahl, D. B.; Sheffield, S. A.; Weinberg, M.; Orler, E. B.; Brown, E. N.; Rae, P. J. *Unpublished report*; Los Alamos National Laboratory, 2007.
115. Roland, C. M.; Hensel-Bielowka, S.; Paluch, M.; Casalini, R. *Rep. Prog. Phys. 68*, 1405, 1995.
116. Dattelbaum, D. M.; Sheffield, S. A.; Weinberg, M.; Stahl, D. B.; Neel, K.; Thadani, N. J. Appl. Phys. 2008 in press.
117. Cady, W. E.; Caley, L. E. Properties of Kel F-800 Polymer, Lawrence Livermore National Laboratory, 1977 UCRL-52301.
118. Clements, B. E.; Maciucescu, L.; Brown, E. N.; Rae, P.; Orler, E. B.; Dattelbaum, D. M.; Sheffield, S. A.; Robbins, D. L.; Gustavsen, R. L.; Velisavljevic, N. Campaign 2, level 2 DOE milestone report: KEL-F 800 Experimental characterization and model development, Los Alamos National Laboratory, 2007.
119. Stevens, L. L.; Orler, E. B.; Dattelbaum, D. M.; Ahart, M.; Hemley, R. J. *J. Chem. Phys. 127*, 104905, 2007.
120. http://en.wikipedia.org/wiki/Diamond.
121. Mao, H. K.; Bell, P. M. *Science 191*, 851, 1976.
122. Jarayman, A. *Rev. Mod. Phys. 55*, 65, 1983.
123. Caminiti, R.; Pandolfi, L.; Ballirano, P. *J. Macromol. Sci., B: Phys. 39*, 481, 2000.
124. Yamamoto, T.; Miyaji, H.; Asai, K. *Jpn. J. Appl. Phys. 16*, 1891, 1977.
125. Fontana, L. e. a. *Phys. Rev. B 75*, 174112, 2007.
126. Wang, H. C.; Aubuchon, S. R.; Thompson, D. G.; Osborn, J. C.; Marsh, A. L.; Nichols, W. R.; Schoonover, J. R.; Palmer, R. A. *Macromolecules 35*, 8794–8801, 2002.

127. Wang, H. C.; Thompson, D. G.; Schoonover, J. R.; Aubuchon, S. R.; Palmer, R. A. *Macromolecules 34*, 7084–7090, 2001.
128. Graff, D. K.; Wang, H. C.; Palmer, R. A.; Schoonover, J. R. *Macromolecules 32*, 7147–7155, 1999.
129. Wang, H. C.; Graff, D. K.; Schoonover, J. R.; Palmer, R. A. *Appl. Spectr. 53*, 687–696, 1999.
130. Dattelbaum, D. M.; Robbins, D. L.; Sheffield, S. A.; Orler, E. B.; Gustavsen, R. L.; Alcon, R. R.; Lloyd, J. M.; Chavez, P. J. *AIP Conf. Proc. 845*, 69–72, 2006.
131. Emmons, E. D.; Velisavljevic, N.; Schoonover, J. R.; Dattelbaum, D. M. *Appl. Spectr.* 2008, 62, p. 142–148.
132. Bunn, C. W.; Howells, E. R. *Nature 174*, 549, 1954.
133. Beecroft, R. I.; Swenson, C. A. *J. Appl. Phys. 30*, 1793, 1959.
134. Bridgman, P. W. *Proc. Amer. Acad. Arts Sci. 76*, 71, 1948.
135. Wu, C. K.; Nicol, M. *Chem. Phys. Lett. 21*, 153, 1973.
136. Flack, H. D. *J. Polym. Sci: Part A-2 10*, 1799, 1972.
137. Koenig, J. L.; Boerio, F. J. *J. Chem. Phys. 50*, 2823, 1969.
138. Boerio, F. J.; Koenig, J. L. *J. Polym. Sci: Part A-2 9*, 1517, 1971.
139. Koenig, J. L.; Boerio, F. J. *J. Chem. Phys. 52*, 4170, 1970.
140. Hannon, M. J.; Boerio, F. J.; Koenig, J. L. *J. Chem. Phys. 50*, 2829, 1969.
141. Rabolt, J. F.; Piermarini, G.; Block, S. *J. Chem. Phys. 69*, 2872, 1978.
142. Brown, R. G.*J. Chem. Phys. 40*, 2900, 1964.
143. Weir, C. E. *J. Res. Natl. Bur. Stand. 50*, 95, 1953.
144. Eby, R. K.; Clark, E. S.; Farmer, B. L.; Piermarini, G. J.; Block, S. *Polymer 31*, 2227–2237, 1990.
145. Nicol, M.; Wiget, J. M.; Wu, C. K. *J. Polym. Sci.: Polymer Phys. Ed. 18*, 1087–1102, 1980.
146. Scheirs, J. In *Modern Fluoropolymers*; J. Scheirs, Ed.; Wiley: Chichester, England, 1997.
147. Christen, D.; Hoffmann, V.; Klaeboe, P. *Z. Naturforsch 31*, 1320, 1979.
148. Duffy, T. S.; Wang, Y. In *Ultrahigh-pressure Mineralogy*, Vol. 37; Hemley, R. J., Ed.; Mineralogical Society of America: Washington, DC, 1998.
149. Flores, J. J.; Chronister, E. L. *J. Raman Spectroscop. 27*, 149, 1996.
150. Marsh, S. (Editor) *LASL Shock Hugoniot Data*, Los Alamos National Laboratory, 1980.
151. J. Fritz; *Upublished results*, Los Alamos National Laboratory.
152. Wunderlich, B. *J. Chem. Phys. 37*, 1207, 1962.
153. Shaner, J. W. *J. Chem. Phys. 89*, 1616–1624, 1988.
154. Cummins, H. Z.; Schoen, P. E. In *Laser Handbook*; Arechhi, F. T., Schulz-Dubois, E. O., Eds.; North-Holland: Amsterdam, 1972.
155. Fabelinskii, I. L. *Molecular Scattering of Light*; Plenum Press: New York, 1968.
156. Krüger, J. K. *Optical Techniques to Characterize Polymer Systems*; Elsevier: New York, 1989.
157. Kruger, J. K.; Marx, A.; Peetz, L.; Roberts, R.; Unruh, H.-G. *Colloid Polymer Sci 264*, 1986.
158. Whitfield, C. H.; Brody, E. M.; Bassett, W. A. *Rev. Sci. Instrum. 47*, 1976.
159. Koski, K. J.; Müller, J.; Hochheimer, H. D.; Yarger, J. L. *Rev. Sci. Instrum. 73*, 2002.
160. Dye, R. C.; Sartwell, J.; Eckhardt, C. J. *Rev. Sci. Instrum. 60*, 1989.
161. Stevens, L. L.; Orler, E. B.; Dattelbaum, D. M.; Ahart, M.; Hemley, R. J. American Physical Society Topical Conference on Shock Compression of Condensed Matter, 2007, accepted.
162. Patterson, G. D. *J. of Polym. Sci: Polym. Phys. Ed. 14*, 1976.
163. Middaugh, R. A.; Goudey, C. A. *Oceans 93 Proceedings vols 1–3: engineering in harmony with the ocean*, A149–A154, 1993.
164. McCluskey, M., Personal communication, 15th American Physical Society Topical Conference on Shock Compression of Condensed Matter, Fairmont Orchid, Hawai'i, July 2007.
165. Mao, H. K., Personal communication, Bethesda, MD, June 2007.
166. Carter, W. J.; Marsh, S. P. *Los Alamos National Laboratory Report* LA-13006-MS, 1995.
167. Sheffield, S. A.; Bloomquist, D. D. *Bull. Am. Phys. Soc. 26*, 660–661, 1981.
168. Barker, L. M. Am. Phys. Soc. Topic. Group Shock Compress. Condens. Mat., Snowbird, UT, 505, 11–17, 1999.
169. Barker, L. M. *Exp. Mech. 12*, 209, 1972.

170. Barker, L. M.; Hollenbach, R. E. *Rev. Sci. Instrum. 36*, 1617, 1965.
171. Barker, L. M.; Hollenbach, R. E. *J. Appl. Phys. 43*, 4669, 1972.
172. Bloomquist, D. D.; Sheffield, S. A. *J. Appl. Phys. 54*, 1717–1722, 1983.
173. Jensen, B. J.; Holtkamp, D. B.; Rigg, P. A.; Dolan, D. H. *J. Appl. Phys. 101*, 13523-1-10, 2007.
174. Mercier, P.; Benier, J.; Azzolina, A.; Lagrange, J. M.; Partouche, D. *J. Phys. IV134*, 805–812, 2006.
175. Strand, O. T. et al. SPIE Proceedings, 2005, p 593.
176. Johnson, J. N.; Dick, J. J.; Hixson , R. S. *J. Appl. Phys. 84*, 2520, 1998.
177. Bourne, N. K.; Gray, G. T. *J. Appl. Phys. 98*, 123503, 2005.
178. Morris, C. E.; Fritz, J. N.; McQueen, R. G. *J. Chem. Phys. 80*, 5203, 1984.
179. Kalashnikov, N. G.; Kuleshova, L. V.; Pavlovshii, M. N. *Zh. Prikl. Mekhan. Tekh. Fiz.*187, 1972.
180. Champion, A. R. *J. Appl. Phys. 42*, 5546, 1971.
181. Robbins, D. L.; Sheffield, S. A.; Alcon, R. R. Proceedings of the 13th American Physical Society Topical Conference on Shock Compression of Condensed Matter, Portland, Oregon, July 2003, Vol. 706, p. 675.
182. Weir, C. E. *J. Res. Nat. Bur. Stand. 50*, 95, 1953.
183. Rabolt, J. F.; Piermarini, G.; Block, S. *J. Chem. Phys. 69*, 2872, 1978.
184. Wu, C.-K.; Nicol, M. *Chem. Phys. Lett. 21*, 153, 1973.
185. Rae, P. J.; Dattelbaum, D. M. *Polymer 45*, 7615, 2004.
186. Nagao, H.; Matsuda, A.; Nakamura, K. G.; Kondo, K. *Appl. Phys. Lett. 83*, 249, 2003.
187. Bourne, N. K.; Gray, G. T. *J. Appl. Phys. 93*, 8966, 2003.
188. Sheffield, S. A.; Gustavsen, R. L.; Alcon, R. R. Proceedings of the American Physical Society Topical Conference on Shock Compression of Condensed Matter, 2000, Vol. 505, p. 1043.
189. Dattelbaum, D. M.; Sheffield, S. A.; Stahl, D. B.; Neel, K.; Thadani, N. *in preparation*.
190. Millet, J. C. F.; Bourne, N. K.; Gray, G. T. *J. Appl. Phys. 96*, 5500, 2004.
191. Wackerle, J.; Johnson, J. O.; Halleck, P. M. Projectile Velocity Measurements and Quartz-and Manganin-Gauge Pressure Determinations in Gas Gun Experiments, Los Alamos National Laboratory, LA-5844, 1975.
192. Anderson, M. U. American Physical Society Topical Conference on Shock Compression of Condensed Matter, 1991, p. 875.
193. Sheffield, S. A.; Alcon, R. R. American Physical Society Topical Conference on Shock Compression of Condensed Matter, 1991, p. 909.
194. Weinberg, M., Master Thesis, New Mexico Institute of Mining and Technology, 2006.
195. Gupta, S. C.; Gupta, Y. M. *High Press. Res. 19*, 785, 1992.
196. Hall, C. A. *Phys. Plasmas 7*, 2000.
197. Reisman, D. B.; Toor, A.; Cauble, R. C.; Hall, C. A.; Asay, J. R.; Knudson, M. D.; Furnish, M. D. *J. Appl. Phys. 89*, 2001.
198. Hall, C. A.; Asay, J. R.; Knudson, M. D.; Stygar, W. A.; Spielman, R. B.; Pointon, T. D.; Reisman, D. B.; Toor, A.; Cauble, R. C. *Rev. Sci. Instrum. 72*, 2001.
199. Duvall, G. E.; Fowles, G. R. In *High Pressure Physics and Chemistry*, Vol. II; Bradley, R. S., Ed.; Academic Press: New York, 1963.
200. Aidun, J. B.; Gupta, Y. M. *J. Appl. Phys. 69*, 1991.
201. Sandia Z-pinch Machine, Sandia National Laboratory.
202. Hall, C. A.; Baer, M. R.; Gustavsen, R. L.; Hooks, D. E.; Orler, E. B.; Dattelbaum, D. M.; Sheffield, S. A.; Sutherland, G. T. *AIP Conf. Proc. 845*, 1311–1314, 2006.
203. Gustavsen, R. L.; Dattelbaum, D. M.; Orler, E. B.; Hooks, D. E.; Alcon, R. R.; Sheffield, S. A.; Hall, C. E.; Baer, M. R. American Physical Society Conference on Shock Compression of Condensed Matter, 2005, 845, pp. 149–152.
204. Baer, M.; Hall, C.; Hobbs, M.; Gustavsen, R.; Hooks, D.; Dattelbaum, D.; Sheffield, S. American Physical Society Conference on Shock Compression of Condensed Matter, 2007, in press.
205. Baer, M. R.; Hall, C. A.; Gustavsen, R. L.; Hooks, D. E.; Sheffield, S. A. *J. Appl. Phys. 101*, 034906, 2007.

206. Sutherland, G. T.; unpublished results.
207. Kober, E. M.; Menikoff, R. American Physical Society Topical Conference on Shock Compression of Condensed Matter,Snowbird, UT, 1999, 505, p. 21.
208. Higuchi, H.; Jamieson, A. M.; Simha, R. *J. Polym. Sci. Part B: Polym. Phys. 34*, 1423, 1996.
209. Williams, M. L.; Landel, R. F.; Ferry, J. D. *J. Am. Chem. Soc. 77*, 3701, 1955.
210. Doolittle, A. K. *J. Appl. Phys. 22*, 1471, 1951.
211. Fox, T. G.; Flory, P. J. *J. Am. Chem. Soc. 70*, 2384, 1948.
212. Fox, T. G.; Flory, P. J. *J. Appl. Phys. 21*, 581, 1950.
213. Fox, T. G.; Flory, P. J. *J. Phys. Chem. 55*, 221, 1951.
214. Fox, T. G.; Flory, P. J. *J. Polym. Sci. 14*, 315, 1954.
215. Fox, T. G.; Flory, P. J. *J. Appl. Phys. 21*, 5, 1950.
216. Simha, R.; Boyer, R. F. *J. Chem. Phys. 37*, 1003, 1962.
217. Hagiwara, K.; Ougizawa, T.; Inoue, T.; Hirata, K.; Kobayashi, Y. *Rad. Phys. Chem. 58*, 525, 2000.
218. Bondi, A. *J. Phys. Chem. 68*, 441, 1964.
219. Simha, R.; Carri, G. *J. Polym. Sci. Part B: Polym. Phys. 32*, 2645, 1994.
220. Dow Corning, "Sylgard 184 Silicone Elastomer"product information sheet.
221. Menikoff, R.; Kober, E. M.; Dattelbaum, D. M. *Unpublished results*; 2005.
222. Jean, Y.; Dattelbaum, D. M. *unpublished results.*
223. Morris, C. E.; Loughran, E. D.; Mortensen, G. F.; Gray, G. T. I.; Shaw, M. S. American Physical Society Topical Conference on Shock Compression of Condensed Matter, 1989.
224. Sander, R. K.; Sheffield, S. A.; Blais, N.; Engelke, R.; Dattelbaum, D. M.; McInroy, R. American Physical Society Topical Conference on Shock Compression of Condensed Matter, 2005, Vol. 845, p. 1165.
225. Graham, R. A.; Richards, P. M.; Shrouf, R. D. *J. Chem. Phys. 72*, 3421, 1980.
226. Eichelberger, R. J.; Hauver, G. E., 1962 p. 363.
227. Hauver, G. E. *J. Appl. Phys. 36*, 2113, 1965.
228. de Icaza Herrera, M.; Migault, A.; Jacquesson, J. *C. R. Acad. Sci. Paris 284*, 503, 1977.
229. de Icaza Herrera, M.; Migault, A.; Jacquesson, J. *C. R. Acad. Sci. Paris 284*, 531, 1977.
230. Novitskii, E. Z.; Ivanov, A. G.; Khokhlov, N. P. Third All-Union Symposium on Combustion and Explosion, Nauka, Moscow, 1971, p. 579.
231. de Icaza Herrera, M.; Migault, A.; Jacquesson, J. In *High Pressure Science and Technology*; Timmerhaus, K. D., Barber, M. S., Eds.; Plenum Press: New York, 1979.
232. Graham, R. A. *J. Phys. Chem. 83*, 3048, 1979.
233. Brish, A. A. *Soviet Physics JETP 11*, 15, 1960.
234. Graham, R. A. VIIth International AIRAPT Conference, Le Creusot, France, 1979, p. 1032–1039.
235. Champion, A. R. *J. Appl. Phys. 43*, 2216, 1972.
236. Kuleshova, L. V. *Sov. Phys. Solid State 11*, 886, 1969.
237. Hauver, G. E. 5th Symposium (International) on Detonation, 1970, p. 387.
238. Lovinger, A. J. *Jpn. J. Appl. Phys. 24*, 18, 1985.
239. Bauer, F.; Graham, R. A.; Lee, L. M. ISAF 92, 1992, pp. 273–276.
240. Bauer, F.; Graham, R. A.; Lee, L. M. ISAF 90, 1990, pp. 288–291.
241. Bauer, F. American Physical Society Topical Conference on Shock Compression of Condensed Matter, 1983 p. 225.
242. Bauer, F. *Ferroelectrics 49*, 231, 1983.
243. Sheffield, S. A.; Dattelbaum, D. M.; Robbins, D. L.; Alcon, R. R.; Gustavsen, R. L. Proceedings of the American Physical Society Topical Conference on Shock Compression of Condensed Matter, 2005, Vol. 845, p. 921.
244. Johnson, J. N. APS/AIRAPT Conference, 1993.
245. Smith, J. H. *ASTM Tech. Pub. 336*, 264, 1963.
246. Staudhammer, K. P.; Johnson, K. A. *International Symposium on Intense Dynamic Loading and its Effects*, 1986.
247. Chang, S. N.; Chung, D. T.; Li, Y. F.; Nemat-Nasser, S. *J. Appl. Mech. Trans. ASME 59*, 305, 1992.

248. Chang, S. N.; Chung, D. T.; Ravichandran, G.; Nemat-Nassar, S. American Physical Society Topical Conference on Shock Compression of Condensed Matter, 1989, p. 389.
249. Kumar, P.; Clifton, R. J. *J. Appl. Phys. 48*, 4850, 1977.
250. Gray, G. T. American Physical Society Topical Conference on Shock Compression of Condensed Matter, 1989.
251. Gray, G. T. The Materials Society Symposium on Modeling the Deformation of Crystalline Solids, 1991.
252. Bourne, N. K.; Gray, G. T. *Proc. R. Soc.* 2005.
253. Clifton, R. J.; Raiser, G.; Ortiz, M.; Espinosa, H. American Physical Society Topical Conference on Shock Compression of Condensed Matter - 1989, 1990, p. 437–440.
254. Gray, G. T. In *High-Pressure Shock Compression of Solides*; Asay, J. R., Shahinpoor, M., Eds.; Springer: Berlin, 1993.
255. Gray, G. T. In *ASM Handbook Vol. 8: Mechanical Testing and Evaluation*; Kuhn, H., Medlin, D., Eds.; ASM International Materials Park, OH, 2000.
256. Gray, G. T.; Follansbee, P. S.; Frantz, C. E. *Mater. Sci. Eng. A. 111*, 9, 1989.
257. Koller, D. D.; Hixson, R. S.; Gray, G. T.; Rigg, P. A.; Addessio L. B.; Cerreta, E. K.; Maestas, J. D.; Yablinksy, C. A. *J. Appl. Phys. 98*, 103518, 2005.
258. Bourne, N. K.; Gray, G. T. *J. Phys. D.: Appl. Phys. 38*, 3690, 2005.
259. Millett, J. C. F.; Bourne, N. K.; Akhavan, J. *J. Appl. Phys. 95*, 4722, 2004.
260. Brown, E. N.; Trujillo, C. P.; Gray, G. T.; Rae, P. J.; Bourne, N. K. *J. Appl. Phys., 101*, 024916, 2007.
261. Antoun, T.; Seaman, L.; Curran, D. R.; Kanel, G. I.; Razorenov, S. V.; Utkin, A. V. *Spall Fracture*; Springer: New York, 2003.
262. Johnson, J. N.; Dick, J. J. American Physical Society Topical Congerence on Shock Compression of Condensed Matter - 1999, 2000, p. 543.
263. "Basic Research Needs for Materials under Extreme Environments," Report of the Basic Energy Sciences Workshop on Materials under Extreme Environments, June 11–13, 2007, Department of Energy, Office of Science, Bethesda, MD.

Chapter 5
Reaction Kinetics

Suhithi M. Peiris

5.1 Introduction

Detonation, which is the result of a reacting high explosive, is not a simple chemical reaction. Instead it is a complex process of rapid chemical changes that give rise to equally rapid mechanical and physical changes. Initially, as the explosive is initiated or subjected to a stimulus such as a shock wave, energy is transferred to the material causing mechanical deformation and heating. The transferred energy and heating produce chemically excited species, leading to initial bond-breaking or reaction initiation. These processes are thought to occur in a Hot-Molecule Zone (HMZ). Once reaction is initiated, sequences of chemical reactions follow resulting in the liberation of chemical energy and the production of gaseous reaction products.

To understand and predict the behavior of energetic materials, reaction rates and kinetic measurements of reactions in the HMZ need to be measured and understood. The initiation of chemical reactions and energy release from the HMZ or "hot spots" has been the subject of a great deal of research. However, dynamic measurements such as shock studies are limited to a few analytical techniques, simply because equipping such experiments with spectrometers and detectors that can capture fast chemistry during shock initiation is extremely challenging. Therefore, the parameters that characterize reaction kinetics, the reaction mechanism, and initial bond-breaking transitions states, are often evaluated at lower pressures and slower heating rates than exist in the HMZ and extrapolated to the probable conditions in the HMZ. Accurate prediction of reaction products and detonation modeling critically requires these parameters at high temperature and high pressure of the hot-molecule zone.

Compared with fast dynamic measurements, static high pressure, such as that maintained in a diamond anvil cell, allows time for detailed chemical analysis at the pressures usually found in the HMZ. When energetic materials are initiated by heat pulses or laser radiation, their global reactions show the slow induction, fast growth, and slow deceleration features of typical sigmoid-type curves. Various kinetic models are available to copy the induction, growth, and deceleration features

S.M. Peiris, G.J. Piermarini (eds.), *Static Compression of Energetic Materials,*
Shock Wave and High Pressure Phenomena,

of reactions occurring in the solid phase. Therefore, this chapter will start with an introduction to the most relevant solid-state sigmoid kinetic models, followed by descriptions of data and measurements on the reaction or thermal decomposition of a number of energetic materials under static high pressure.

5.2 Kinetic Models

5.2.1 Background

A decomposition reaction is the chemical process that signifies the breakdown of the reactant to produce simpler, more stable products. In the case of solid high explosives (HEs), the solid HE decomposes to form final products that are gases such as CO, NO_2, CO_2. or N_2, with perhaps some remaining solid C in the form of soot. Usually the reactions of HEs are easily initiated with the use of heat, shear energy, friction forces, or shock waves. Since all these methods of reaction initiation lead to the decomposition of HEs, most of the discussions in this chapter will apply to all of them. However, the decomposition pathway or mechanism by which the decomposition occurs might depend on the initiation mechanism and may start with initial bond breaking, or with melting, or with evaporation followed by initial bond breaking, or other yet unknown mechanisms. In addition, under high pressure when the solid HE is compressed, the solid-state lattice is held together to higher temperature and intermolecular reactions become more important than at ambient or low pressure where volatilization/evaporation is seen prior to chemical reaction. Therefore, the method of initiation, the rate of heating, etc. will most likely change the first few mechanistic steps of HE decomposition reactions, and this chapter will focus on solid-state reaction kinetics as relevant to higher pressure studies and fast heating rates.

Considering the decomposition of a solid HE where the initial reaction occurs within the lattice of the solid, the reaction is spatially limited to the location where it was initiated. Unlike with gas-phase or even liquid-phase reactions, solid-phase reactions propagate in a ever-widening volume within the limits of the three-dimensional solid [1]. In such a case, there is a "boundary" or a surface, where the reaction is occurring, with reacted material on one side and unreacted material on the other side. This boundary or surface is usually referred to as the "reaction interface." Because reactions occur in the reaction interface, concentration of species, usually defined as the total number of moles of that species in the volume, and the commonly known kinetic rate laws that use concentration, cannot be easily applied to solid-state reactions. Instead, the kinetics of a solid-state reaction is determined by the velocity with which the reaction interface advances into unreacted or unchanged material [1]. This is a process necessarily measured by both the numbers of species within the reaction interface, and the spatial distribution or surface area of the interface.

Considering reaction as a growing interface, reactions propagating within a solid can be considered to consist of several processes. The first process of reaction initiation is *Nucleation*, where the reaction interface is initially established [1]. The initially generated reaction nucleus may be a small collection of molecules that are reacting and creating high strain in a localized spot. In current detonation theory, such localized spots have been popularly called "hot spots," and many hot-spot initiation theories about why or how energy was localized in that spot etc., abound. Some theories describe energy localization due to an imperfect lattice such as a few layers of displaced lattice planes or due to defect sites such as lattice vacancies or molecules substituted by impurities. Other theories explain the creation of hot spots by discussing the density differences between crystalline HE and the surrounding amorphous polymeric binder. Chapter 8 of this book takes an extensive look at such reaction initiation theories so they will not be discussed further in this chapter.

Beyond the initial nucleation, more areas are nucleated, extending the reaction interface. In most materials, because the initial steps of the reaction require energy to overcome an activation barrier, initially, the reaction interface grows slowly [1]. This time during which little reaction may be observed is known as the *Induction Period*. In HEs, the decomposition of a few initial molecules during nucleation produces heat. This heat then rapidly heats the next layer of molecules, etc., reducing the induction period to a few microseconds. As the reaction interface continues to expand, the reaction is said to undergo a *Growth* process. As most of the reacting material gets consumed, the continued expansion of the reaction is no longer possible, leading to *Deceleration* ending in reaction completion [1]. Figure 5.1 shows the s-shaped or sigmoid curves typically seen as the reaction interface grows from almost nothing to consume the whole material.

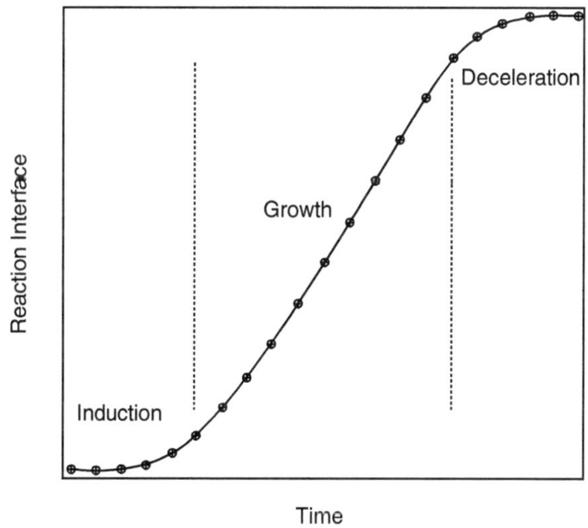

Fig. 5.1 Sketch depicting the decomposition reaction of a solid material. The dotted lines draw attention to the three common 'regions' of the plot

The decomposition rate or the velocity of advance of the reaction interface within a solid, has been described from two different viewpoints [1]. One viewpoint is simply that the increasing nucleation sites geometrically advance the reaction interface and the other viewpoint is that of the nucleated reactions yield chemical intermediates or products that activate more reactants in a chain-type manner. From the first viewpoint, deceleration is thought to be the result of two scenarios, one known as *Coalescence* where the reaction interface of one growing nucleus meets the reaction interface of another growing nucleus, and the other known as *Ingestion* where unreacted material is eliminated and no new nucleation can take place. From the second viewpoint, deceleration is thought to be the result of the consumption of the reactant causing a decrease in the reaction rate. The most commonly used sigmoid kinetic models (described below) for decomposition of a solid such as a high explosive were developed starting from these two different viewpoints [1].

5.2.2 Decomposition Kinetics

The kinetic models most commonly used to understand decomposition of a solid material assume that the reaction is occurring under isothermal conditions. No heating or heating rate dependencies are built into the models, and therefore, temperature-dependent phenomena such as melting are not included in the model.

Modeling decomposition kinetics requires an estimation of the reaction interface at any given time during the reaction. Considering that the reaction is 50% complete when the interface has consumed 50% of the reactant volume, the fraction of reaction that has occurred (α) is usually quantified as a means of estimating reaction progress. Experimental kinetic studies evaluate α by measuring a parameter that changes with the fraction of reaction that has occurred. Then plots of α versus time show typical sigmoid-type curves, and fitting such data curves with the different forms of the different models help determine the kinetic model that prevails.

5.2.2.1 Advancing Reaction Interface Models

Advancing interface models developed by Avrami, Erofe'ev, and others, is based on the viewpoint that increasing nucleation sites geometrically advance the interface with time, t [1]. In these models deceleration would occur via coalescence and ingestion. Reaction deceleration can also occur because the products formed at the interface have to diffuse through an increasing thick layer of intermediates or products and then, these models are sometimes considered diffusion-controlled reactions. In either case, the expression found can be written in the general form:

$$-[\ln(1-\alpha)]^{1/n} = [k(t-t_0)]$$

5.2.2.2 Chain-Type Nucleation Models

These models developed by Prout, Tompkins, and others, is based on the viewpoint that increasing nucleation yield chemical intermediates or products that activate more reactants in a chain-type manner, similar to an auto-catalytic process [1]. The same chain-type behavior would also occur if lattice strain developed at the interface caused crack propagation in different directions to become branching reaction zones. Considering that sooner or later the reaction must slow down due to consumption of reactant, and including an integration constant c, the following equation is found to apply:

$$\ln[\alpha/(1-\alpha)] = kt + c$$

5.2.2.3 Lee-Tarver Ignition and Growth Model

For energetic materials, through the years, it is the Lee-Tarver model [2] that has been successfully used to relate an energetic materials' reaction rate with its chemical energy and expanding gaseous products. Similar to chain-type nucleation, this model assumes that ignition involves a small fraction of the explosive (such as a few hot spots) that is initially ignited, and that these micro-sized reacting regions grow and interact to consume the whole material. However, in this model reaction growth is thought to be controlled by the pressure and surface area, similar to an advancing interface. Their generalized model is:

$$\frac{\partial \alpha}{\partial t} = I(1-\alpha)^x \left(\frac{V_0}{V} - 1\right)^r + G(1-\alpha)^x \alpha^y P^z$$

where V_0 is the initial specific volume of the unreacted material, V is the specific volume of the unreacted material at pressure P, and I, G, r, x, y, and z are constants. In this equation, the first part on the right-hand side using the constant I represents reaction ignition and accounts for the induction time with "r" usually having a value of 4. From empirical fits of this equation to various materials, "x" is evaluated to be 2/9. The second part on the right-hand side using the constant G represents reaction growth and deceleration to consumption. For spherically expanding reaction interfaces, "y" would be set to 2/3 to correspond with the surface area of a sphere reacting outward. The pressure term represents the pressure dependence of the reaction rate. For deflagrating or fast-burning materials the pressure exponent "z" has been measured to be between 1 and 2.

Later, this initial Lee-Tarver model was modified to include two growth terms, to accommodate high-intensity short-duration shock pulse initiation [3]. The third term is necessary because when highly intense short-duration shocks are used for initiation, a larger fraction of the explosive is rapidly ignited with a pressure-dependent reaction growth. In this three-term model, the first step is still initiation via formation of hot spots created by the various mechanisms such as void closure, viscous heating, shear banding, etc. The second step is a relatively slow growth of reaction modeled by inward burning of a spherical grain, and the third step is rapid completion of the reaction as hot spots begin to coalesce. Similar to the second step, the third is also modeled by a pressure-dependent growth rate with an exponent large enough for self-sustained detonation.

$$\frac{\partial \alpha}{\partial t} = I(1-\alpha)^x \left(\frac{V_0}{V} - 1 - a\right)^r + G_1(1-\alpha)^x \alpha^s P^y + G_2(1-\alpha)^{(1-x)} \alpha^u P^z$$

In this equation, minimum and maximum values of α are used to turn on or off terms at appropriate values. Most kinetic data collected for HEs were analyzed according to one of the three models described above. However, as described in [1], there are many other models that can be used to model kinetics in solids.

5.2.3 Method

Typical experimental methods (further detailed in the rest of this chapter) consist of loading pressure cells with the sample and perhaps a material used for monitoring the pressure. Pressure media, specially organic liquids, may affect the chemical reaction of the sample under pressure and temperature, and are most often not employed.

As described in the rest of the chapter, most studies focus on heating the sample at some set high pressure inside the pressure cell. While slower heating rates popularly known as “cook off” mimic situations where an EM is a hazard that may cause accidents, it is fast heating at the heating rates found in shock studies that mimic detonation dynamics. Therefore, the more recent studies described herein employ very fast heating with lasers initiating a single spot within the sample. The early studies described here, used visual methods such as microscopy to “see” the sample reacting. Others used temperature run away as an indication of reaction. These studies result in burn rates or time-to-explosion measurements, and measure the effect of pressure on such values. Later studies that combined laser heating with time-resolved spectroscopy provides a glimpse into the reaction mechanisms occurring at detonation. Additional specific details of the methods used for the various studies are described in the rest of this chapter.

5.3 Data

The earliest static-pressure reaction research focused on measuring the “time to explosion” (t_x) of nitromethane (NM), TNT, PETN, and HMX at 1 and 5 GPa and various temperatures [4]. These samples were loaded in high-pressure cells made of gold and compressed between BN anvils. The cell is heated at rates measured by seconds, and resultant t_x values range around 100 s. Both PETN and HMX showed increased t_x with increased pressure, while the t_x of TNT was independent of pressure. NM showed an opposing trend where t_x was lower at higher pressure. These results were interpreted to mean that the rate-controlling step of the reactions of PETN and HMX involved formation of gaseous products which could then be suppressed by pressure. The TNT decomposition rate-controlling step was interpreted to include TNT melting and the formation of solid "coke," thereby making that reaction independent of externally applied pressure.

Through the years, nitromethane (NM) kinetics at static high pressures have been studied utilizing several different methods. A detailed study of the decomposition kinetics of NM between 2 and 7 GPa and 393 and 453 K was performed and is described in Chapter 1 of this book. That study reports that pressure accelerates decomposition kinetics and that at least two different kinetic pathways of chemical decomposition are seen depending on the static pressure and temperature [5]. Decomposition of NM was also studied over the range of 115–180°C and 0.6–8.5 GPa, where samples were compressed in Merrill-Basset DACs and then heated with external heaters to the point at which reaction starts occurring [6]. Decomposition products were collected and analyzed using infrared absorption microscopy. The volatile products observed were N_2O, CO_2, and H_2O, with the products remaining as fluids below 2.5 GPa, and solids above that pressure [6]. Considering the complicated phase diagram of NM now discovered, it is not surprising that several different mechanisms and kinetic rates have been reported at various pressures and temperatures [7].

Another study, that measured Combustion Front Propagation Rates (CFPR) found that NM did not ignite below 2.5 GPa when initiated with a 10 ns pulse from a frequency-doubled Nd:YAG (532 nm) pulse laser [8]. The authors listed in Ref. [8] loaded samples in DACs and used the difference in light transmitted through the reacted part of the sample in comparison to the light transmitted through unreacted sample, to monitor the combustion front. They measured linearly increasing CFPR from $5\,\mathrm{m\,s^{-1}}$ at ambient pressure to $100\,\mathrm{m\,s^{-1}}$ at 30 GPa. From 30 to 40 GPa, CFPR reduces from 100 to $40\,\mathrm{m\,s^{-1}}$. One explanation for this change in rate around 30 GPa is that the reaction products change from carbon-based dark solid to a clear and transparent material inside the DAC [8].

Ammonium perchlorate (NH_4ClO_4, AP) was studied from ambient to 26 GPa and 700 K. At almost ambient pressure (~0.5 GPa) heating to 603–620 K decomposed AP to its gaseous components:

$$2NH_4ClO_4 \rightarrow Cl_2 + 2NO + O_2 + 4H_2O$$

When pressure is increased to 10 GPa the decomposition temperature increases to 670 K or so [9].

A faster heating rate comparable to explosive decomposition and time-resolved spectroscopic measurements of chemical changes were used to study ammonium perchlorate at static high pressures [10]. In these experiments samples were loaded in Merrill-Bassett cells and heated by a pulsed dye laser (514 nm, 6 μs FWHM) with a measured laser fluence of 2.7, 11, or $22\,\mathrm{J\,cm^{-2}}$. Reaction products were spectroscopically identified with no AP remaining unreacted in the cell. To monitor the reaction progress, simultaneous with pulsed-laser heating, an UV-Vis source light is transmitted through the sample. Transmitted light is collected and sent to a spectrometer (SPEX 270M) for wavelength dispersion, and then to a streak camera (Hamamatsu 1487) for time dispersion. The two-dimensional output is then recorded in a CCD. A delay generator (Stanford Research DG535) controls synchronization for the complete data-collection system. The absorbance is calculated from $A(t,\lambda) = \log_{10}(I_0/I)$, where I_0 is collected prior to pulsed laser heating and I is collected during and immediately following the laser pulse. Then temporal change in absorption during decomposition can be used to determine the extent of reaction $(\partial\alpha/\partial t)$ [10].

Figure 5.2 shows experimental data obtained for AP together with the Lee-Tarver model adapted for laser-initiation experiments [10]. In this adapted model, the laser fluence L, which is a time-dependent function, was included in the Ignition term and the first Growth term. The parameters h and w are used to contain the effect of L on the final result. A new parameter "i" was added to the first Growth term so that it would not be turned off when the laser pulse died off. Therefore, $i = 4e^{-5}\,\mathrm{J\,cm^{-2}}$ was used simply to yield a non-zero value for the first Growth term, even when laser fluence L was zero [10].

$$\frac{\partial\alpha}{\partial t} = I_g(1-\alpha)^{2/9}\left(\frac{V_0}{V}-1\right)^4 L^h + G_1(1-\alpha)^{2/9}\alpha^{2/3}P(i+L)^w + G_2(1-\alpha)^{2/3}\alpha^{2/9}P$$

During data fits to this adapted model, the value of $h = 0.66$ was arbitrarily chosen. The value of $w = 0.1$ was chosen such that $w \ll h$, because the growth term would

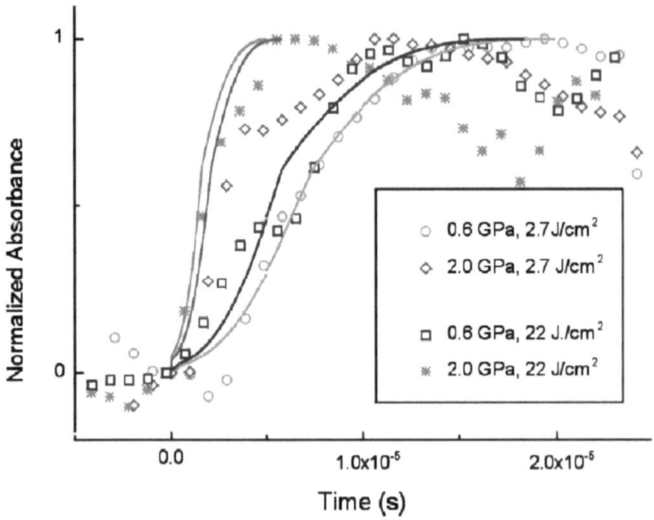

Fig. 5.2 Decomposition of AP monitored using time-resolved absorbance changes at 0.6 and 2.0 GPa when initiated by laser fluence of 2.7 and 22 $\mathrm{J\,cm^{-2}}$. The solid lines are from an adapted form of the Lee-Tarver model, fit with a single set of parameters [10]

be less affected by the initiation laser. The values chosen for the other parameters are described in the Lee-Tarver reference [3]. A single set of values for I_g, G_1, and G_2 were found that described the observed rates for 16 experiments performed with different initial pressures and laser fluence.

As shown in Fig. 5.2, the model captures the slow rise (or induction period) at lower pressures of 0.6 GPa, and the growth rates at all the pressures. Although the calculated deceleration happens faster than is observed at higher pressures, the deceleration at low-pressure measurements do fit well. The good correlation between the data points and the predictions of the Lee-Tarver-based model indicates that the static-high-pressure, high-temperature reactions probed in these experiments have the potential to describe real detonation chemistry.

Similar experiments have been performed on β-HMX, also laser-initiated at high pressure approaching the condition of shock or detonation studies [11]. The experimental setup is shown in Fig. 5.3. A single pulse from a flashlamp-pumped dye laser was used to initiate the samples. Simultaneously, a source lamp for incident absorption provides light covering the wavelength region 250–900 nm. The pulse duration of this absorption source lamp is 25 μs. The transmitted light is dispersed in wavelength by a spectrometer and is streaked in time by a streak camera. A CCD detector then measures the intensities of the wavelength- and time-resolved light. In each CCD image collected, the x-axis is wavelength, and the y-axis is time, with $t = 0$ being at the start of the initiating laser pulse.

More than 20 samples of β-HMX were compressed to various initial pressures between 0.6 and 3.7 GPa and initiated with a laser pulse with various fluence between 1.4 and 8.5 J cm^{-2}, reacting each sample completely to product gases. For each sample, first, a "pre-initiation" transmission image was recorded through the pressurized sample before initiating reaction with the laser. This transmitted image is considered a reference intensity $I_0(t,\lambda)$. Then a transmitted image is recorded during reaction to yield $I(t,\lambda)$. Then the "change in absorbance" (ΔA) with respect to the reference intensity is obtained according to: $\Delta A(t,\lambda) = \log_{10}[I_0(t,\lambda)/I(t,\lambda)]$ [11].

The ΔA intensity within 20 pixels around a chosen wavelength is averaged to obtain ΔA versus time, at that chosen wavelength. Considering absorbance to be

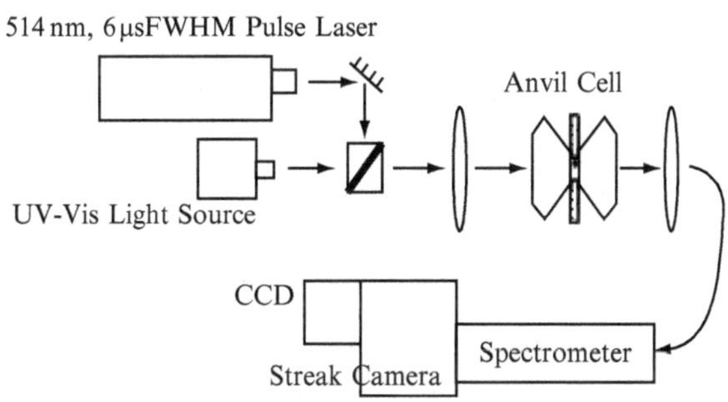

Fig. 5.3 Experimental setup for single-shot laser initiation and reaction propagation measurements using time- and wavelength-resolved spectroscopy [11]

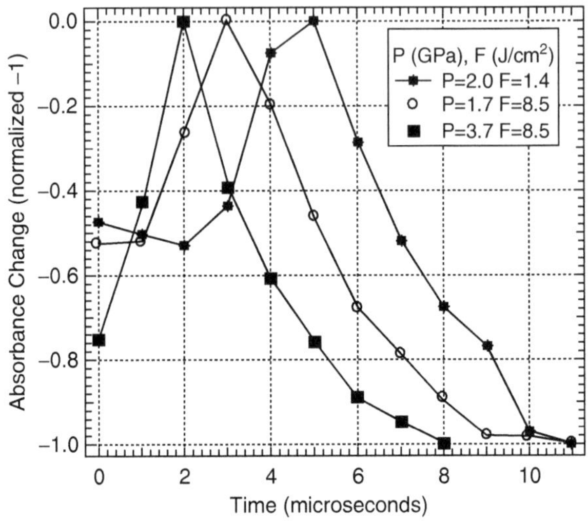

Fig. 5.4 Change in absorbance (at 470 nm) normalized to −1, after laser initiation of β-HMX at time = 0. The three curves (solid lines are to guide the eye) show the pressure (P) dependence and laser fluence (F) dependence

proportional to the quantity of HMX through which the UV-Vis light source is transmitted (similar to Beer's Law), allows $.A$ to be an estimate of change in concentration of HMX with time, or the reaction rate [11].

Figure 5.4 shows the typical absorbance change that occurs during HMX reaction. The data in solid circles shows an initial s-shaped rise to 5 μs, and then a later s-shaped drop to 11 μs. Unlike the reactions of AP shown in Fig. 5.2, these two sigmoid curves seen for HMX reactions reveal an initial increased absorbance due to the production of some darker intermediates or two-phase interactions, and a later decreased absorbance producing the clear final-product gases [11]. The lower pressure, lower fluence data (solid circles) has an induction period of almost 2 μs. As fluence is increased to 8.5 J cm^{-2} as shown in the data marked by open diamonds, the induction period is reduced to 1 μs, and as pressure is increased at that high fluence, as demonstrated by the data in closed squares, the period is further reduced to the beginning of laser initiation, $t = 0$. Initiating laser fluence therefore, comparable to the initial temperature of the reaction, decreases the total time for production of clear gases or enhances the global reaction rate of HMX.

The latter part of the global reaction of HMX described by the second sigmoid accounts for a larger part of the total reaction time. To determine the kinetic model that would account for a larger part of the HMX reaction, the change in absorbance (considered the extent of reaction) of this second sigmoid was fitted to the two kinetic models of Advancing Reaction Interface Model and the Chain-Type Nucleation Model. Because both models have linear equations, plotting the data as $\ln\alpha$ and $\ln(1-\alpha)$ versus time according to each model, should result in straight lines. A sample of the HMX data, plotted according to each model, are shown in Fig. 5.5.

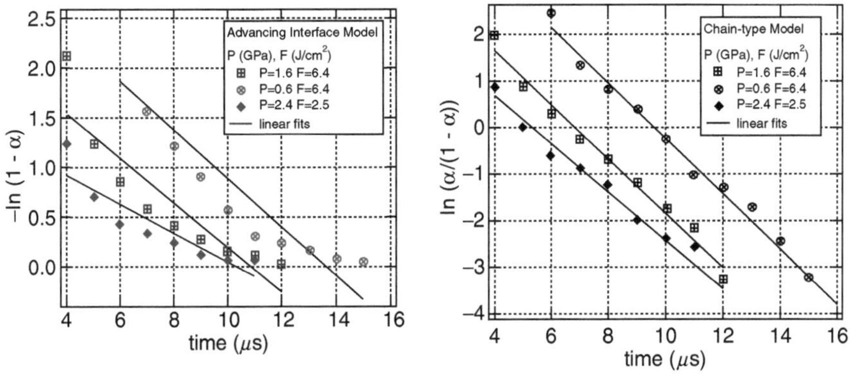

Fig. 5.5 Chain-type nucleation kinetic model and Advancing-interface kinetic model fits to HMX data at different pressure (P) and laser fluence (F). The better linear fits to the chain-type model indicates HMX reactions proceed through chain nucleation

Note that the advancing-interface model data is far from linear. However, the chain-type model does result in well-characterized lines. This indicates that HMX reaction growth is best modeled by a chain-type nucleation model.

An increase of reaction rate with increasing temperature was also seen in the previous work on HMX by Piermarini et al. [12]. They loaded the sample in the DAC between layers of NaCl to provide a clear window for FTIR spectroscopy. Samples were pressed to an initial pressure, then heated at a rate of about $5^{\circ}\mathrm{C\,h^{-1}}$ at that pressure to an initial temperature and held at that temperature until decomposition was seen by decreasing intensity of IR-active HMX vibrational modes. It was assumed that the quantity of HMX in the DAC is proportional to the intensity of the spectral peaks. Peak intensity before initiation (I_0) and at various times during thermal decomposition (I_t), was ratioed (I_t/I_0), and the ratios from nine peaks were averaged to estimate the fraction of HMX decomposed at various times. Their study shows a single sigmoid for the decomposition of HMX above 3.6 GPa, at initial temperatures of 280°C, 285°C, 290°C, 295°C, and 300°C [12]. They reported that data below 3.6 GPa was inconsistent, citing reasons such as a more complex reaction mechanism at lower pressures or leaking product gases that caused pressure fluctuations during the thousands of seconds over which decomposition took place in their study.

Their HMX data was also modeled by applying various forms of kinetic rates. They also found that the chain theory of nucleation gave remarkably linear fits at several temperatures and pressures over 3.6 GPa. These linear fits then resulted in values for the rate constant (k) from the slope of the line.

$$\ln[\alpha/(1-\alpha)] = kt + c$$

At each pressure a plot of ln k versus $1/T$, which is the typical Arrhenius plot, revealed a linear dependence or Arrhenius behavior with temperature. This allows the calculation of activation energies of HMX at various pressures. In addition, plotting ln k versus P yields the activation volume at various temperatures. The positive activation volume they obtained indicates that the intermediate or transient state of HMX during decomposition is volumetrically larger than the reactant HMX molecule at these pressures and temperatures. Therefore, they concluded that the decomposition of HMX involved a unimolecular transition state [12]. If two individual reactant molecules came in close proximity to make a bimolecular transition state, the activation volume would be negative because the transition state would be volumetrically smaller than the two molecules in their usual lattice spacing.

There are many publications of the thermal decomposition reaction mechanism and transition states of HMX at nominal pressures and high temperatures. These are well reviewed by Thomas Brill in a 1995 paper on multiphase chemistry considerations of nitramines [13]. In that paper, the accepted chemical scheme for the initial steps of HMX decomposition (and RDX decomposition) is described as two competing mechanistic paths, as follows:

$$\text{HMX} \rightarrow 4CH_2O + 4N_2O \quad \text{(a)}$$
$$\text{HMX} \rightarrow 4HCN + 4NO_2 + 2H_2 \quad \text{(b)}$$

The rate-determining step of path (b) is speculated to be $N-NO_2$ homolysis involving a single molecule of HMX, while the exact steps for path (a) are complex. In addition, path (a) is reported to be faster at temperatures below 600 K and path (b) faster at temperatures above 600 K [13]. Therefore, the dominant mechanistic path and the initial concentrations of individual species depends on the temperature. Subsequently, the secondary reactions that follow the above initial mechanistic paths to give the final gaseous products, also depend on the temperature. To discern the pressure dependence of the decomposition mechanism, detailed high-pressure, time-resolved FTIR or Raman spectroscopy that result in the concentrations of individual chemical species is necessary for future work.

There is also a laser-initiated reaction propagation rate study of HMX at static high pressure, similar to the NM and AP studies discussed previously in Refs. [8] and [9, 15]. This study simply followed the "flame front" or spread of light when HMX was pressed to initial pressures between 0.7 and 35 GPa and then initiated with a 532 nm Nd:YAG laser of 9 ns pulse width. They found the reaction propagation rate increased with pressure according to power law expressions, for both small ($\sim$3 μm) and large ($\sim$10 μm) grain samples [15].

Decomposition studies of RDX global reactions at high pressure and temperature have also been performed [14]. These are detailed in Chapter 1. In contrast to HMX, they found the lower pressure (to 2 GPa) and lower temperature (to 215°C) RDX decomposition to be interface advancing rather than chain-type. In this lower P–T range α-RDX decomposes directly, while at higher P–T α- and γ-RDX convert to the β-phase before decomposition. The decomposition of the β-phase was found to be autocatalytic or chain-type similar to β-HMX decomposition. Further, at lower P–T α-RDX decomposition demonstrated a negative activation volume indicating a bimolecular transition state, while at higher P–T β-RDX with a positive activation volume demonstrated a unimolecular reaction [14].

The chemical kinetics of TATB have also been investigated at high P and T. The reaction propagation rate of TATB was measured at pressures from 3 to 40 GPa, similar to other propagation rate studies [16]. The propagation rate was found to increase linearly with pressure, other than the discontinuities observed at 19, 30, and 40 GPa. A progressive shift of color was also observed, from yellow at ambient pressure to red and then black above 30 GPa [16].

Laser-initiation reaction rates of TATB were also measured similar to those described in [11] above for HMX [17]. In these experiments, an additional dichroatic beam splitter was used to separate the transmitted light between two wavelength regions of 270–420 nm and 550–780 nm. Unreacted TATB exhibited a broad absorption band centered at 370 nm, probably accounting for the yellow color of TATB. As shown in Fig. 5.6 this band loses intensity during reaction, and absorbance measurements of the clear solid product/s formed (post-laser) does not show this feature. Therefore, absorbance at 370 nm was used to probe the global reaction rate of TATB. After laser pulse initiation, the 370 nm absorption band was seen to decrease continuously and reach a minimum by 9 μs. However, none of the samples to pressures of 2 GPa with laser fluence of up to 17 J cm^{-2} reacted completely to product gases [17].

Under the P–T and laser fluence used, since TATB did not react completely to product gases, global reaction kinetics could not be obtained. Therefore, the extent of reaction at each P–T was estimated to note any trends. The percent of reacted TATB was calculated by dividing the post-laser absorbance spectrum by the pre-laser absorbance spectrum (both centered at 370 nm). The effect of pressure on the extent of reaction is shown in Fig. 5.7. Clearly higher pressure allows a higher percent of TATB to be reacted.

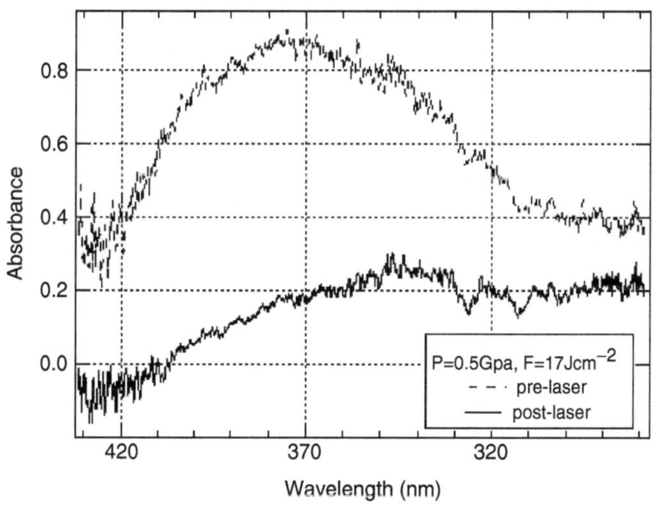

Fig. 5.6 Absorbance centered at 370 nm (against a standard NACl sample of the same thickness in the same DAC at the same pressure) of TATB at 0.5 GPa before laser initiation (pre-laser) and its disappearance after laser initiation (post-laser)

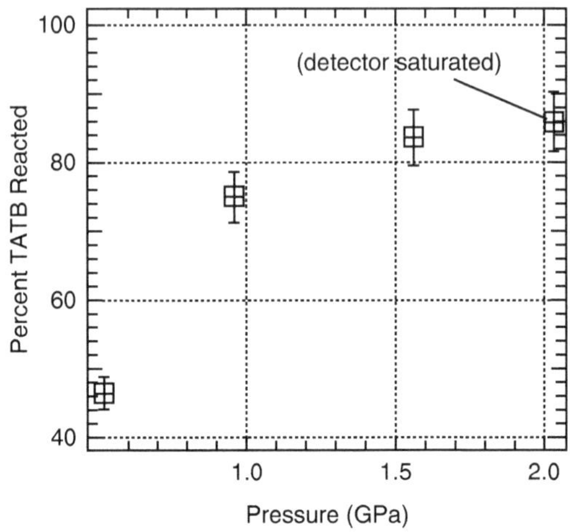

Fig. 5.7 Percent TATB reacted estimated from the change in absorbance after laser initiation. Higher pressure allows more TATB to be reacted

In addition, the photolysis of TATB has been studied using 568.2, 514.5, 488.0, and 457.9 nm wavelengths of light [18]. Prolonged exposure to 200–300 mW of radiation at 568.2, 514.5, and 488.0 nm wavelengths at pressures in excess of 1.5 GPa resulted in no photolytic degradation. However, 457.9 nm wavelength light of greater than 250 mW resulted in the formation of a brown spot, interpreted to be thermal degradation of the sample [18].

A single study of the reactions of CL-20 has been reported [19]. Samples of γ CL-20 were used and compressed to 2.7 GPa to give ζ CL-20. Reaction was initiated with a 8 ns, 532 nm, Nd:YAG laser pulse and used matrix-isolation techniques together with IR vibrational spectroscopy for analyzing the products. Even though the gem anvil cells used were cooled to 50 K to isolate the products, the laser fluence of $5\,\mathrm{J\,cm^{-2}}$ heated the samples above 300°C to a possible 740°C. Infrared absorption spectra of the starting material and the reaction products after laser initiation were measured. They observed mostly CO_2, with some NO_2, CO, and HNCO. NO_2 is reported to be the major final product of a previous T-jump IR spectroscopy study performed at ambient pressure. Therefore, the authors conclude that in their study when products are cooled or matrix-isolated, the reaction of CL-20 is arrested along the path to final detonation products.

5.4 Conclusions

This chapter describes available kinetic data and measurements of the reaction or thermal decomposition of energetic materials under static high pressure. As is, there are very few such studies performed, and work on very popular materials such as

TNT is seriously lacking. Hydrocodes, and other models of detonation of various formulations involving explosives, require data with burn rates, reaction rates, and the effect of pressure on the burn rate or reaction rate, and rate mechanisms. Therefore, there remains much scope for future studies of the reactions and kinetics of energetic materials at static high pressure.

Acknowledgments I gratefully acknowledge support from the Defense Threat Reduction Agency's Advanced Energetics Program, during the compilation of this manuscript.

References

1. Comprehensive Chemical Kinetics (1980) Volume 22 "Reactions in the Solid State" edited by C. H. Bramford and C. F. H. Tipper, Elsevier, Amsterdam.
2. E. L. Lee and C. M. Tarver (1980) "Phenomenological model of shock initiation in heterogeneous explosives", Phys. Fluids 23, 2362–2372.
3. C. M. Tarver and J. O. Hallquist (1985) "Modeling short-pulse duration shock initiation of solid explosives", Eighth International Detonation Symposium, Albuquerque, NM, 1985, Proceedings published by the Office of Naval Research, Arlington, VA, pp. 951–961.
4. E. L. Lee, R. H. Sanborn and H. D. Stromberg (1970) "Thermal decomposition of high explosives at static pressures to 50 kilobars", Fifth International Detonation Symposium, Pasadena, CA,1970, Proceedings published by the Office of Naval Research, Arlington, VA, pp. 331–337.
5. G. J. Piermarini, S. Block and P. J. Miller (1989) "Effects of Pressure on the Thermal Decomposition Kinetics and Chemical Reactivity of Nitromethane", J Phys. Chem. 93, 457–462.
6. S. F. Angew, B. I. Swanson, J. Kenney and I. Kenney (1989) "Chemistry of nitromethane at very high pressure", Ninth International Detonation Symposium, Portland, OR, 1989, Proceedings published by the Office of Naval Research, Arlington, VA, pp. 1019–1025.
7. S. Courtecuisse, F. Cansell, D. Fabre and J. P. Petitet (1995) "A Raman spectroscopic study of nitromethane up to 350°C and 35 GPa", J. Phys. IV (Paris), 5, 359–363.
8. S. F. Rice and M. F. Foltz (1991) "Very high pressure combustion: reaction propagation rates of nitromethane within a diamond anvil cell", Combust. Flame 87, 109–122.
9. M. Frances Foltz and Jon L. Maienschein (1995) "Ammonium perchlorate phase transitions to 26 GPa and 700K in a diamond anvil cell", Mater. Lett. 24, 407–414.
10. G. I. Pangilinan and T. P. Russell (1998) "Global Reaction Rates of the Laser-Induced Decomposition of Ammonium Perchlorate at static high pressures", Eleventh International Detonation Symposium, Snowmass, CO, 1998, Proceedings published by the Office of Naval Research, Arlington, VA, pp. 847–851.
11. J. Gump, L. Parker and S. M. Peiris (2003) "HMX (beta phase): laser-ignited reaction kinetics and isothermal equations of state", 13th APS Topical Conference on Shock Compression of Condensed Matter, Portland, OR, July 20–25, 2003. Proceedings published by the American Institute of Physics, USA, 967–972.
12. G. J. Piermarini, S. Block and P. J. Miller (1987) "Effects of pressure and temperature on the thermal decomposition of rate and reaction mechanism of β-octahydro-1,3,5,7-tetranitro-1,3,5,7-tetrazocine", J. Phys. Chem. 91, 3872–3878.
13. T. B. Brill (1995) "Multiphase chemistry considerations at the surface of burning nitramine monopropellants", J. Propul. Power, 11, 740–751.
14. P. J. Miller, S. Block and G. J. Piermarini (1991) "Effects of pressure on the thermal decomposition kinetics, chemical reactions and phase behavior of RDX" Combust.Flame, 83, 174–184.
15. Anthony P. Esposito, Daniel L. Faber, John E. Reaugh and Joseph M. Zaug (2003) "Reaction propagation rate in HMX at high pressure", Propell. Explos. Pyrot. 28, 83–88.

16. M. France Foltz (1993) "Pressure dependence of the reaction propagation rate of TATB at high pressure" Propell. Explos. Pyrot. 18, 210–216.
17. S. M. Peiris, G. I. Pangilinan and T. P. Russell (1999) "The laser-induced decomposition of TATB at static high pressure", 11th APS Topical Conference on Shock Compression of Condensed Matter, Snowbird, UT, 27 June–2 July, 1999, Proceedings published by the American Institute of Physics, USA, pp. 849–852.
18. Sushil K. Satija, Basil Swanson, Juergen Echkert and J. A. Goldstone (1991) "High-pressure Raman scattering and inelastic neutron scattering studies of triaminotrinitrobenzene", J. Phys. Chem. 95, 10103–10109.
19. J. K. Rice and T. P. Russell (1995) "High-pressure matrix isolation of hetrogenous condensed phase reactions under extreme conditions", Chem. Phys. Lett. 234, 195–202.

Chapter 6
Understanding Shock-Induced Changes in Molecular Crystals

Zbigniew A. Dreger

6.1 Introduction

Understanding the response of molecular crystals to shock-wave compression is of considerable interest to shock-wave research. This is, in part, because most energetic materials (EM) are made of molecular crystals (MC). Because of their high compressibility, low threshold for inelastic deformation, and low symmetry, molecular crystals tend to undergo a variety of physical and chemical changes under shock compression. The changes largely depend on the stress history, and involve deformation, increased temperature, and are of short duration. The microscopic processes occurring in molecular solids under these conditions are rather complex. They can involve the molecular and crystal transformations, and formation or dissociation of chemical bonds, all in a very short time. In energetic materials, an extensive decomposition can occur. To reveal and unravel these effects from a single event of the shock experiment is scientifically very challenging. Because of the complexity of the problem and experimental and theoretical limitations, our understanding of molecular-level processes in shocked molecular crystals/energetic crystals is limited. The governing microscopic mechanisms, both in the unreacted and reacted molecular crystals, remain largely unidentified.

Over the last 20 years, various approaches have been undertaken to increase the level of understanding of physical and chemical processes occurring in shocked energetic materials. They include experimental as well as theoretical and computational efforts. Good recent reviews of these efforts can be found in [1–3], and other chapters in this book. Synergy between these fields has been a challenging requirement, because the time and length scales of current experiments are many orders of magnitude larger than those in atomistic models. Despite this, the advances are being made in developing experimental approaches that provide more detailed information on molecular processes in compressed energetic materials.

S.M. Peiris, G.J. Piermarini (eds.), *Static Compression of Energetic Materials*, Shock Wave and High Pressure Phenomena,

In this chapter, we discuss how and to what extent static compression research can feed and guide shock compression research. Specific examples on a model molecular crystal and on energetic crystals will be provided, followed by a brief review of current approaches and advancements in shock compression of energetic materials.

The main body of research on shock-compressed energetic materials has been directed toward assessing their performance [4]. Such studies are made on actual high explosives (HE) that are inhomogeneous, multicomponent systems. Further, most of the work has been concerned with continuum and thermodynamic properties, including detonation characteristics. While these studies provide important information on the mechanics, thermodynamics, and hydrodynamics of EM under shock compression, they do not provide a necessary insight into molecular-level processes. Here, we present studies on well-defined energetic materials in the form of homogeneous single crystals to avoid additional complexity involved in heterogeneous materials (particle size, defects, grain boundary, interface microstructure, heterogeneity, etc.). Further, we focus on diagnostic techniques suitable for both the shock and static experimentation, and on techniques for examining molecular-level structural and chemical processes.

The obvious similarity between static and shock compression is that in both cases stress is imposed on the object under the investigation. However, loading paths are very different in these two methods, leading to qualitatively different responses (see, Table. 6.1). In particular, shock-wave compression causes deformation and heating, and all this is happening on a very short timescale. Nonetheless, attempts have been made to relate the shock and static compression approaches as much as possible (see, e.g. [3]). This is often accomplished by performing the static experiment at elevated temperatures and under non-hydrostatic compression. The development of a diamond anvil cell (DAC) (see, Chapter 1) greatly advanced these type of experiments. Despite other differences in static and shock compression effects (e.g., timescale) the static high-pressure approach can complement shock-wave experimentation. In addition to this complementary role, there are obvious technical and practical benefits of employing a static high-pressure approach to understanding processes in shocked energetic materials, particularly in an unreactive regime.

In this chapter, we emphasize an integrated approach to the study of energetic materials by: (i) employing both types of compression on the same material, (ii) controlling the experimental conditions, and (iii) using similar diagnostic methods.

Table 6.1 Comparison of shock and static loading characteristics

Feature	Shock	Static
Event	Single	Repeatable
Duration	Short (ns–μs)	Long (>s)
Stress	Uniaxial strain	Hydrostatic
Strain rate	High (manageable)	Low (uncontrollable)
Temperature	Elevated (jump)	Constant
Measurable	Time dependent	Steady state

Our purpose is not to make an extensive review but to point out some interesting features leading to a general comprehension of the response of molecular crystals, in general, and energetic materials, in particular, to dynamic compression using static compression for guidance.

6.2 Energetic Materials Under Shock Compression

6.2.1 Microscopic Mechanisms

Shock-compressed energetic materials undergo a very complex evolution from shock initiation to steady-state detonation. The characteristics of this evolution depend on the mechanical, thermodynamic, and chemical properties of EM. Importantly, these properties vary strongly with time and are often transient in time. Therefore, time-dependent data on mechanical, thermal, and chemical properties are needed to obtain in-depth insights into the microscopic mechanisms shock-induced change.

As proposed by Gupta [5], the process of shock-wave induced chemical decomposition of energetic material can be conveniently divided into three stages as illustrated in Fig. 6.1. The three stages have been related to different timescales of sequential events. The first stage involves energy deposition from the shock wave to the energetic material and consequently to the molecule. This process is believed to take place on a femtosecond–picosecond scale. Depending on the peak stress it can lead to a reversible or irreversible process. A reversible process involves only structural changes. The irreversible process is thought of as a preconditioning ("excitation") of the molecule to dissociation. The second stage is associated with the onset of chemical decomposition resulting in the release of chemical energy. Usually, this process expands over hundreds of nanosecond range. Finally, coupling of the liberated energy to the shock wave can result in eventual detonation. Below, we briefly review recent progress in understanding the shock initiation of homogeneous energetic materials.

The mechanism of the first step, i.e., energy transfer to the molecule, has been a subject of extensive theoretical interest: see recent reviews [7–12]. An understanding of this mechanism is essential for predictive assessment of energetic materials decomposition. Several microscopic models have been put forward. They can be classified into two models: vibrational and electronic. Dlott and Fayer [13, 14] proposed a multi-phonon up-pumping model, suggesting that the shock wave produces a bath of excited phonons which are absorbed by the lowest molecular vibrational modes (vibrational doorway modes). Increased phonon absorption and intramolecular vibrational energy redistribution lead to excitation of higher energy modes. This up-pumping mechanism heats the molecules enough to break chemical bonds. This mechanism requires hundreds of picoseconds for a ladder of vibrational states to reach thermal equilibrium. In contrast, several other models have been proposed

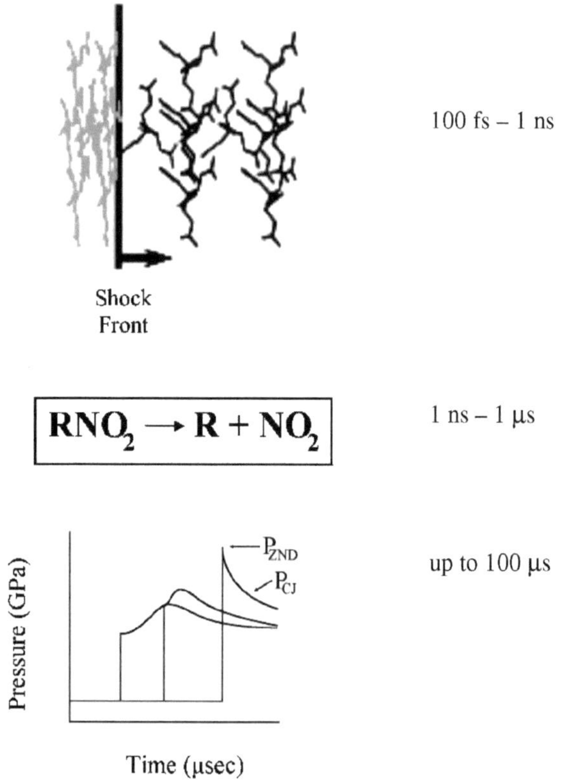

Fig. 6.1 Diagram of processes and their timescales in shocked EM. Top box illustrates a deformation of molecules due to the passage of the shock front. Middle box represents a hypothetical decomposition step. Bottom box [6] illustrates a pressure vs. time plots leading to a detonation wave (P_{ZND} and P_{CJ}correspond to pressures at the Zel'dovich-von-Neuman-Doring (ZND) and Chapman-Jouguet (CJ) points, respectively

that favor microscopic electronic mechanisms (see, e.g. [10]). Gilman [15, 16] suggested that compression from the wave front bends the covalent bonds reducing the energy gap between the highest occupied and lowest unoccupied molecular orbital (HOMO–LUMO gap) producing local metallization. In addition, Kuklja and Kunz showed that the edge dislocations can significantly reduce the optical gap [17, 18], leading to initiation chemistry. Recent calculations by Manaa [9] and Reed [10, 19] show that even highest pressures are insufficient for lowering a band-gap to produce a significant population of excited states in nitromethane crystal, contradicting the previous calculations [18]. Further work by Luty [20] and Munn [21] indicate that the HOMO–LUMO band-gap closure is not needed for mechanically induced chemical reactions. It was suggested that compression reduces the gap between the neutral ground state of the crystal and an ionized state in which an electron is transferred from one molecule to another adjacent molecule generating a charge transfer pair (CT) exciton. This model emphasizes the role of CT pairs where, once they recombine release energy sufficient to break bonds. Despite these efforts, the mechanism of the first step in shock-induced decomposition of EM remains an open question mostly due to a lack of experimental data. On the other hand, the second stage, chemical decomposition on the nanosecond scale, has been more accessible experimentally (see, e.g. [22]). This step is discussed later in the chapter.

6.2.2 Experimental Methods

6.2.2.1 Shock-Wave Compression

Shock-wave/high-amplitude dynamic compression is produced by a rapid deposition of energy on the object. The methods for generation of shock waves can be in general divided into two categories: flyer drivers (powder, gas, laser, and electromagnetic guns) and irradiation (lasers, magnetic accelerators, etc.). Each method has own advantages and limitations [23–25]. For example, electric foil accelerators are unable to launch high-impedance metal flyers. In the direct laser irradiation technique, extensive tailoring of the laser temporal and spatial profile is required to avoid the production of ill-conditioned shock waves. Any viable experimental system for real-time examination of shock-induced physical and chemical changes in materials should combine techniques for well-controlled and well-characterized shock loading with suitable high-speed diagnostic methods. Therefore, to approach optimal shock loading conditions, a large gas-gun driver is often used to generate high-speed plate impact where the geometry is analytically tractable (e.g., planar).

Energetic material can dynamically be compressed by single shock, multiple shocks, or ramp wave (shockless). The pressure–temperature state of the material under the investigation will depend on these methods. The examples presented in this chapter concern the experiments where shock-wave compression was produced in the form of stepwise loading (multiple shocks). This type of shock-wave compressing prevented the excessive heating of the sample. This method is very efficient in tailoring the desired pressure–temperature state [22]. Below we discuss the principles of this type of shock-wave loading. Note that the pressure–volume end states in the shock experiments are determined by the Hugoniot (irreversible adiabat) and in the static experiments by isotherm (e.g., [26]).

A schematic diagram of stepwise (or ring-up) loading is shown in Fig. 6.2. Upon impact, a shock wave of pressure P_1, propagates into the front window. When this wave reaches the window/crystal interface, a partial reflection occurs due to the impedance mismatch between the crystal and the front window. Therefore, the shock wave with reduced pressure of P_2 is transmitted to the crystal. An impedance mismatch also occurs at the back window. A reflection occurs there, which sends a higher-pressure shock wave. The propagation of these multiply reflected waves is illustrated schematically in Fig. 6.2A. In this way, the shock wave reverberates in the crystal between two windows until pressure equilibrium is reached. The pressure

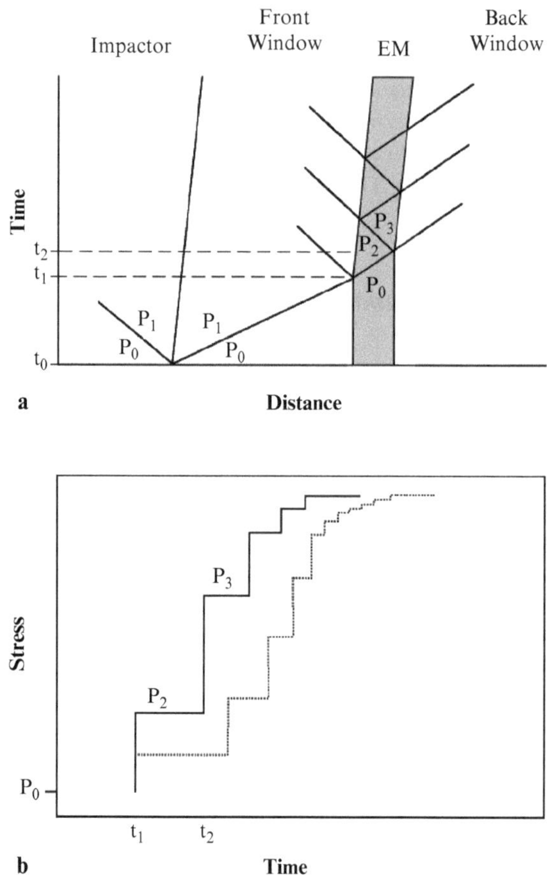

Fig. 6.2 Diagram of shock-wave stepwise loading. (**a**) Propagation of shock wave in energetic crystal. (**b**) A sketch of stepwise loading profiles for different impactor materials. The final temperature for the stress profile represented by the gray line is lower than temperature for the profile represented by the dark line

of the final state is determined only by the projectile velocity and the properties of the impactor and window material. However, the profile can be calculated using a one-dimensional wave propagation code [27] and material model for the crystal. By varying the impactor velocity, type of widows, and thickness, shock-wave amplitudes and duration can be controlled precisely. In addition, by proper selection of material for the front window, different temperatures could be produced for the same peak stresses. An increased number of reverberations will decrease the final temperature (Fig. 6.2B). Sample temperature can be calculated based on the known material model.

6.2.2.2 Optical Spectroscopy Under Shock Compression

To date the most efficient methods of obtaining molecular-level information from shocked energetic materials are vibrational and electronic spectroscopies. The combination of time-resolved absorption, fluorescence, emission, and Raman spectroscopies have been used to examine the microscopic mechanisms in shocked materials including energetic, plate impact [28–37], and laser shocks [38–42].

Coupling spectroscopic methods to shock-wave experiments requires careful synchronization between the shock event, excitation, and detection. In plate-impact experiments, using appropriate trigger pins, the pulsed laser and the streak camera are carefully synchronized with both the impact event and the arrival of a shock wave at the sample (Fig. 6.3). Because of the planar impact, the resulting shock compression produces a state of uniaxial strain in the sample and in the impactor. This state is maintained in the central region of the sample and in the impactor until the arrival of wave reflections from the edges of the sample; the time of arrival defines the duration of the experiment. The short duration of the shock event requires fast and sensitive detection systems. The evolution of processes in the shocked sample can be captured by time-resolved measurements.

Time resolution in shock experiments is limited by several factors. The most prominent are: duration of the experiment, probe pulse duration, timing uncertainty in shock front arrival, and the inherent coupling of temporal and spatial resolution by the shock-wave velocity. Currently, in plate-impact shock experiments the spectroscopic measurements are attainable with nanosecond and subnanosecond resolutions [43–45]. Picosecond resolution has been demonstrated with laser-induced shocks, using a Coherent Anti-stokes Raman spectroscopy (CARS) and infrared spectroscopy (e.g., [42,46–48]).

Vibrational spectroscopy (Spontaneous Raman, CARS, infrared absorption) is one of the most powerful methods for investigating the dynamics and stability of

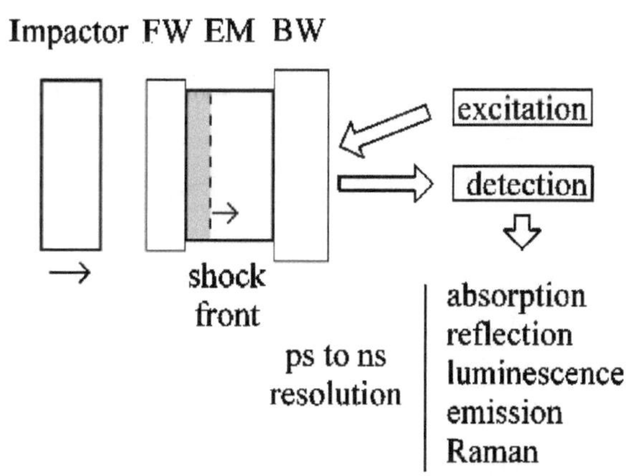

Fig. 6.3 Generic experimental setup for time-resolved spectroscopic measurements under shock-wave compression. The changes in shocked material are probed behind the shock front. FW – front window, EM – energetic material (single crystal), BW – back window

molecules in shocked energetic materials. Furthermore, vibrational spectroscopy can detect molecular and crystal symmetry changes due to deformation, phase transitions, or chemical reaction. In principle, it can also be used to determine the temperature, pressure, and composition of a shocked material with high-time resolution.

Below, we briefly describe two Raman systems that were developed at the Institute for Shock Physics (ISP) and have been used in studies presented here. These two systems allow performing: (i) time-resolved Raman spectroscopy to monitor the evolution of molecular changes as the shock wave reverberates through the material, and (ii) high-resolution Raman spectroscopy to probe the material at peak stress.

For time-resolved Raman measurements [22,32,33] a tunable flash lamp pumped dye laser, tuned to 514.5 nm with pulse width 2 μs was employed. Around 65 mJ pulse energies were used and coupled to the target with a 400 μm optical fiber. Backscattered light was collected by a second 400 μm optical fiber, passed through a holographic notch filter to remove scattered 514.5 nm light, and sent to an imaging spectrometer with a holographic grating. The spectrally dispersed light was coupled to a streak camera followed by an intensified CCD to provide a time-resolved record of spectral changes. The spectral resolution of this system was $\sim 25\,\mathrm{cm}^{-1}$, but Raman data could be collected continuously for ~2 μs during the experiment. The time resolution, determined by the spot size on the detector, was ~40 ns.

High-spectral resolution (short-pulse) Raman spectra were obtained by illuminating the sample with a single laser pulse from a Q-switched frequency doubled Nd:YAG laser (532 nm) [49, 50], see Fig. 6.4. Typical laser pulse energies were ~10 mJ with a pulse duration of 20 ns. A mechanical shutter allowed a single pulse from the 5 Hz pulse train to couple into an optical fiber connected to the target assembly. The excitation light was focused to a size of 600 μm on the sample. Backscattered light was collected with a 600 μm core optical fiber and delivered to a 0.5 m spectrometer with a holographic notch filter to reject scattered 532 nm light.

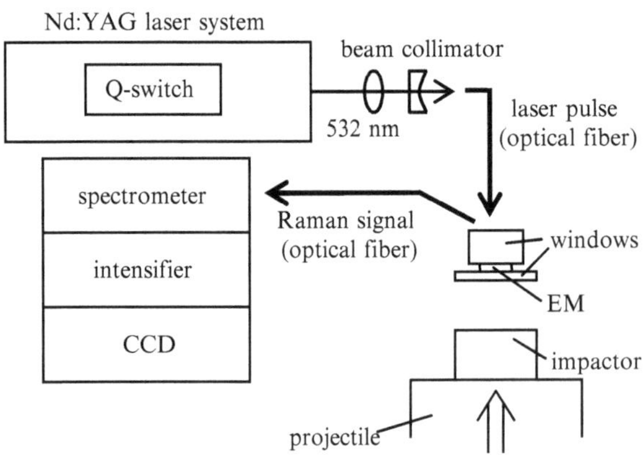

Fig. 6.4 Schematic view of the experimental configuration for high resolution Raman measurements. A single 532 nm pulse from a Nd:YAG laser is used for Raman excitation. The Raman scattered light is recorded by the spectrometer–intensifier–CCD detection system

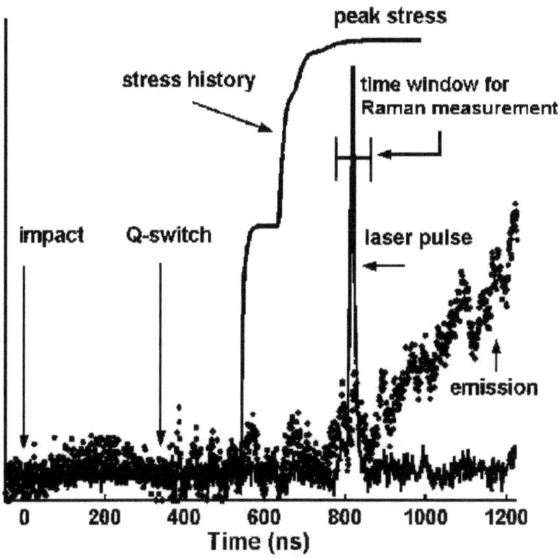

Fig. 6.5 Time synchronization for acquisition of high spectral resolution Raman spectra. The laser pulse is controlled by a Q-switch triggered by the projectile impact. The Raman spectrum is acquired after the crystal reaches the desired peak stress. The stress history curve illustrated by the solid line was calculated using a finite difference wave code [27]. The points represent the time profile for shock-induced light emission (Reprinted with permission from Ref. [49])

The output of the spectrometer was amplified with a gated image intensifier and detected by a back-illuminated CCD. The overall spectral resolution of the detection system was $\sim 3.5\,\mathrm{cm}^{-1}$. The intensifier could be gated between 60 and 150 ns, depending on the need, to reject any extraneous light due to emission by the sample. Synchronization of the shock event, the firing of the laser, and the gating of the image intensifier was accomplished with an electronic delay control system. Pulse-to-pulse timing jitter in the firing of the laser pulse was typically less than 20 ns.

The sequence of events after impact is illustrated in Fig. 6.5. Reverberation of the shock wave between the front and back windows brings the sample to the peak stress through a stepwise loading process. To minimize background light due to shock-induced emission, the CCD was gated for 60 ns and the laser pulse was synchronized to arrive at the sample during this time window, after the sample reached the desired peak stress.

6.3 Effects of Non-hydrostaticity

Non-hydrostaticity or non-hydrostatic/non-homogeneous stresses can be encountered under both shock and static compression. The presence of non-hydrostaticity can significantly change the response of a solid under compression. For instance, it has long been recognized that the existence of non-hydrostatic stresses can bias equation of states of metals and ionic crystal determined under static compression, e.g., [51, 52]. As mentioned above, due to the weakness and anisotropy of intermolecular forces, molecular crystals are particularly sensitive to even small variations in the homogeneity of the applied stresses. Below, we show that non-hydrostaticity plays a major role in producing microscopic changes in the molecular crystals. An example of non-hydrostatic effects on electronic structure of molecular crystal of anthracene is presented.

6.3.1 Non-hydrostaticity Under Shock Compression

Non-hydrostaticity is inherent in the shock-wave compression of solids. As shown in the cartoon in Fig. 6.6, under the uniaxial strain conditions imposed by a planar shock wave, the solid experiences highly non-hydrostatic stresses ($\sigma_3 \neq \sigma_1$). Note that because of the symmetry of the one-dimensional planar waves, $\sigma_1 = \sigma_2$ (no motion parallel to wave front is considered). The only pressure component measured in the experiment with planar shock wave is σ_3, which can be looked upon as composed of mean stress (pressure), $\bar{p}$, and shear stress, τ:

$$\sigma_3 = \bar{p} + (4/3)\tau$$

where $\tau = (\sigma_3 - \sigma_1)/2$ and is often called "maximum resolved shear stress." Furthermore, $\sigma_3 - \sigma_1$ is the uniaxial stress component (a measure of the deviatoric stress). Increasing the shock-wave amplitude or longitudinal stress results in a larger mean stress and a larger uniaxial stress component in the solid. An increase in the uniaxial stress component increase can result in plastic deformations. Under the

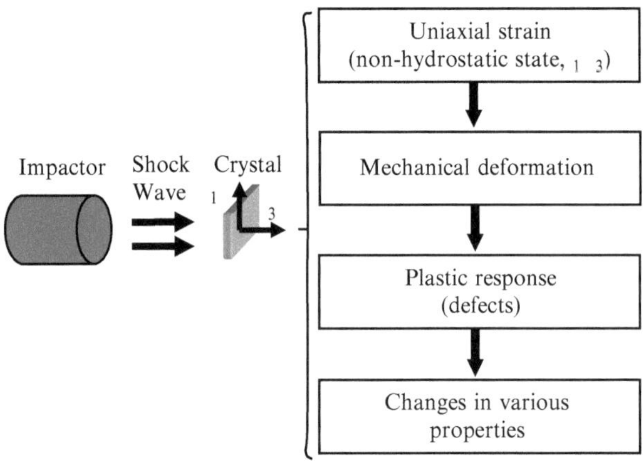

Fig. 6.6 Uniaxial strain effects in shocked crystal, $\sigma_3 \neq \sigma_1$. σ_3-longitudinal stress, σ_1-radial stress

uniaxial strain conditions imposed by a shock wave, the crystal can accommodate plastic deformation through dislocation motion on preferred slip systems. In molecular crystals, including energetic ones, the formation of such extended structural defects may further enhance the reactivity of these solids. Therefore, processes occurring due to uniaxial stress component can be very different from those under mean stress.

6.3.2 Non-hydrostaticity Under Static Compression

Diamond anvil cell experiments are usually performed under hydrostatic compression. Since the DAC is a uniaxial stress device, maintaining hydrostatic conditions in a compressed solid requires containing the sample within a fluid or gas medium. However, at room temperature a completely hydrostatic environment cannot be sustained at elevated pressures due to the solidification of all known pressure media. Recently, development of third-generation synchrotrons made it possible to use x-ray sources to examine hydrostaticity/non-hydrostaticity in compressed solids, e.g., [52–59]. Access to these methods is limited however.

In general, in static high-pressure experiments, the hydrostaticity (negligible shear stress) of a pressure medium is the most conveniently evaluated using the ruby fluorescence technique [60, 61]. Fig. 6.7 illustrates three basic cases of stress states exerted on the sample in a DAC chamber, together with the corresponding response in ruby fluorescence. Let us take the stress component σ_3 to be parallel to the load axis of DAC. The other two components σ_1 and σ_2 (normal to the load axis) are assumed to be equal because of cylindrical symmetry of the sample chamber.

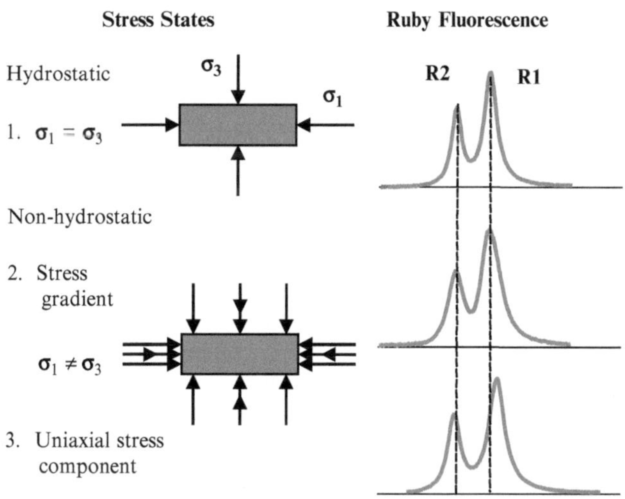

Fig. 6.7 Diagram of ruby fluorescence response to different state stresses in DAC experiments

Under ideal hydrostatic conditions $\sigma_1 = \sigma_3$ in any position in the chamber. This gives narrow ruby lines (R_1,R_2) with constant separation (splitting) between them over the entire pressure range. The second case can be characterized by a stress gradient, where $\sigma_1 = \sigma_3$ in any position, but they are spatially distributed. In this case ruby peaks become broad but they should maintain the same position as under hydrostatic compression. The third case shows uniaxial or deviatoric stress, where σ_1 and σ_3 are different by the same amount throughout the sample chamber. In this case, the separation between ruby lines changes, usually it increases. In practice, in diamond anvil cell experiments non-hydrostatic stress states exist usually as a convolution of the two last cases. In other words, the splitting of R_1–R_2 ruby peaks and their width are generally correlated, although they have a different physical meaning. It was shown in shock-wave experiments that deviatoric stress itself changes the splitting between ruby lines, but not the ruby line width [62–64]. Furthermore, it was demonstrated that the magnitude of splitting between R_1 and R_2 lines depends on the ruby orientation with respect to the applied stress tensor [62–64].

Building on these findings, we have recently introduced oriented ruby crystals for DAC measurements to better assess and control stress state in the pressure-transmitting media [65]. This approach allowed us to obtain more precise information on the non-hydrostatic stress field imposed on a crystal surrounded by pressure transmitting medium. In the studies, we used both *a*-axis and *c*-axis oriented ruby crystals (35 μm in diameter and 20 μm thick) to determine the onset and extent of non-hydrostaticity in various pressure-transmitting media. We determined the onset and extent of the uniaxial stress component by measuring changes in splitting between the R_1 and R_2 ruby lines. The mean stress in the transmitting media was established from the shift of R_2 line because its shift is not affected by non-hydrostaticity [62].

In Fig. 6.8, we present an example of data obtained with two ruby crystals in poly(dimethylsiloxane) (PDMS). As shown, the *a*-axis is more sensitive to the uni-

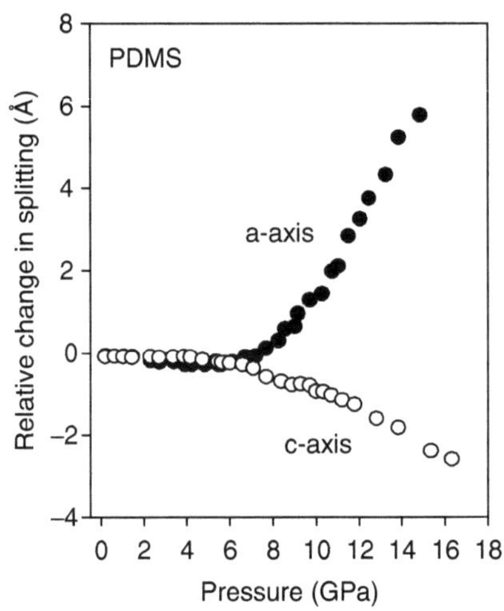

Fig. 6.8 Relative change in splitting between R_1 and R_2 ruby lines as a function of mean stress. PDMS was used as the pressure transmitting medium. Solid symbols correspond to *a*-axis oriented ruby and open symbols to *c*-axis oriented ruby

axial stress component (σ_3–σ_1) than the *c*-axis. The two crystals also show opposite splitting on compression. The onset of this splitting indicates the onset of non-hydrostaticity, and the extent of the splitting is related to the deviatoric stress [63, 64]. Therefore, by selecting the proper pressure-transmitting medium a desired stress field can be created in the DAC experiment. In particular, the uniaxial stress component could be increased relative to the hydrostatic stress by choosing pressure media of increasing shear strength [65].

6.3.3 Electronic Structure of Anthracene Under High Pressure

Anthracene ($C_{14}H_{10}$) represents a large class of polycyclic hydrocarbons that have served as models for organic molecular crystals. The interest in anthracene was mainly stimulated by its attractive electronic and optical properties, and by advances in purification and crystal growth techniques, see e.g., [66–68]. Although anthracene has been studied extensively under static high pressure compression in the past [69–74] the results reported from different laboratories vary significantly. Recently, it has been recognized and shown that observed changes in the electronic structure are driven by non-hydrostaticity [75–77]. Below we present main findings from these studies to emphasize the importance of static studies under non-hydrostatic compression and to offer insight into microscopic changes in shocked molecular crystals.

At ambient conditions, the fluorescence spectrum of anthracene crystals displays a profound vibrational progression from the first excited state $-{}^1L_a$ band. Under hydrostatic compression, this vibrational structure is well preserved; essentially only the red shift of vibrational peaks is observed. However, non-hydrostatic compression transforms the fluorescence spectrum to a broad, featureless band occurring at much lower energies than the spectrum under hydrostatic compression. Fig. 6.9 shows the above differences, together with the significant changes in the optical images under these two conditions.

6.3.3.1 Non-hydrostatic Compression

To examine further the role of non-hydrostaticity on the electronic structure of anthracene, both absorption and fluorescence spectra were measured under static and shock compression. The static compression was applied in a DAC with water as the pressure-transmitting medium. This medium was chemically inert and introduced the non-hydrostatic stresses above 5 GPa. The shock compression was generated by plate impact onto the single crystal sandwiched between two optical windows. This way the sample was loaded in a stepwise fashion, limiting temperature increase. Despite some differences in these two loading conditions (timescale and temperature) the electronic spectra show many similarities [76, 77]. Some details are shown in Fig. 6.10. The absorption band edge, in addition to the monotonic redshift, reveals

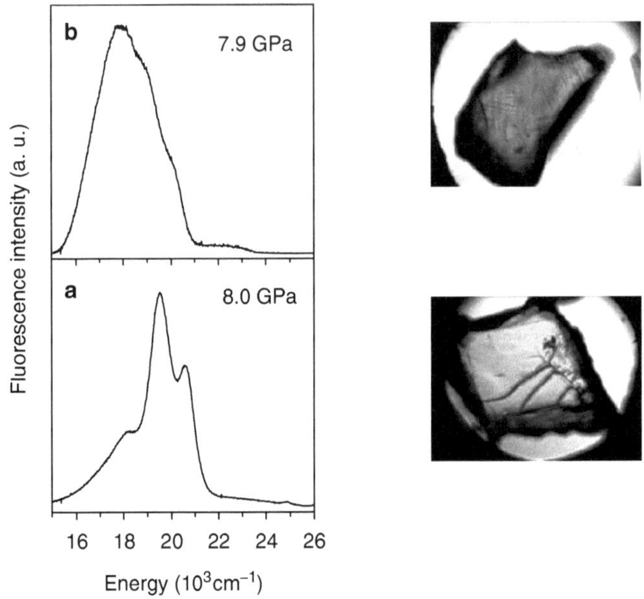

Fig. 6.9 Fluorescence spectra and optical images of anthracene single crystal under hydrostatic (**a**) and non-hydrostatic (**b**) compressions. Fluorescence was excited with a 337 nm line of nitrogen laser. The pressure given in the graphs corresponds to mean stress as determined from the position of R_2 ruby line

also a new band developed as a shoulder on the redshifted absorption-band edge. This new band occurs around 5 GPa under both loading conditions. The new band has a broad energy distribution, which seems to consist of multiple contributions. The monotonic redshift observed in the absorption experiments indicates a change in the electronic structure, i.e., a reduction in the energy gap between the ground and first excited state of anthracene singlet. The appearance of the new absorption band observed above 5 GPa indicates additional changes in the electronic structure at high pressures.

To explore the origin of the new absorption band, we excited the compressed crystals with a 514.5 nm wavelength ($\sim 19.4 \times 10^3\,\text{cm}^{-1}$). This excitation matched the energy of the new absorption band, but it was below the energy of the singlet state absorption. As shown in Fig. 6.10C, D, this excitation induces fluorescence with spectra that are very different from this at ambient pressure. Both static and shock compression produce a very similar, broad, and structureless spectra. Because fluorescence occurs upon excitation with the energy matching the energy of the compression-induced band; the same electronic origin for both, the absorption and fluorescence is implied. Furthermore, the close resemblance of new features in the electronic spectra of non-hydrostatically and shock-compressed anthracene strongly indicate the similar microscopic cause of the observed changes in the spectra. A brief discussion of the possible origin of changes in the electronic structure of anthracene is presented below.

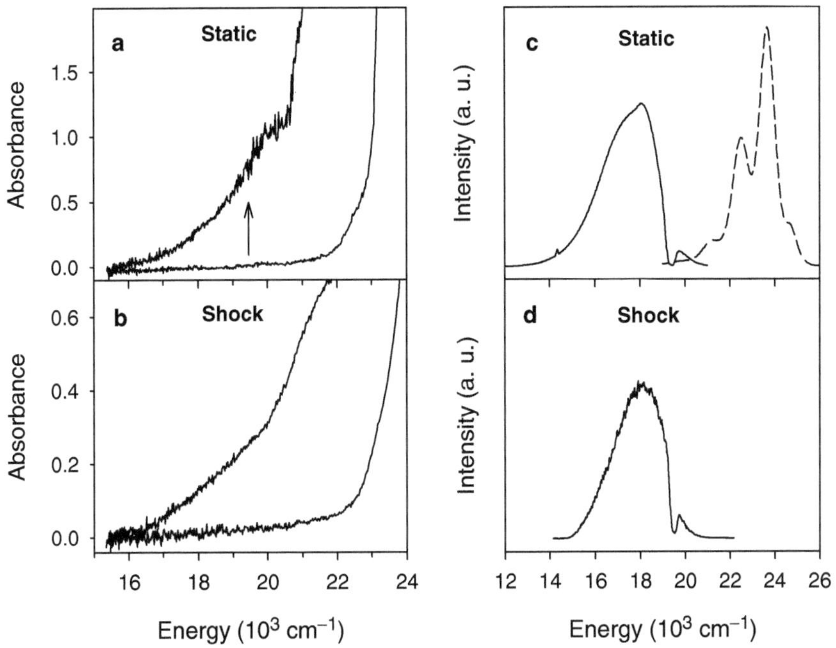

Fig. 6.10 Left panel: Absorption spectra (absorption edges) under static and shock compression: (**a**) spectra under static compression at 3.0 and 6.1 GPa, (**b**) spectra under shock compression at peak stresses of 2.5 and 6.0 GPa. Right panel: Fluorescence spectra upon excitation of new absorption band: (**c**) spectrum under ambient pressure (dashed line) and under static compression at 6.8 GPa, (**d**) spectrum under shock compression at peak stress of 6.0 GPa

6.3.3.2 Microscopic Origin of Changes in Electronic Structure

In this study, static non-hydrostatic and shock loading impose nonhomogeneous stresses on the crystal. Providing that the stress applied exceeds the elastic limit for anthracene [78], it should lead to plastic deformation and ultimately to the formation of defects. The appearance of the new band in the absorption spectrum could be associated with stress-induced defects, likely in the form of dislocations. Indeed, the analyses of optical images of non-hydrostatically compressed crystal in DAC indicate formation of macroscopic defects [76]. As shown in Fig. 6.11, characteristic lines are observed on the surface of the [001] plane. The lines intersect at the angle of $73^\circ \pm 5^\circ$. This angle matches quite well the angle between the [110] and $[1\bar{1}0]$ directions in the compressed anthracene crystal [79]. Therefore, it can be concluded that non-hydrostatic compression, either static or shock, can introduce specific sets of dislocations in the crystal. Of particular interest are $1/2$ dislocations, e.g., $1/2$ [110], which can form sites of van der Waals dimers in the crystal by bringing closer parallel molecules. These sites can be responsible for formation of new absorption bands and structureless, excimer-like fluorescence spectra [76, 77].

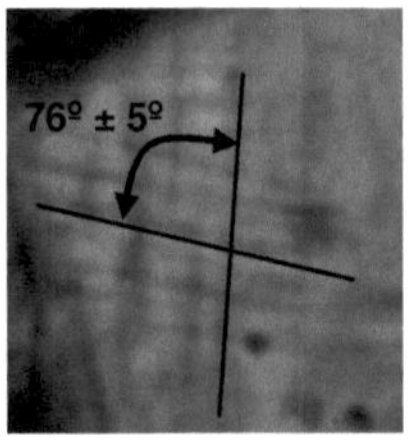

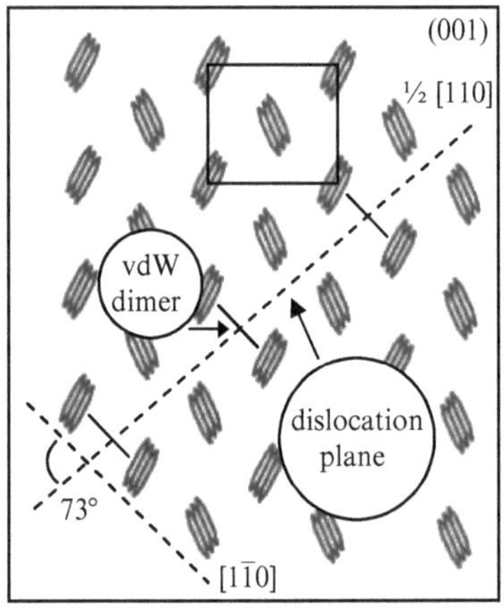

Fig. 6.11 The effect of non-hydrostatic compression on microscopic changes in anthracene crystal. On the left, the optical image of the (001) plane of anthracene crystal compressed to 6.3 GPa. On the right, diagram of the [001] plane with indication of dislocation planes and van der Waals dimers

6.4 High-Pressure Polymorphism of Energetic Crystals

Polymorphism (Greek: poly = many, morph = form) by definition is the property of crystalline solids to exist in different forms. These forms correspond to different arrangements of the constitutive elements of the solid. In molecular crystals, polymorphism can originate due to changes in the arrangement between molecules as well as within the molecules. The latter is called conformational polymorphism. Polymorphic modifications of a compound can exhibit significant differences in its physical and/or chemical properties. Many examples of polymorphism have been reported in the literature. Further, many studies have attempted to understand the mechanisms of transformation between different polymorphs and to relate polymorphic structure with the properties in question [80].

Polymorphism is common among energetic materials and has been a subject of extensive studies. Various polymorphs have been found to result from changes in crystallization conditions at ambient temperature and pressure. Remarkable examples of this include HMX (octahydro-1,3,5,7-tetranitro-1,3,5,7-tetrazocine) and HNIW (or CL-20, hexanitrohexaazaisowurtzitane). HMX can exist in four polymorphs, designated as α, β, γ, and δ [81, 82]. The γ-phase was however found to be a hydrated form of HMX [83]. The stabilities of these polymorphs are known to be in the order of $\beta > \alpha > \gamma > \delta$ at ambient conditions [84]. HNIW was also found in four polymorphs: $\varepsilon, \alpha, \beta$, and γ at ambient conditions [85–88]. The α-polymorph may exist as a hydrate or solvate. Two different stability ranks among these polymorphs have been proposed: α-hydrate $> \varepsilon > \alpha$-anhydrous $> \beta > \gamma$ [87], and $\varepsilon > \gamma > \alpha$-hydrate $> \beta$ [85]. Further information on the kinetics and activation energies between different phases for these two compounds can be found in Refs. [89–94] for HMX and [85–88] for HNIW.

These two examples show the richness and complexity of polymorphic phases of energetic materials, even at ambient conditions. However, from the viewpoint of shock-wave initiation, an understanding of the high-pressure/high-temperature polymorphic phases of energetic materials is often more important than an understanding of ambient phases. In particular, an understanding of high-pressure polymorphism of EM is required for proper characterization of their reactive behavior. Very often, a relatively insensitive EM may have a high-pressure polymorph that is responsible for its performance. High-pressure polymorphism of EM is usually investigated under static compression and elevated temperature (for review see other chapters in this book). However, the ultimate verification of the occurrence of high-pressure polymorphs requires measurements under shock-wave compression. Here we present the observation of polymorphic transformation in an energetic crystal of RDX (hexahydro-1,3,5-trinitro-1,3,5-triazine) under shock compression. To our knowledge, it is one of two cases where polymorphism was investigated on homogeneous EM at the molecular level under shock compression [49,95].

6.4.1 α–γ Phase Transition in RDX Crystal

RDX, the cyclic trimer of methylenenitramine, is one of the most important crystalline energetic materials. It is used extensively in explosives and monopropellants. The RDX molecule consists of three NO_2 groups bonded to the nitrogen atoms of a triazine ring. In the solid state, RDX is known to exist in three polymorphs: α, β, and γ [96–107]. The α–polymorph exists at ambient conditions and has an orthorhombic structure belonging to the *Pbca* space group with eight molecules per unit cell [96, 97]. All molecules occupy sites of C_1 symmetry, but they possess C_s pseudosymmetry with two of the nitro groups in the axial (A) position (with respect to the s-triazine ring) and the third in equatorial (E) position. This molecular conformation is usually referred to as the chair AAE conformer. The application of high temperature and pressure is one means of obtaining the β-polymorph [101, 102]. The crystal structure of this polymorph is unknown. However, the molecular symmetry was suggested to be C_{3v} [100, 108]. The up γ-polymorph was observed under pressures greater than 3.8–4.0 GPa, using x-ray diffraction [98, 103, 105], Raman [101], and IR spectroscopy [102, 104, 106]. X-ray diffraction results yielded a volume change of $\sim$1.6% at 4.0 GPa, suggesting that the γ-polymorph, like the α-polymorph, may have an orthorhombic structure. It was also suggested that the RDX molecule in the γ-polymorph can adopt the AEE conformation [104], i.e., one nitro group in the axial position and two nitro groups in the equatorial position. Despite these studies our understanding of the α–γ transformation and the structure of γ-polymorph of RDX is far from complete.

Our recent spectroscopic studies have provided more detailed information on the response of RDX to high compression, in particular, on the symmetry of the γ-polymorph and the α–γ transformation under shock-wave compression [95, 107]. Static high-pressure experiments were used to determine the main feature of the γ-polymorph and to guide the shock compression experiments.

6.4.1.1 Static Compression

Raman spectroscopy was used to detect changes in interactions and symmetries in the crystal by probing modifications in lattice dynamics, molecular arrangements, and conformations. To mitigate problems that are potentially encountered in studies under high pressure of molecular solids (random morphology of samples and uncertain compression conditions, see section above), the experiments reported here were performed with good quality single crystals and under well-controlled pressure conditions.

Polymorphism can be investigated through changes in the external (lattice) modes. For RDX, group theory predicts 24 lattice Raman active modes at ambient conditions [107, 109, 110]. Twelve of these modes are librational and the other 12 are translational. However, not all modes are seen in the static high-pressure experiment. As seen in Fig. 6.12, only nine modes are detected for the α-polymorph. The modes exhibit large pressure dependence, as can be expected with molecular crystals. Above 4 GPa, clear discontinuities in pressure dependence of Raman shifts are observed, confirming the onset of the α–γ transformation. The number of detected peaks for the γ-polymorph apparently increases slightly from 9 to 12. An extra couple of peaks which show up above the phase transition could be ones that were not well separated in the low-pressure phase. The pressure dependencies of the Raman

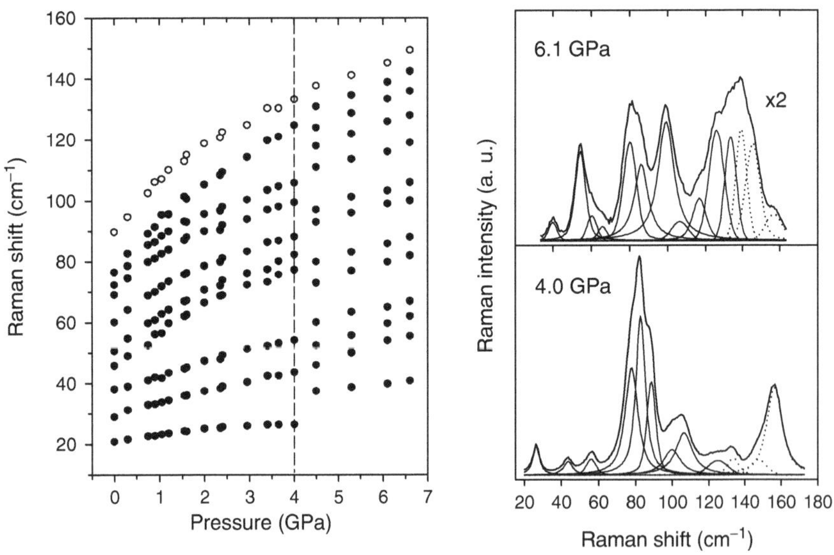

Fig. 6.12 The effect of pressure on Raman spectra of external modes of RDX crystal. Left panel: Pressure-induced Raman shifts of low frequency modes. Open symbols denote the internal mode. The vertical dashed line marks the onset of discontinuity. Right panel: Example of separation of vibrational modes from the low frequency spectra of α-polymorph (4 GPa) and γ-polymorph (6.1 GPa). The spectrum at 6.1 GPa was magnified by a factor of 2. Internal modes are denoted by dotted lines (Reprinted with permission from Ref. [107])

shifts of external modes in the γ-phase are noticeably smaller than at ambient conditions, but to some extent higher than those just before the phase transition. This would indicate the hardening of the structure when pressure approaches the phase transition and some softening of the structure above the phase transition.

Given that we observe 12 external Raman modes in the γ-polymorph, only crystal structure with four or eight molecules per unit cell are allowed. This limits the possible structure to $D_{2h}[C_1(8)], C_{2v}[C_1(4)]$ and $D_2[C_1(4)]$. It is expected that the Raman and infrared (IR) vibrations should be coincidental for the $C_{2v}[C_1(4)]$ and $D_2[C_1(4)]$ factor groups and non-coincidental for the $D_{2h}[C_1(8)]$ group. Therefore, the comparison between Raman and IR modes in the γ-polymorph could be useful in discriminating between the three proposed factor groups. Unfortunately, the available IR spectral data at 9.8 GPa [104] do not have adequate resolution to make an effective comparison with our Raman data. However, the available x-ray data demonstrate that the dimensions of the unit cell hardly change at the α–γ transition [103, 105]. It would indicate that the number of molecules in the γ-polymorph remains the same as in the α-phase. Consequently, the only structure that preserves eight molecules in the unit cell is $D_{2h}[C_1(8)]$. Although our Raman data cannot be considered conclusive, we propose that the γ-polymorph structure is isomorphous with the orthorhombic $D_{2h}[C_1(8)]$ group.

Another clue regarding the possible nature of changes in the RDX structure can be derived from the behavior of internal modes. Fig. 6.13 shows a clear increase in the number of modes in the γ-polymorph. In principle, two factors can contribute to an increase in the number of internal molecular vibrations in the crystal. The first arises from the effect of site symmetry, which lower the effective molecular

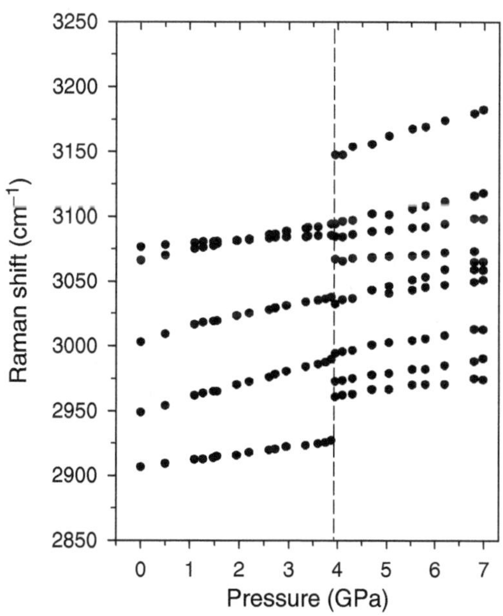

Fig. 6.13 Pressure-induced shifts for C–H stretching modes. The vertical dashed line marks the onset of discontinuity (Reprinted with permission from Ref. [107])

symmetry and remove degeneracies. The second is the effect of factor group coupling which can split vibrations due to interactions between different molecules in the unit cell. If the γ-polymorph is isomorphous with the $D_{2h}[C_1(8)]$ point group, then RDX molecules occupy the C_1 site symmetry. Therefore, all selection rules and degeneracies for the molecule under its intrinsic symmetry are lost [111]. Consequently, only factor group coupling can be responsible for the observed splitting of the internal vibrational modes. Examination of the correlation diagram for the D_{2h} factor group indicates that this group would cause fourfold splitting of each internal Raman active mode [112]. Actually, twofold splittings are observed in the experiments. This could be because the magnitude of splitting is too small to permit resolution of all four peaks or because the increase in intermolecular interactions between pairs of RDX molecules is larger than those within the pairs.

6.4.1.2 Shock Compression

Static compression studies clearly indicate that changes in the CH-stretching modes are the fingerprints of the α–γ transition [107]. This finding was used to guide the shock-compression studies to address the following questions: (i) does the α–γ transition occur under shock loading, (ii) does shocked RDX exhibit anisotropy at the molecular level at stresses close to the phase transition, and (iii) Is the phase transition time dependent? Investigating the structural response of RDX is an important first step because changes in molecular structure and orientation are expected to be important for shock initiation.

All shock experiments were performed on RDX single crystals oriented along one of three crystallographic orientations: [111], [210], or [100]. Typical lateral dimensions of the samples were 7–8 mm and thickness of ~400 μm. The crystals were sandwiched between two windows of z-cut quartz. Planar shock waves were produced by impacting the sample assembly with another z-cut quartz window. Upon impact, a shock wave propagated through the front quartz window and into the RDX crystal. Because of the impedance mismatch between z-cut quartz and RDX, the sample was loaded in a stepwise or ring-up fashion [95].

To investigate the stress dependence of the molecular response of RDX to shock compression, a series of high-resolution Raman experiments was performed on [111] oriented crystals. Selected spectra from this series are shown in Fig. 6.14 together with static pressure Raman spectra acquired at similar stresses. Despite the different loading states in the two cases (the shock state is one of uniaxial strain whereas the static pressure state is hydrostatic), such comparisons are valuable because both sets of data are at comparable densities. The Raman spectra of shocked RDX [111] show a marked similarity to the Raman spectra of RDX under static pressure. From this finding, we infer that the RDX crystal undergoes a shock-induced transition to the γ-polymorph above ~4 GPa. The phase transition appears to occur between 3.5 and 4.5 GPa, but the exact pressure was not determined. This range agrees well with static pressure results that show the phase transition to occur between 3.8 and 4.0 GPa [101, 102, 107].

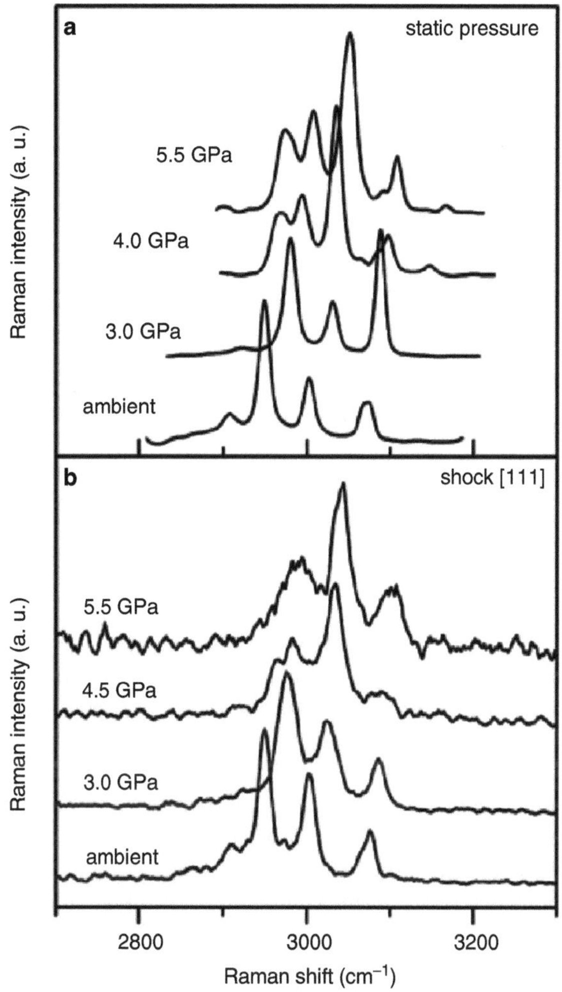

Fig. 6.14 (**a**) Raman spectra of C–H stretching modes under static compression; (**b**) Raman spectra of the C–H stretching modes of single crystals of [111] orientation under shock compression to several peak stresses (Reprinted with permission from Ref. [95])

Experiments performed on three different orientations show no measurable orientation dependence in the Raman spectra, either below or above phase transformation [95]. This result seems surprising because compressibility of RDX is different for the three crystallographic directions [98]. Also, recent wave profiles measurements at 2.5 GPa show differences for the orientations used in our experiments [113]. Finally, these results for RDX are different from those on energetic crystal of PETN, where crystal orientation had a noticeable effect on the shifting of the C–H stretching modes ([49] and sections below). Thus, the molecular-level

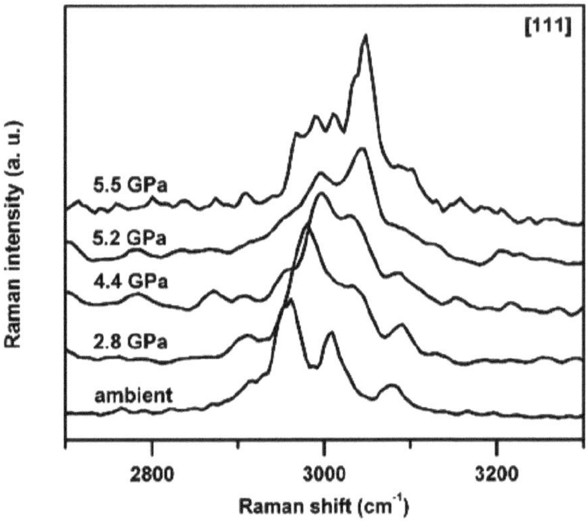

Fig. 6.15 Time-resolved spectra taken during stepwise loading of the [111] oriented crystal to peak stress of 5.5 GPa (Reprinted with permission from Ref. [95])

response of the two materials appears to be different and indicates the need for further examination of the factors governing anisotropy in shocked energetic crystals.

Time-resolved Raman spectroscopy gives us some insight into the dynamics of the phase transition. Selected spectra from a time-resolved experiment are shown in Fig. 6.15. The stresses were calculated for the stepwise loading path of RDX confined between two z-cut quartz windows. Comparison of spectra in Fig. 6.14 with those in Fig. 6.15 indicates that some finite time is required for the phase transition to occur. We can make a prediction of the time-dependent Raman shifts based on the assumptions that (i) the Raman shifts under shock are identical to those under hydrostatic compression and (ii) there is no time dependence in the shift. The $2{,}949\,\text{cm}^{-1}$ mode is used for this comparison.

Figure 6.16 presents the results of this comparison, together with the calculated stress history computed at the midpoint of the RDX crystal under stepwise loading [27, 95, 114]. As the stress increases during the first step of the ring-up process, the shift of the Raman frequency is predicted to increase.

However, at the second step, the predicted shift decreases slightly because of the phase transition. As the stress continues to increase, this peak shifts further. The observed shifts, however, exhibit different behavior than the prediction. At the second and third steps in the loading path, the observed shift overshoots the prediction. The shift does decrease, apparently once the phase transition occurs, but this takes place at least 100 ns after the 4.0 GPa threshold was crossed. Therefore, it indicates that the α–γ phase transition requires an incubation time of at least 100 ns when RDX is shocked to a peak stress of 5.5 GPa.

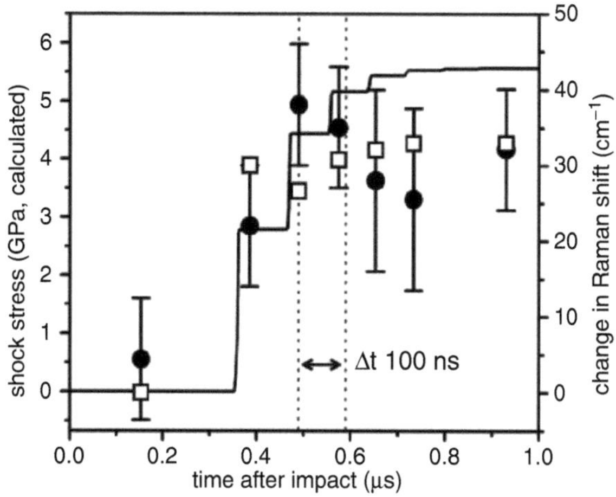

Fig. 6.16 Comparison between simulated stress history for 5.5 GPa shock wave experiment (solid line). Predicted Raman shifts for 2,949 cm^{-1} mode under shock indicated by open squares. Observed Raman shift indicated by closed circles (Reprinted with permission from Ref. [95])

6.5 Shock Initiation of PETN Crystals

In previous sections we presented examples of how static high-pressure experiments can further our knowledge of molecular-level processes in shocked unreacted molecular and energetic crystals. However, shock compression often leads to initiation of chemical reactions in shocked solids. In particular, energetic materials decompose under suitable shock conditions. A good understanding of the microscopic processes in shock-initiated energetic materials is needed for the development of a predictive capability regarding issues of explosive sensitivity, safety, performance, etc.

Although, the role of molecular structure parameters on sensitivity of EM has long been acknowledged [115], the work by J. J. Dick brought a new dimension to this problem. He found that under plane-wave loading the time and run distance to detonation in PETN (pentaerythritol tetranitrate) depends strongly on the direction of shock propagation relative to the crystal axes [116]. Subsequently, Dick and coworkers put forward a steric hindrance model to explain the observed anisotropic behavior [117–119]. They proposed that the mechanical anisotropy is due to differences in the dislocation slip systems that are activated in response to shock-wave compression along different orientations. Furthermore, they suggested that shear deformation on the slip system depended on steric hindrance or resistance to molecular motion on either side of the slip interface. Shock compression along the [110] (sensitive) orientation resulted in significant steric hindrance, while compression along the [100] (insensitive) orientation did not [119]. The steric hindrance was correlated to the observed anisotropic, elastic–plastic response arising from activation of different slip systems under uniaxial strain deformation along the different crystal

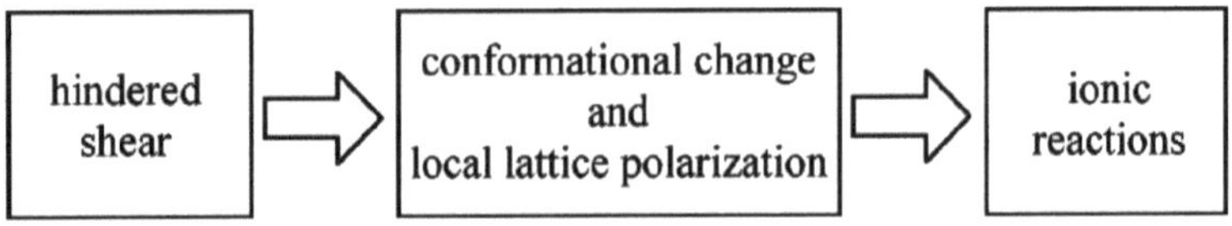

Fig. 6.17 Conceptual diagram of molecular model of PETN anisotropic sensitivity

orientations. Although the model correctly predicted the dependence of the elastic wave amplitude on crystal orientation, it remained largely correlational from the chemical viewpoint.

6.5.1 Mechanism of Anisotropic Sensitivity

In response, Gruzdkov and Gupta proposed a model that linked mechanical anisotropy to initiation chemistry [120]. In this model, the steric hindrance concept was related through shear deformation at the lattice level, to rotational conformations of PETN molecules. Moreover, they proposed that conformational changes could lead to local lattice polarization and consequently to ionic reactions. The concept of the model is presented in Fig. 6.17.

According to the molecular hypothesis in Ref. [120], the difference between sensitive and insensitive orientations is that in the former the hindered shear produces local lattice polarization close to the shear planes, while in the latter the unhindered shear does not. Thus, changes in rotational conformations of PETN molecules are central to this mechanism.

6.5.2 Conformational Changes

The propensity of PETN molecule to conformational changes was examined theoretically [121]. A number of stable conformers, corresponding to different molecular point-symmetries, were identified using semiempirical calculations. Harmonic vibrational frequencies for the conformers were determined through normal-mode analyses. Simulated Raman spectra in the region of C–H stretching modes for several molecular conformers are shown in Fig. 6.18A. The calculations indicate that upon symmetry change from initial, S_4, to lower symmetries all peaks tend to split. For example for the $C_2(\alpha)$ conformer, the initial A (2,984 cm^{-1}), E(3,041 cm^{-1}), and B(3,042 cm^{-1}) peaks transform into two A peaks (2,977 and 2,993 cm^{-1}), two B peaks (3,041 and 3,061 cm^{-1}), and two A peaks (3,041 and 3,061 cm^{-1}), respectively. Spectral variations predicted for symmetry change from S_4 to C_1 are significantly more complex and produce combinations of peaks that are less regular than C_2.

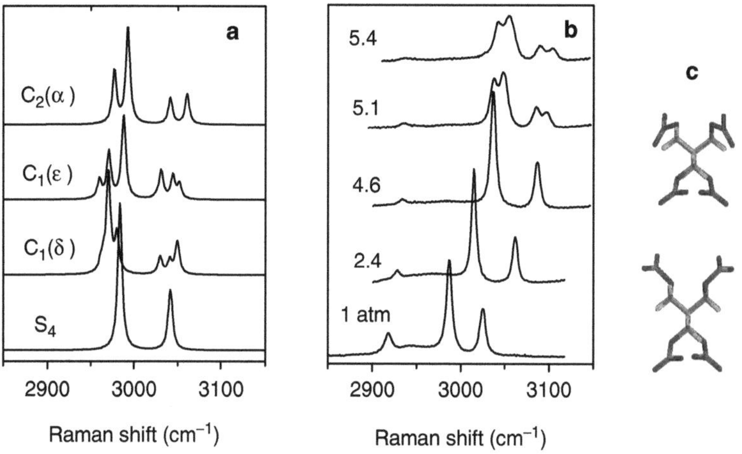

Fig. 6.18 Raman spectra of PETN in the region of C–H stretching modes: (**a**) simulated spectra for several molecular conformers, (**b**) experimental spectra for single crystal under static compression (pressure values are given in gigapascals), (**c**) molecular structures of S_4 (ambient) and C_2 (high pressure) conformers

The theoretical predictions on conformational changes of PETN molecule were examined experimentally using Raman spectroscopy. Static high-pressure experiments performed on single crystals in DAC with glycerol as quasi-hydrostatic medium-indicated splitting of several peaks above 5 GPa. For instance as shown in Fig. 6.18B, the two stronger C–H stretching modes split into doublets. This suggests structural changes in the PETN crystal taking place under static compression. The similarity of experimental spectra above 5 GPa to those of the C_2 conformer suggests conformational changes in compressed PETN from S_4 to C_2 symmetry. The plausible conformational symmetry at high compression is shown in Fig. 6.18C. This conformational change likely involves the transformation from a tetragonal to orthorhombic crystal structure [121]. Recent high-pressure experimental studies on polycrystalline samples [122, 123], and computations [124] suggest that the PETN structural transition is sensitive to non-hydrostatic compression.

The static compression results were instructive in examination of the feasibility of conformational changes in shocked PETN. To test the anisotropic response, the single-pulse Raman spectra were measured on crystals shocked along the [100] (insensitive) and [110] (sensitive) orientations. High-resolution Raman spectra in the region of C–H stretching modes for several pressures are presented in Fig. 6.19. These results reveal measurable differences in the response of the two orientations. In particular, the slopes of the stress-dependent frequency shifts for the [100, 110] orientations differ, suggesting that the PETN molecules respond differently for uniaxial strain along the two crystal orientations [49]. Also, there is substantial broadening of peaks for [110] orientation above ~4 GPa.

This broadening was analyzed in comparison with static high-pressure data. Voigt functions were used in fitting the spectra at 5.4 GPa under static compression

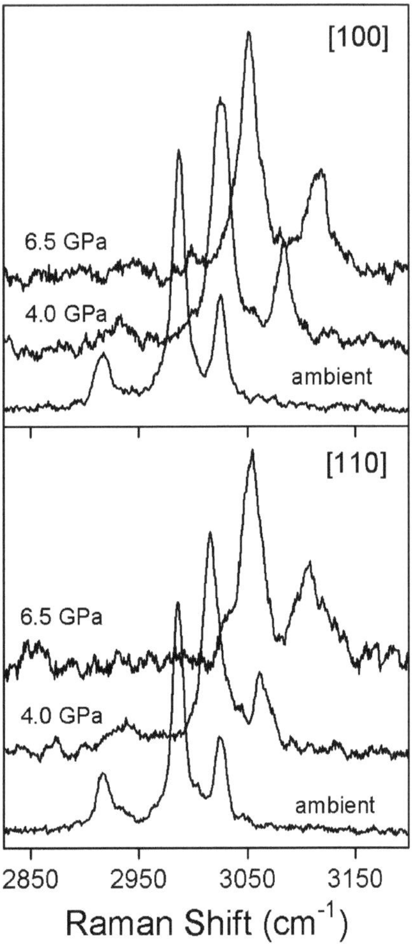

Fig. 6.19 Raman spectra of PETN crystal in the region of C–H stretching modes at several peak stresses for the [100, 110] orientations (Reprinted with permission from Ref. [49])

and at 6.5 GPa under shock compression along the [110] orientation. Results are shown in Fig. 6.20. The bottom panel of this figure shows that additional two or four spectral components are needed to properly fit the shock data. The need for additional components suggests that the anomalous broadening of the C–H stretching modes for the [110] orientation is due to the appearance of new spectral contributions, likely caused by the splitting of existing peaks. Comparison with static high-pressure spectra suggests that qualitatively similar changes in the molecular structure occur during shock compression, i.e., lowering the molecular symmetry in the crystal. However, as manifested by the differences in the spectra, the conformer that is formed under shock compression can be different from that produced under static high pressure. It is also possible that a mixture of conformers, rather than a single one, is present in the shocked PETN along the [110] direction.

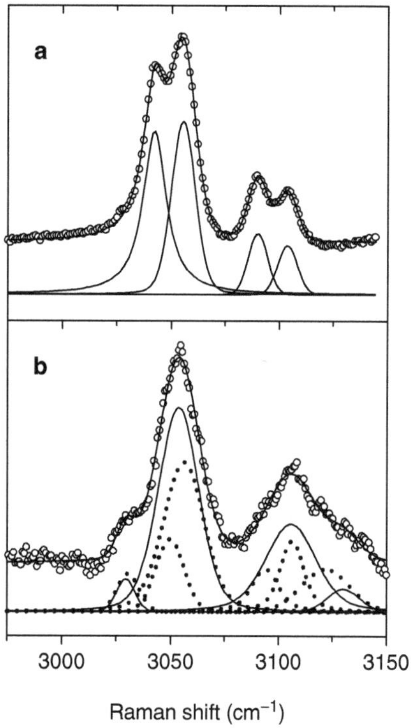

Fig. 6.20 Peak fitting analysis for the C–H stretching modes. The open circle are the experimental data. Solid lines represent Voigt functions for fitting peaks. (**a**) Spectrum at 5.4 GPa under static compression; (**b**) Spectrum at 6.5 GPa under shock compression for the [110] orientation, two different fittings were used (solid and dotted lines)

6.5.3 Shock-Induced Decomposition

During shock-wave propagation in PETN crystal, significant tangential forces are imposed on the molecules across the shear plane. Because of these forces, the molecules can deform and change conformations as proposed by [120] and as suggested by spectroscopic experiments. Shear along different slip planes will likely result in different conformations. Hindered shear could transform the PETN from a non-polar to a polar crystal. It is important to note that the polarization is microscopic in nature. Because the dominant initial reaction of nitrate esters at elevated pressures appears to be ionic [125–127], the formation of polar crystal would also promote an ionic reaction in shocked PETN.

Investigation of light emitted from shocked PETN provided further insight into reaction mechanisms in shocked EM [37]. Analysis of the emission spectra revealed two bands: a high-energy band (HEB) at ~3.0 eV and the low-energy band (LEB) at ~2.4 eV (Fig. 6.21). Both the HEB and LEB were observed in every experiment regardless of stress or crystal orientation. However, their relative and absolute intensities and kinetics revealed stress and orientation dependence.

To find the source of the emission, photoluminescence experiments were performed under static compression. A 280 nm wavelength from a second harmonic of dye laser was used for the excitation. A significant difference observed between

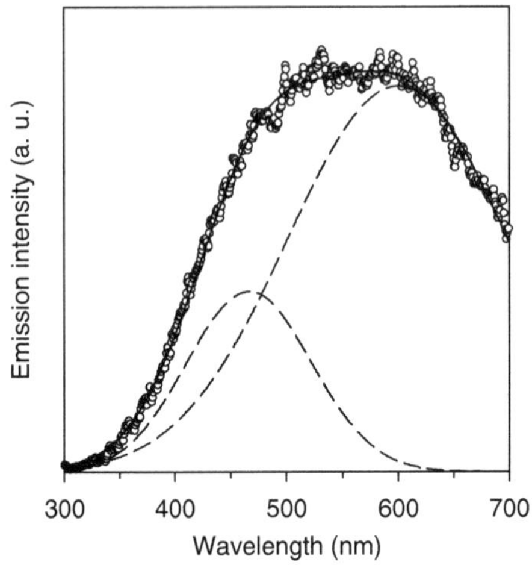

Fig. 6.21 Typical emission spectrum of PETN shocked to peak stress of 9.6 GPa (open circles). The spectrum is deconvoluted into the low- and high energy band (dashed lines)

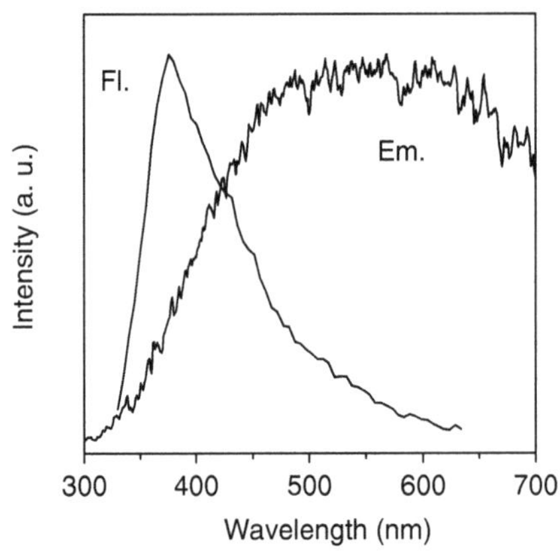

Fig. 6.22 Comparison between the shock-induced emission spectra and laser-induced fluorescence spectra. Both spectra are corrected for instrumental responses. Experiments were performed at the same pressure of 7.5 GPa

photoluminescence and emission spectra (Fig. 6.22) ruled out electronically excited PETN as a possible source. Therefore, the emission was identified as chemiluminescence from excited intermediates. The nitronium ion, NO^{2+}, was proposed as an emitting intermediate, on the basis of the electronic structure and its properties. The nitronium ion was analyzed using ab initio calculations; these revealed transition energies matching those of the experimentally observed emitting intermediate. Several chemical reactions compatible with the formation of NO^{2+} were considered and evaluated. Furthermore, a four-step chemical initiation mechanism was proposed in shocked crystalline PETN [37].

6.6 Concluding Remarks

Shock compression produces an unusual realm of pressure, temperature, and deformation, generated on a short timescale that can induce a wide variety of physical and chemical changes in molecular and energetic crystals. The recent years showed several advancements in understanding the response of molecular crystals/energetic crystals under shock-wave conditions. It was possible due to the progress that has been made in modeling as well as in the experiments. Very promising prospects emerge from the experimental developments, including: ultrafast vibrational spectroscopy of the laser-shocked molecular materials, high-spectral resolution of Raman spectroscopy of the plate impacted energetic materials, and controlled deformation experiments in a diamond anvil cell.

The static high-pressure approach already plays an important role in advancing our knowledge of microscopic processes in shocked energetic crystals. It provides an experimental tool that can separate the effects (pressure/stress, temperature, and time) that are inherently tangled in shock compression. As pointed in this chapter a further coupling of static and shock approaches is needed and necessary for scientific benefits of both fields. To prove this case we have presented selected examples from our laboratory, Institute for Shock Physics at Washington State University. We illustrated the interplay between static and shock experiments in the cases of: (i) the role of deformation on formation of new electronic states in anthracene crystal, (ii) polymorphism in energetic crystal of RDX, and (iii) conformational changes and decomposition in the shocked energetic crystal of PETN. Our aim was to provide within the framework of a single chapter the examples which demonstrate the usefulness of static high-pressure approach to improve the understanding of microscopic processes in shocked molecular and energetic crystals.

Acknowledgments Professor Yogendra M. Gupta is sincerely acknowledged for many stimulating discussions, encouragement, and enthusiastic support. Also, I would like to acknowledge several colleagues for their contributions over the years: Y.A. Gruzdkov, N. Hemmi, J.M. Winey, J.E. Patterson, and K.A. Zimmerman. This work was supported by funding from the Office of Naval Research and the Department of Energy.

References

1. P. Politzer and J.S. Murray, eds., *Energetic Materials: Part 1. Decomposition, Crystal and Molecular Properties*, Elsevier: Amsterdam (2003).
2. P. Politzer and J.S. Murray, eds., *Energetic Materials: Part 2. Detonation, Combustion*, Elsevier: Amsterdam (2003).
3. M.R. Manaa, ed., *Chemistry at Extreme Conditions*, Elsevier: Amsterdam (2005).
4. J.A. Zukas and W.P. Walters, eds., *Explosive Effects and Applications*, Springer: New York (2002).
5. Y.M. Gupta, Recent developments to understand molecular changes in shocked energetic materials, *J. De Phys. IV Colloque* **C4**, 345–356 (1995).

6. J.B. Bdzil and D.S. Stewart, The dynamics of detonation in explosive systems, *Ann. Rev. Fluid Mech.* **39**, 263–292 (2007).
7. D.D. Dlott, Multi-phonon up-pumping in energetic materials, in: Overviews of recent research on energetic materials, D.L. Thomson, T.B. Brill, and R.W. Shaw, eds., *Adv. Ser. Phys. Chem.* **16**, 303–333 (2005).
8. D.D. Dlott, Fast molecular processes in energetic materials, in: *Energetic Materials, Part 2: Detonation, Combustion*, P.A. Politzer and J.S. Murray, eds., Elsevier: Amsterdam (2003), pp. 125–191.
9. M.R. Manaa, Initiation and decomposition mechanisms of energetic materials, in: *Energetic Materials, Part 2: Detonation, Combustion*, P.A. Politzer and J.S. Murray, eds., Elsevier: Amsterdam (2003), pp. 71–100.
10. L.E. Fried, M.R. Manaa, P.F. Pagoria, and R.L. Simpson, Design and synthesis of energetic materials, *Ann. Rev. Mater. Res.* **31**, 291–321 (2001).
11. M.R. Manaa, L.E. Fried, and E.J. Reed, Explosive chemistry: simulating the chemistry of energetic materials at extreme conditions, *J. Comp.-Aided Mat. Design* **10**, 75–97 (2003).
12. T. Luty, Explosive molecular crystals: on the mechanism of detonation, *Mol. Phys. Rep.* **14**, 157–167 (1996).
13. D.D. Dlott and M.D. Fayer, Shocked molecular solids: vibrational up pumping, defect hot spot formation, and the onset of chemistry, *J. Chem. Phys.* **92**, 3798–3812 (1990).
14. A. Tokmakoff, M.D. Fayer, and D.D. Dlott, Chemical reaction initiation and hot-spot formation in shocked energetic molecular materials, *J. Phys. Chem.* **97**, 1901–1913 (1993).
15. J.J. Gilman, Shear-induced metallization, *Phil. Mag. B* **67**, 207–214 (1993).
16. J.J. Gilman, Chemical reactions at detonation fronts in solids, *Phil. Mag. B* **71**, 1057–1068 (1995).
17. M.M. Kuklja and A.B. Kunz, Simulation of defects in energetic materials. 3. The structure and properties of RDX crystals with vacancy complexes, *J. Phys. Chem. B* **103**, 8427–8431 (1999).
18. M.M. Kuklja, E.V. Stefanovich, and A.B. Kunz, An excitonic mechanism of detonation initiation in explosives, *J. Chem. Phys.* **112**, 3417–3423 (2000).
19. E.J. Reed, J.D. Joannopoulos, and L.E. Fried, *Phys. Rev. B* **62**, 19500–19509 (2000).
20. T. Luty, P. Ordon, and C.J. Eckhardt, A model for mechanochemical transformations: applications to molecular hardness, instabilities, and shock initiation of reaction, *J. Chem. Phys.* **117**, 1775–1785 (2002).
21. D. Tsiaousis and R.W. Munn, Energy of charged states in the RDX crystal: trapping of charge-transfer pairs as a possible mechanism for initiating detonation, *J. Chem. Phys.* **122**, 184708-1–184708-9 (2005).
22. G.I. Pangilinan and Y.M. Gupta, Molecular processes in a shocked explosive: time resolved spectroscopy of liquid nitromethane, in: *High Pressure Shock Compression of Solids III*, M. Shahinpoor and L.W. Davison, eds., Springer: New York (1998), pp. 81–100.
23. W.J. Nellis, Dynamic compression of materials: metallization of fluid hydrogen at high pressures, *Rep. Prog. Phys.* **69**, 1479–1580 (2006).
24. L.M. Barker, M. Shahinpoor, and L.C. Chhabildas, Experimental and diagnostic techniques, in: *High Pressure Shock Compression of Solids*, J.R. Asay and M. Shahinpoor, eds., Springer: New York (1993), pp. 43–74.
25. L.C. Chhabildas, L. Davison, and Y. Horie, eds., *High Pressure Shock Compression of Solids VIII*, Springer: New York (2005).
26. J.R. Asay and M. Shahinpoor, *High Pressure Shock Compression of Solids*, Springer: New York (1993).
27. Y.M. Gupta, *COPS Code*, Stanford Research Institute: Menlo Park, CA, unpublished (1976).
28. D.S. Moore, S.C. Schmidt, J.W. Shaner, D.L. Shampine, and W.T. Holt, Coherent anti-Stokes Raman scattering in benzene and nitromethane shock-compressed to 11 GPa, in: *Shock Waves in Condensed Matter-1985*, Y.M. Gupta, ed., Plenum: New York (1986), pp. 207–211.
29. C.S. Yoo, Y.M. Gupta, and P.D. Horn, Pressure-induced resonance Raman effect in shocked carbon disulfide, *Chem. Phys. Lett.* **159**, 178–183 (1989).

30. N.C. Holmes, Dual-beam, double-pass absorption spectroscopy of shocked materials, *Rev. Sci. Instrum.* **64**, 357–362 (1993).
31. D.S. Moore and S.C. Schmidt, Vibrational spectroscopy of materials under extreme pressure and temperature, *J. Mol. Struct.* **347**, 101–112 (1995).
32. R.L. Gustavsen and Y.M. Gupta, Time resolved Raman measurements in α-quartz shocked to 60 kbar, *J. Appl. Phys.* **75**, 2837–2844 (1994).
33. G.I. Pangilinan and Y.M. Gupta, Time-resolved Raman measurements in nitromethane shocked to 140 kbar, *J. Phys. Chem.* **98**, 4522–4529 (1994).
34. J.M. Winey and Y.M. Gupta, UV-visible absorption spectroscopy to examine shock-induced decomposition in neat nitromethane, *J. Phys. Chem.* **49**, 9333–9340 (1997).
35. Y.A. Gruzdkov and Y.M. Gupta, Emission and fluorescence spectroscopy to examine shock-induced decomposition in nitromethane, *J. Phys. Chem. A* **102**, 8325–8332 (1998).
36. Y.M. Gruzdkov and Y.M. Gupta, Mechanism of amine sensitization in shocked nitromethane, *J. Phys. Chem. A* **102**, 2322–2331 (1998).
37. Z.A. Dreger, Y.A. Gruzdkov, Y.M. Gupta, and J.J. Dick, Shock wave induced decomposition chemistry of pentaerythritol tetranitrate single crystals: time-resolved emission spectroscopy, *J. Phys. Chem. B* **106**, 247–256 (2002).
38. X. Hong, S. Cheu, and D.D. Dlott, Ultrafast made-specific intermolecular vibrational energy transfer to liquid nitromethane, *J. Phys. Chem.* **99**, 9102–9109 (1995).
39. X. Hong and D.D. Dlott, Two-dimensional vibrational spectroscopy and its application to high explosives, in: *Time-Resolved Vibrational Spectroscopy VII*, W. Woodruff, Ed. (Springer Proceedings in Physics), Santa Fe, pp. 269–270 (1995).
40. D.E. Hare, I.Y.S. Lee, J.R. Hill, J. Franken, H. Suzuki, B.J. Baer, and E.L. Chronister, Ultrafast dynamics of shock waves and shocked energetic materials, *Proc. Mat. Res. Symp.* **418**, 357–362 (1996).
41. D.D. Dlott, Nanoshocks in molecular materials, *Acc. Chem. Res.* **33**, 37–45 (2000).
42. S.D. McGrane, D.S. Moore, and D.J. Funk, Shock induced reaction observed via ultrafast infrared absorption in poly(vinyl nitrate) films, *J. Phys. Chem. A* **108**, 9342–9347 (2004).
43. M.D. Knudson and Y.M. Gupta, Real-time observation of a metastable state during the phase transition in shocked cadmium sulfide, *Phys. Rev. Lett.* **81**, 2938–2941 (1998).
44. M.D. Knudson, K.A. Zimmerman, and Y.M. Gupta, Picosecond time-resolved electronic spectroscopy in plate impact shock experiments: experimental development, *Rev. Sci. Instrum.* **70**, 1743–1750 (1999).
45. N.C. Holmes and R. Chau, Fast time-resolved spectroscopy in shock compressed matter, *J. Chem. Phys.* **119**, 3316–3319 (2003).
46. J. Franken, S. Hambir, D.E. Hare, and D.D. Dlott, Shock waves in molecular solids: ultrafast vibrational spectroscopy of the first nanosecond, *Shock Waves* **7**, 135–145 (1997).
47. J. Franken, S.A. Hambir, and D.D. Dlott, Picosecond vibrational spectroscopy of shocked energetic materials, in: *Shock Compression of Condensed Matter-1997*, S.C. Schmidt, D.P. Dandekar, and J.W. Forbes, eds., AIP: New York (1998), pp. 819–822.
48. D.D. Dlott, H. Yu, S. Wang, Y. Yang, and S.A. Hambir, Nanotechnology energetic materials dynamics studied with nanometer spatial resolution and picosecond temporal resolution, in: *Advances in Computational & Experimental Engineering Science-'04*, S.N. Atlurl and A Tadeu, eds., pp. 1427–1432 (2004).
49. N. Hemmi, Z.A. Dreger, Y.A. Gruzdkov, J.M. Winey, and Y.M. Gupta, Raman spectra of compressed pentaerythritol tetranitrate single crystals: anisotropic response, *J. Phys. Chem. B* **110**, 20948–20953 (2006).
50. N. Hemmi, Y.A. Gruzdkov, K.A. Zimmerman, Z.A. Dreger, and Y.M. Gupta, in preparation.
51. Y. Ma, H.K. Mao, and R.J. Hemley, Lattice strains in gold and rhenium under nonhydrostatic compression to 37 GPa, *Phys. Rev. B* **60**, 15063–15073 (1999-II).
52. J.W. Otto, J.K. Vassiliou, and G Frommeyer, Nonhydrostatic compression of elastically anisotropic polycrystals. I. Hydrostatic limits of 4:1 methanol-ethanol and paraffin oil, *Phys. Rev. B.* **57**, 3253–3263 (1998-II).

53. J.W. Otto, J.K. Vassiliou, and G. Frommeyer, Nonhydrostatic compression of elastically anisotropic polycrystals. II. Direct compression and plastic deformation, *Phys. Rev. B* **57**, 3264–3272 (1998-II).
54. A.K. Singh, The lattice strains in a specimen (cubic system) compressed nonhydrostatically in an opposed anvil device, *J. Appl. Phys.* **73**, 4278–4286 (1993).
55. A.K. Singh, C. Balasingh, H. Mao, R.J. Hemley, and J. Shu, Analysis of lattice strains measured under nonhydrostatic pressure, *J. Appl. Phys.* **83**, 7567–7575 (1998).
56. L. Dubrovinsky and N. Dubrovinskaia, Angle-dispersive diffraction under non-hydrostatic stress in diamond anvil cells, *J. Alloys Comp.* **375**, 86–92 (2004).
57. N. Funamori, T. Yagi, and T. Uchida, Deviatoric stress measurement under uniaxial compression by a powder x-ray diffraction method, *J. Appl. Phys.* **75**, 4327–4331 (1994).
58. T. Kenichi, Evaluation of the hydrostaticity of a helium-pressure medium with powder x-ray diffraction techniques, *J. Appl. Phys.* **89**, 662–668 (2001).
59. T.S. Duffy, G. Shen, J. Shu, H. Mao, R.J. Hemley, and A.K. Singh, Elasticity, shear strength, and equation of state of molybdenum and gold from x-ray diffraction under nonhydrostatic compression to 24 GPa, *J. Appl. Phys.* **86**, 6729–6735 (1999).
60. G.J. Piermarini, S. Block, and J.D. Barnett, Hydrostatic limits in liquids and solids to 100 kbar, *J. Appl. Phys.* **44**, 5377–5382 (1973).
61. D.M. Adams, R. Appleby, and S.K. Sharma, Spectroscopy at very high pressures: Part X. Use of ruby R-lines in the estimation of pressure at ambient and at low temperatures, *J. Phys. E: Sci. Instrum.* **9**, 1140–1144 (1976).
62. Y.M. Gupta and X.A. Shen, Potential use of the ruby R_2 line shift for static high-pressure calibration, *Appl. Phys. Lett.* **58**, pp. 583–585 (1991).
63. S.M. Sharma and Y.M. Gupta, Theoretical analysis of *R*-line shifts of ruby subjected to different deformation conditions, *Phys. Rev. B* **43**, 879–893 (1991).
64. X.A. Shen and Y.M. Gupta, Effect of crystal orientation on ruby *R*-line shifts under shock compression and tension, *Phys. Rev. B* **48**, 2929–2940 (1993).
65. Z.A. Dreger, N. Trotman, and Y.M. Gupta, in preparation.
66. M. Pope and E.Ch. Swenberg, *Electronic Processes in Organic Crystals and Polymers*, Oxford University Press: New York (1999).
67. E.A. Silinsh and V. Capek, *Organic Molecular Crystals: Interaction, Localization and Transport Phenomena*, Springer: New York (2000).
68. N. Karl, High purity organic molecular crystals, in: *Crystals*, H.C. Freyhardt, ed., Springer: New York (1980), pp. 1–100.
69. S. Wiederhorn and H.G. Drickamer, The effect of pressure on the near-ultra-violet spectra of some fused-ring aromatic crystals, *J. Phys. Chem. Solids* **9**, pp. 330–334 (1959).
70. P.F. Jones and M. Nicol, Excimer fluorescence of crystalline anthracene and naphthalene produced by high pressure, *J. Chem. Phys.* **43**, 3759–3760 (1965).
71. H.W. Offen, Fluorescence spectra of several aromatic crystals under high pressures, *J. Chem. Phys.* **44**, 699–703 (1966).
72. P.F. Jones, Excimer emission of naphthalene, anthracene, and phenanthrene crystals produced by very high pressures, *J. Chem. Phys.* **48**, 5440–5447 (1968).
73. B.Y. Okamoto and H.G. Drickamer, Evaluation of configuration coordinate parameters from high pressure optical data. I. phenanthrene, anthracene, and tetracene, *J. Chem. Phys.* **61**, 2870–2877 (1974).
74. M. Nicol, M. Vernon, and J.T. Woo, Raman spectra and defect fluorescence of anthracene and naphthalene crystals at high pressures and low temperatures, *J. Chem. Phys.* **63**, 1992–1999 (1975).
75. Z.A. Dreger, H. Lucas, and Y.M. Gupta, High-pressure effects on fluorescence of anthracene crystals, *J. Phys. Chem. B* **107**, 9268–9274 (2003).
76. Z.A. Dreger, E. Balasubramaniam, Y.M. Gupta, and A.G. Joly, in preparation.
77. N. Hemmi, Z.A. Dreger, and Y.M. Gupta, Time-resolved electronic spectroscopy to examine shock-wave-induced changes in anthracene single crystals, *J. Phys. Chem.* C **112**, 7761–7766 (2008).

78. P.M. Robinson and H.G. Scott, Plastic deformation of anthracene single crystals, *Acta. Metall.* **15**, 1581–1590 (1967).
79. M. Oehzelt, R. Resel, and A. Nakayama, High-pressure structural properties on anthracene up to 10 GPa, *Phys. Rev. B* **66**, 174104–174104-5 (2002).
80. J. Bernstein, *Polymorphism in Molecular Crystals*, Clarendon Press: Oxford (2002).
81. H.H. Cady and L.C. Smith, Studies on the polymorphs of HMX, *LAMS-2652 (TID-4500 17th ed.)* (1962).
82. Holston Defense Corporation (Eastman Kodak, Kingsport, TN), Physical and chemical properties of RDX and HMX, *Control No.20-P-26 Series B* (1962).
83. P. Main, R.E. Cobbledick, and R.W.H. Small, Structure of the fourth form of 1,3,5,7-tetranitro-1,3,5,7-tetraazacyclooctane (γ-HMX), $2C_4H_8N_8O_8.0.5H_2O$, *Acta. Cryst.* *C***41**, 1351–1354 (1985).
84. F. Goetz, T.B. Brill, and J.F. Ferraro, Pressure dependence of the Raman and infrared spectra of α-, β-, γ-, and δ-octahydro-1,3,5,7-tetranitro-1,3,5,7-tetrazocine *J. Phys. Chem.* **82**, pp. 1912–1917 (1978).
85. M.F. Foltz, C.L. Coon, F. Garcia, and A.L. Nichols III, The thermal stability of the polymorphs of hexanitrohexaazaisowurtzitane, Part I and Part II, *Propel. Explosiv. Pyrotech.* **19**, 19–25; 133–144 (1994).
86. T.P. Russell, P.J. Miller, G.J. Piermarini, and S. Block, High-pressure phase transition in γ-hexanitrohexaazaisowurtzitane, *J. Phys. Chem.* **96**, 5509–5512 (1992).
87. T.P. Russell, P.J. Miller, G.J. Piermarini, and S. Block, Pressure/temperature phase diagram of hexanitrohexaazaisowurtzitane, *J. Phys. Chem.* **97**, 1993–1997 (1993).
88. J. Lin and T.B. Brill, Kinetics of Solid Polymorphic Phase Transitions of CL-20, *Propellants, Explosives, Pyrotechnics,* **32**, 326–330 (2007).
89. T.B. Brill and R.J. Karpowicz, Solid phase transition kinetics. The role of intermolecular forces in the condensed-phase decomposition of octahydro-1,3,5,7-tetranitro-1,3,5,7-tetrazocine, *J. Phys. Chem.* **86**, 4260–4265 (1982).
90. B.F. Henson, B.W. Asay, R.K. Sander, S.F. Son, J.M. Robinson, and P.M. Dickson, Dynamic Measurement of the HMX β- δ Phase Transition by Second Harmonic Generation, *Phys. Rev. Lett.* 82, 1213–1216 (1999).
91. B.F. Henson, L. Smilowitz, B.W. Asay, and P.M. Dickson, The $\beta - \delta$ phase transition in the energetic nitramine octahydro-1,3,5,7-tetranitro-1,3,5,7-tetrazocine: Thermodynamics, *J. Chem. Phys.* **117**, 3780–3788 (2002).
92. L. Smilowitz, B.F. Henson, B.W. Asay, and P.M. Dickson, The β–δ phase transition in the energetic nitramine-octahydro-1,3,5,7-tetranitro-1,3,5,7-tetrazocine: Kinetics, *J. Chem. Phys.* **117**, 3789–3798 (2002).
93. L. Smilowitz, B.F. Henson, M. Greenfield, A. Sas, B.W. Asay, and P.M. Dickson, On the nucleation mechanism of the β- δ phase transition in the energetic nitramine octahydro-1,3,5,7-tetranitro-1,3,5,7-tetrazocine, *J. Chem. Phys.* **121**, 5550–5552 (2004).
94. A.K. Burnham, R.K. Weese, and B.L. Weeks, A distributed activation energy model of thermodynamically inhibited nucleation and growth reactions and its application to the β - δ phase transition of HMX, *J. Phys. Chem. B* **108**, 19432–19441 (2004).
95. J.E. Patterson, Z.A. Dreger, and Y.M. Gupta, Shock wave-induced phase transition in RDX crystals, *J. Phys. Chem. B* **111**, 10897–10904 (2007).
96. W.C. McCrone, Crystallographic data 32. RDX cyclotrimethylenetrinitramine *Anal. Chem.* **22**, 954–955 (1950).
97. C.S. Choi and E. Prince, The crystal structure of cyclotrimethylenetrinitramine, *Acta Cryst. B* **28**, 2857–2862 (1972).
98. B. Olinger, B. Roof, and H. Cady, in Proceedings of international symposium on high dynamic pressures, Paris, France, (1978), pp. 3–8.
99. R.J. Karpowicz, S.T. Sergio, and T.B. Brill, β-polymorph of hexahydro-1,3,5-trinitro-s-triazine. A Fourier transform infrared spectroscopy study of an energetic material, *Ind. Eng. Chem. Prod. Res. Dev.* **22**, 363–365 (1983).
100. R.J. Karpowicz and T.B. Brill, Comparison of the molecular structure of hexahydro-1,3,5-trinitro-s-triazine in the vapor, solution and solid phases, *J. Phys. Chem.* **88**, 348–352 (1984).

101. B.J. Baer, J. Oxley, and M. Nicol, The phase diagram of RDX (hexahydro-1,3,5-trinitro-s-triazine) under hydrostatic pressure, *High Pressure Res.* **2**, 99–108 (1990).
102. P.J. Miller, S. Block, and G.J. Piermarini, Effects of pressure on the thermal decomposition kinetics, chemical reactivity and phase behavior of RDX, *Combust. Flame* **83**, 174–184 (1991).
103. C.S. Yoo, H. Cynn, W.M. Howard, and N. Holmes, Equations of state of unreacted high explosives at high pressures, in: *Eleventh International Detonation Symposium*, pp. 951–957 (1998).
104. N. Goto, H. Yamawaki, K. Wakabayashi, Y. Nakayama, M. Yoshida, and M. Koshi, High pressure phase of RDX, *Science and Technology of Energetic Materials* **66**, 291–300 (2005).
105. N. Goto, H. Fujihisa, H. Yamawaki, K. Wakabayashi, Y. Nakayama, M. Yoshida, and M. Koshi, Crystal structure of the high-pressure phase of hexahydro-1,3,5-trinitro-1,3,5-triazine (γ-RDX), *J. Phys. Chem. B* **110**, 23655–23659 (2006).
106. J.A. Ciezak, T.A. Jenkins, Z. Liu, and R.J. Hemley, High-Pressure Vibrational Spectroscopy of Energetic Materials: Hexahydro-1,3,5-trinitro-1,3,5-triazine, *J. Phys. Chem. A* **111**, 59–63 (2007).
107. Z.A. Dreger and Y.M. Gupta, High pressure Raman spectroscopy of single crystals of hexahydro-1,3,5-triazine (RDX), *J. Phys. Chem. B* **111**, 3893–3903 (2007).
108. A. Filhol, C. Clement, M. Forel, J. Paviot, M. Rey-Lafon, G. Richoux, C. Trinquecoste, and J. Cherville, Molecular conformation of 1,3,5-trinitrohexahydro-s-triazine (RDX) in solution, *J. Phys. Chem.* **75**, 2056–2060 (1971).
109. M. Rey-Lafon, R. Cavagnat, C. Trinquecoste, and M.T. Forel, Etude des specters de vibration de la trinitro-1,3,5 hexahydro s-triazine *J. Chem. Phys. Physicochim. Biol.* **68**, 1575–1577 (1971).
110. J.J. Haycraft, L.L. Stevens, and C.J. Eckhardt, Single-crystal, polarized, Raman scattering study of the molecular and lattice vibrations for the energetic material cyclotrimethylene trinitramine *J. Appl. Phys.* **100**, 053508–053817 (2006).
111. J.C. Decius and R.M. Hexter, *Molecular Vibrations in Crystals*, McGraw-Hill: New York (1977).
112. W.G. Fateley, F.R. Dillish, N.T. McDevitt, and F.F. Bentley, *Infrared and Raman Selection Rules for Molecular and Lattice Vibrations: The Correlation method*, Wiley: New York (1972).
113. D.E. Hooks, K.J. Ramos, and A.R. Martinez, Elastic-plastic shock wave profiles in oriented single crystals of cyclotrimethylene trinitramine (RDX) at 2.25 GPa, *J. Appl. Phys.* **100**, 024908-1–024908-7 (2006).
114. J.M. Winey, private communications.
115. R. Cheret, *Detonation of Condensed Explosives*, Springer: New York (1993).
116. J.J. Dick, Effect of crystal orientation on shock initiation sensitivity of pentaerythritol tetranitrate explosive, *Appl. Phys. Lett.* **44**, 859–861 (1984).
117. J.J. Dick, R.N. Mulford, W.J. Spencer, D.R. Pettit, E. Garcia, and D.C. Shaw, Shock response of pentaerythritol tetranitrate single crystals, *J. Appl. Phys.* **70**, 3572–3587 (1991).
118. J.J. Dick, Anomalous shock initiation of detonation in pentaerythritol tetranitrate crystals, *J. Appl. Phys.* **81**, 601–612 (1997).
119. J.J. Dick and J.P. Ritchie, Molecular mechanics modeling of shear and the crystal orientation dependence of the elastic precursor shock strength in pentaerythritol tetranitrate, *J. Appl. Phys.* **76**, 2726–2737 (1994).
120. Y.A. Gruzdkov and Y.M. Gupta, Shock wave initiation of pentaerythritol tetranitrate single crystals: mechanism of anisotropic sensitivity, *J. Phys. Chem.* **100**, 11169–11176 (2000).
121. Y.A. Gruzdkov, Z.A. Dreger, and Y.M. Gupta, Experimental and theoretical study of pentaerythritol tetranitrate conformers, *J. Phys. Chem. A* **108**, 6216–6221 (2006).
122. K.E. Lipinska-Kalita, M.G. Pravica, and M.F. Nicol, Raman scattering studies of the high-pressure stability of pentaerythritol tetranitrate $C(CH_2ONO_2)_4$, *J. Phys. Chem. B* **109**, 19223–19227 (2005).

123. M.G. Pravica, K.E. Lipinska-Kalita, Z. Quine, E. Romano, Y. Shen, M.F. Nicol, and W.J. Pravica, Studies of phase transitions in PETN at high pressures, *J. Phys. Chem. Solids* **67**, 2159–2163 (2006).
124. B.M. Rice and E.F.C. Byrd, Theoretical chemical characterization of energetic materials, *J. Mat. Res.* **21**, 2444–2452 (2006).
125. D.L. Naud and K.R. Brower, Pressure effects on the thermal decomposition of nitramines, nitrosamines, and nitrate esters, *J. Org. Chem.* **57**, 3303–3308 (1992).
126. L.L. Davis and K.R. Brower, Reactions of organic compounds in explosive-driven shock waves, *J. Phys. Chem.* **100**, 18775–18783 (1996).
127. L.L. Davis, Reactions of organic compounds in explosive-driven shock waves, *Ph.D. Dissertation*, New Mexico Institute of Mining and Technology, (1996).

Chapter 7
Equilibrium Molecular Dynamics Simulations

Betsy M. Rice and Thomas D. Sewell

7.1 Introduction

Molecular dynamics (MD) is a widely used atomistic simulation method due to the detailed information it can provide, often with a relatively small computational investment. The most distinguishing attribute of MD among molecular simulation methods is that it provides a means to monitor the time evolution of a system of particles (usually atoms) in phase space, thus allowing for an atomic-level view of the dynamics of a material in a given equilibrium or nonequilibrium thermodynamic state. This is particularly appealing for those in the energetic materials (EM) community since such a detailed description could reveal the fundamental mechanisms controlling the initiation of an energetic material to detonation, a phenomenon for which direct experimental measurement is in short supply due to the small time and spatial scales involved and the accompanying large rates of chemical energy release. MD is not affected by any of these factors; rather, its main limitations are the description of interatomic interactions (potential energy functions) used in the simulations and the viability of using classical mechanics to study molecular-scale phenomena. MD is receiving increased use in condensed-phase EM research as interaction potentials emerge that "realistically" describe the chemistry associated with initiation of an EM. However, MD is not limited to studying nonequilibrium dynamic events only; it has proven to be extremely useful for predicting thermodynamic equilibrium properties in the condensed phase.

Often a complete mapping of the equation of state (EOS) or the shock Hugoniot locus for an EM is extremely difficult to accomplish using traditional experimental methods of diamond anvil cells or shock waves [1–3]. Further, off-Hugoniot data can be measured only through the use of specialized equipment designed to study quasi-isentropic compression or using multiple-shock methods in which experimental and analytic uncertainties multiply quickly [4, 5]. MD simulations of any of these states, on the other hand, are straightforward and can readily provide a description of the EM under conditions not amenable to experimentation. Note

S.M. Peiris, G.J. Piermarini (eds.), *Static Compression of Energetic Materials,*
Shock Wave and High Pressure Phenomena,

that in many cases, particularly ones involving dynamic phenomena, the comparison between macroscopic and atomic-based results can be complicated due to the effects of finite simulation domains or slow relaxation phenomena. However, recent increases in time and spatial resolution of experimental diagnostics such as computed microtomography [6, 7] and ultrafast dynamic ellipsometry of laser-driven shocks on thin-film samples [8], which can provide the entire shock Hugoniot based on stress-induced optical effects and low-strain particle motion within a single-shot experiment, enable measurements of properties on scales routinely accessible by molecular simulation methods.

The focus of this chapter will be limited to the use of the MD method to predict thermodynamic, structural, and mechanical properties of EMs under static compression, although it will include references to MD studies of dynamic behavior of EMs. A brief description of the method and design of appropriate MD simulations to study EMs at static high pressures will be given, along with descriptions of interaction potentials developed to study EMs. Since the accuracy of MD results is so strongly dependent on the model of interatomic interactions used, we will describe the performance of recent potential models for predicting the properties of EMs at various thermodynamic conditions, describe capabilities and deficiencies of the models, and identify what we think are the most crucial simulations that should be performed to evaluate emerging methods and models. First, however, we will provide a description of certain aspects of energetic materials on the mesoscale (roughly 0.1–1,000 μ) that will provide a touchstone for the discussions of particular methods, materials, and properties that follow.

7.2 A Mesoscale Perspective on Energetic Materials

Military energetic materials formulations, whether used in conventional munitions, warheads, or propellants, are almost always highly filled composites comprised of ∼1–100 μ-sized crystals (or "grains") of energetic material suspended within a polymeric binder matrix with filler fractions of ∼70–95% [9]. While the binder may or may not contribute to the energetic output of the charge, its principal roles are to provide mechanical integrity and to mediate initiation sensitivity to accidental ignition under abnormal conditions of dynamic stress.

Energetic formulations are multiscale and complex. Spatial scales required for a complete understanding range from Å–nm up through mm, with associated timescales ranging from sub-picosecond for fundamental chemical events to μs or beyond for prompt initiation processes, and much longer for slow processes such as thermal cook-off. Practical engineering applications using Lagrangian or Eulerian hydrodynamic simulation schemes require homogenized (scale-free, continuum) models, known as "constitutive descriptions" with characteristic spatial scales of mm or above. (Here, for convenience, we consider the constitutive description to encompass material equation of state [10], thermomechanical response [11], and chemical kinetics [12].) These range widely in complexity, from fairly simple descriptions to extremely complicated ones involving many physical properties, thermodynamics and, in some cases, kinetics of phase transitions, chemical reaction rates, strain rate dependence, and so on. Practically, it is impossible to include explicitly in such models the fully resolved grain-scale microstructural features of EMs, let alone the "sub-scale" physical and chemical processes that determine the effective continuum response.

Nevertheless, in order to realize the goal of "predictive capability" for continuum codes of energetic materials initiation and performance that goes beyond the dominant "engineering calibration paradigm," it is necessary that whatever homogenization schemes are employed include sufficient physics to capture at least in a statistical sense the salient "sub-grid" features of the underlying material based on physical models that can be connected to directly measurable or computable phenomena. Among the latter are the spatial distributions of material heterogeneities and enormous amount of interfacial area present on the grain scale; these arise due to crystal defects (e.g., dislocation twins, voids, or solvent inclusions), regions of poor binder wetting of the grains, or direct crystal–crystal contacts [13, 14]. Under thermomechanical loading such heterogeneities lead to localization of energy, corresponding to low-probability "tails" of distributions of temperature and stress within the material [15, 16]. These regions of localized energy are generally known as "hot spots."

It has been accepted since the 1940s that the generation of a sufficiently high density of sufficiently large hot spots at sufficiently high temperature is required for the initiation of most energetic material formulations [17, 18] and in some cases for the propagation of detonation waves [19]. The characteristic scale of "effective" hot spots in a typical EM is on the order of 1–10 μ, well below the resolution limit of practical engineering calculations. It should be clear that the number and spatial distributions of potential hot spots in a composite material can be affected by material processing, thermal history, mechanical insult, chemical and physical aging processes, and so on. Coupled with the fact that the simple material properties are generally functions of temperature, pressure, and sometimes strain rate (e.g., polymer viscoelasticity), and that these thermodynamic dependencies can interact in complicated ways (sometimes offsetting, other times reinforcing), the problem of how to describe a material in a way that yields "predictive" capability is nontrivial.

One challenge (among many) that sub-grid localization causes for continuum simulations is the trade-off between the need to use simple chemical reaction rates/models based on an effective, homogenized temperature – in fact more commonly pressure, for technical reasons – and the fact that those effective rates are not the true rates for thermal chemistry in the material on the sub-grid scale. (Further, the effective rates may themselves be mesh dependent [12].) Specifically, no matter how it is achieved (e.g., shock heating, thermal diffusion or convection, or dissipation), the average temperature in a computational cell will necessarily be lower than the hot spot temperature, such that the effective rate must be higher than the true chemical rate in order to match macroscopic experiments (for example a wedge test or "pop plot"). Menikoff has recently compared three different (simple) constitutive descriptions in a study of the initiation and detonation properties of the HMX-based plastic-bonded explosive PBX-9501 [20]. His work elaborates on several outstanding issues for such simulations.

Many of the thermomechanical equation of state properties, transport properties, and fundamental inelastic deformation processes needed for the parameterization or as aids to the formulation of mesoscale simulation models for energetic materials can be computed directly using carefully designed molecular dynamics or Monte Carlo simulations. Among the properties that are needed and that can be predicted as functions of pressure and temperature (stress state) with a high degree of reliability, given an accurate potential energy surface, are polymorphic crystal structures and lattice energies, isothermal compression curves and coefficients of thermal expansion (both volumetric and linear), second-order elastic coefficients and derived isotropic moduli, heats of sublimation and vaporization, and linear transport coefficients for mass, momentum, and energy (self-diffusion, shear viscosity, and thermal conductivity, respectively). The major complication that arises in transferring these quantities directly from the atomic scale to mesoscopic or continuum scales is the current inability to include in a realistic way within the limited framework of a molecular dynamics simulation the kinds of defect structures that exist in real materials. As a consequence, MD predictions for many of the properties listed above based on structurally defect-free samples will represent upper or lower bounds to values that would be measured in the laboratory. Additional properties for which information is required but whose computation is more problematic using purely classical molecular dynamics, typically due to the need to account in some way for quantum effects, include the temperature- and pressure-dependent specific heat and modal Gruneisen coefficients. Large-scale molecular dynamics simulations are increasingly used to study dynamic processes with explicit inclusion of all mechanical degrees of freedom, for example mechanisms of inelastic deformation under shock loading in neat crystals or dissipative mechanisms during pore collapse. Finally, with the advent of reactive force fields – that is, ones which can be used to study chemistry in the condensed phase – it is increasingly possible to make predictions of chemical reaction rates and mechanisms for thermodynamic conditions relevant to understanding energetic materials initiation and detonation phenomena. With these emerging capabilities, and subject to the caveats outlined above, it is reasonable to attempt to construct on the basis of molecular dynamics simulations a substantially complete, internally consistent thermal-mechanical-chemical constitutive description for condensed-phase energetic materials.

7.3 The Method of Molecular Dynamics

Molecular dynamics simulations are routinely employed in many areas of the chemical and physical sciences and numerous reviews exist that describe methodologies and applications specific to the various fields, including that of EMs [21]; therefore, we will provide only a brief overview of the method. A molecular dynamics simulation involves solving equations of motion for a system of classical particles. The equations can be formulated in the Newtonian, Hamiltonian, or Lagrangian approaches [22], all of which correspond to the microcanonical (NVE) statistical mechanical ensemble in which the number of particles, volume, and total energy are conserved. Simulations in other statistical mechanical ensembles can be performed by using modified forms of the equations of motion, and are especially useful for generating EOS information. These include the isothermal-isochoric (NVT, canonical), the isothermal-isobaric (NPT, constant pressure), and isothermal-isostress (N*s*T) ensembles. In all of these, the number of particles and temperature are conserved. The volume is conserved for the NVT ensemble while isotropic pressure is conserved in the NPT ensemble. The N*s*T ensemble is a generalization of the isothermal-isobaric ensemble that allows anisotropic deformation of the simulation cell in response to a prescribed tensorial stress rather than to simple hydrostatic pressure. N*s*T-MD simulations are particularly useful for assessing interaction potentials for EMs since they provide a test of the ability of an interaction potential to accurately predict and maintain the correct crystal space group symmetry (i.e., size and shape of the simulation cell and the arrangement of molecules within the cell) during such a simulation. Other statistical ensembles include isobaric-isoenthalpic (NHT, where H is enthalpy) and grand canonical (μVT, where μ is chemical potential) ensembles, but these are not routinely used in EM research. Alterations to the equations of motion have also been made use of in simulations that drive the system to points on its shock Hugoniot [23, 24] or that sample states of the shocked materials [25–27].

Integration of any of the aforementioned equations of motion will produce a trajectory; that is, a time-ordered sequence of solutions that provide a description of the dynamic behavior of the particles in phase space over the time interval of the simulation. In the simplest cases, physical properties can be obtained by time averaging of simple observables over the duration of the trajectory, for example, crystal lattice parameters or liquid density. In many cases, however, it is necessary to construct the fluctuation or second moment of some quantity in order to compute the desired observable, for example, the elastic tensor, bulk modulus, or specific heat. This generally requires additional computational effort to obtain a given level of statistical precision, and special care must be taken if using extended ensembles to ensure that the set of chosen external coupling parameters (e.g., thermostat and/or barostat) does not affect the results. (Except for pathological cases, this choice of parameters is not usually crucial when calculating simple first moments, but can significantly affect the predicted higher moments.) In other cases, it is necessary to compute integrals over time autocorrelation functions, for instance to predict the shear viscosity or density of states. Even with clever averaging over all time origins in an equilibrium simulation, such quantities can take several nanoseconds to converge and care must be taken to ensure that long-time relaxations do not invalidate the results.

For equilibrium simulations pressure and temperature are thermodynamic state variables; thus, prediction of most quantities as functions of those variables is straightforward and useful [28] since, in most cases, experimental data are available only in the vicinity of room temperature and pressure, or along one or the other axis (i.e., isothermal or isobaric conditions; for an exception, see Ref. [1]). In the absence of such pressure- and temperature-dependent information, most continuum models treat important variables such as specific heat as constants and then "tune" other adjustable parameters to agree with experiment; this level of approximation can lead to difficulties for complicated situations such as materials subjected to multiple shocks [10].

Beyond accuracy of the potential energy surface in the intervals of thermodynamic space of interest, it is necessary to ensure the correctness of the equations of motion and the stability limits of the numerical scheme used to solve them – particularly for properties computed in the microcanonical (NVE) ensemble. In the case of extended ensembles such as NVT or NPT, the simple integral invariants (conservation of energy, linear, and [sometimes] angular momentum) that serve as touchstones for accuracy in the NVE ensemble may not exist. Thus, it is wise to always check the accuracy of the equations of motion in a microcanonical limit prior to performing calculations in some other ensemble. If possible, one should also include error bars on calculated properties to reflect statistical precision. (It is perhaps permissible not to include them if some quantity is computed for several different values of pressure or temperature, in which case the "smoothness" of the calculated points may provide some indication of statistical variation.) Finally, one must assess critically the validity of the approximation of classical mechanics in the context of prediction of physical properties at the atomic scale. This is a reasonable approach for most equilibrium properties, but in some cases the classical approximation is not valid; the most obvious of these is prediction of the specific heat for organic materials (and for low-Z metals such as Li or Be) for which the Debye temperature is $>$2,000 K and thus significant population of high-frequency vibrons such as C–H stretching modes is clearly inaccurate [29]. This "specific heat problem" is basically a manifestation in the condensed phase of the "zero point energy problem" that is well known in gas-phase chemical dynamics [30]. An unfortunate consequence of the error in specific heat is that, in a full dimensional purely classical calculation, the temperature behind a weak to moderately strong shock wave will be underestimated.

7.4 Properties Calculated in Static High-Pressure MD Simulations

The prediction of intensive materials properties using MD is an attractive (and sometimes necessary) alternative to experimental measurement. Measuring common materials properties of EM can be difficult since phase changes [31–34] or changes in chemical reactivity can occur in EMs upon compression [34–36]. Further, many measurements exhibit systematic errors due to sample purity, thermal history, and even particle size (for formally intensive properties!). In such cases, MD predictions of material properties for structurally perfect crystals are particularly useful since they will often provide upper or lower bounds to those properties.

The most basic thermodynamic properties directly calculated from an MD simulation of a bulk EM are temperature, pressure, internal energy, and specific volume. (By specifying two of the preceding properties and predicting the remaining ones one can arrive at an equation of state.) Also, a variety of other thermal or mechanical properties can be obtained from these properties, including thermal expansion coefficients, specific heat, and isotropic or anisotropic elastic properties. Others can be obtained either through time averaging of properties calculated during the MD simulation or derived from the root-mean-square (RMS) fluctuations of thermodynamic properties. In some cases it is necessary to compute convolutions of time correlation functions.

Structural properties of an EM can be obtained simply by time-averaging structural descriptors that use the atomic positions of the material over the course of an MD simulation. One of the most straightforward representations is the radial distribution function, a simple measure of the local structure of the atoms within a material (i.e., for a given atom, the number of its neighbors as a function of distance). For materials that can be described by pair-additive interactions, this function can be used to predict certain thermodynamic properties, such as energy. Structure factors, similar to those obtained from crystallographic measurement can also be calculated, and are particularly useful in characterizing molecular crystals. Another useful exercise is to compare thermally averaged atomic positions of all the molecules in the simulation cell with those in an ideal crystal composed of symmetry equivalents generated using the thermally averaged lattice vectors, thermally averaged atomic coordinates of a single molecule, and space group symmetry operators of the various crystalline space groups. This activity will determine the crystallographic space group of the simulated crystal. The occurrence of a pressure-induced phase transition during an MD simulation can be observed by monitoring structural order parameters, variables defined to represent translational, orientational, or conformational order within a crystal. (Only the first of these is relevant in the often-studied case of simple metals.) These parameters are evaluated for each molecule in the system at each integration step, then averaged and normalized, for instance the second Legendre polynomial. These have been used to identify the melting transitions for various EMs [37–42]. Other order parameters, for instance the Cremer-Pople ring puckering coordinates [43] or the convex hull which defines first Vornoi polyhedra (a sophisticated "nearest neighbor" method that is not invalidated by uniaxial or hydrostatic compression) [44], can be used to identify transitions among solid polymorphs; these have been applied to stress-induced phase transitions and creation of stacking faults in RDX (1,3,5-trinitro-1,3,5-*s*-triazine) crystals under compression [45].

7.5 Design of the Simulation

A representation of bulk EM is often required for the evaluation of nearly all materials properties of practical interest. Advances in both architectures and algorithms for scalable parallel computing now allow billion-atom MD simulations [46, 47].

However, even such large system sizes are too small to represent bulk material as a discrete object in vacuum, since the surface-to-volume ratio will be large. (Particles on the faces of a crystal or droplet are subject to forces that differ from those for particles in the interior of the simulation cell. Thus, any thermal averages obtained from such a system might be affected by edge effects.) Fortunately, infinitely large system sizes are not necessary in order to study many bulk materials properties and processes in an EM. Instead, the bulk material can be emulated using simulation cells containing comparative atoms ($\sim 10^3$–10^6) by using periodic boundary conditions. In this procedure, Cartesian space is completely filled with replicas of the primary simulation cell along each of the three Cartesian axes **x**, **y**, and **z**. The only formal constraint on the shape of the cell is that it comprises a space filling, three-dimensional tessellation. In this way, particles within the simulation cell located at a position **r** have identical images located at $\mathbf{r} + (i\mathbf{x} + j\mathbf{y} + k\mathbf{z})$, where i, j, and k are integers. If a particle leaves the simulation cell through one of its faces, its image enters the simulation cell through the opposite face. Each particle within the simulation cell can interact with other particles in the simulation cell, or with images outside the simulation cell. Further (and necessary) computational efficiency is obtained by limiting the range of the interatomic interactions (i.e., the potential energy is set to zero beyond a specified cutoff distance) and restricting the simulation cell to have dimensions that are at least twice the range of interaction. By doing this, any particle within the simulation cell will interact only with the nearest periodic images of particles within the interaction range; interactions with all other images do not need to be calculated, except in the case of non-truncated long-range electrostatics for which specialized lattice summation techniques are usually employed.

It is vitally important to reiterate that, even with periodic boundary conditions, accurate evaluation of long-range forces, and an assumed "perfect" atomic force field, developing predictive tools for many properties and processes needed for simulations of energetic materials on mesoscopic and macroscopic scales will remain a difficult proposition due to complicated defect structures and complex interactions among them on spatial scales that even the largest molecular simulation tools cannot yet approach, or due to physical relaxation phenomena whose mathematical scaling properties do not extend down to the spatial scales currently accessible to even the most determined molecular dynamicist (i.e., for organic materials with electrostatic interactions, space scales below a few hundred nanometers in a 3-D cell and timescales of less than a microsecond).

Numerical solution of the equations of motion is an initial value problem. Thus, at the beginning of any condensed phase MD simulation, an initial configuration (i.e., atomic positions, velocities, and lattice vectors [in the case of a periodic system]) must be given. For those systems for which experimental crystallographic information is available, the initial simulation cell can be composed of replicas of unit cells of the material arranged in the experimental configuration at the conditions of interest, and atomic velocities are assigned randomly from either a uniform or Maxwell–Boltzmann distribution. Frenkel [48] notes that assigning atomic velocities uniformly and subjecting the simulation to an equilibration trajectory will produce a Maxwell–Boltzmann distribution of velocities; however, care must be taken to ensure complete equilibration, as will be discussed below. For those systems for which a priori knowledge of the crystal structure is not available, probable candidate structures can be generated using a classical molecular simulation method known as ab initio crystal prediction. Various implementations of the method generally follow a three-step computational approach [49–79]. The first step corresponds to generating a 3-D model of the molecule that will be used to construct candidate crystals of different symmetries. In the next step hypothetical crystal structures using the molecular models are created. Finally, in the third step the energy of each hypothetical crystal is minimized, leading to the selection of the best candidates for crystal topology. Each of these steps requires a description of interatomic interactions. The creation of hypothetical crystal structures using the molecular model can be accomplished in a variety of ways. The most general method of generating a hypothetical crystal structure is to place Z images of the molecular model randomly into a cell [80]. However, it is unlikely that the global energy minimum will be identified during the energy minimization step if the random insertion of particles into the cell occurs without imposing space group symmetry restrictions. Thus, space group symmetry for the cell and its contents is usually assumed at this stage of the procedure. Once the choice of space group has been made, a reasonable starting structure for the cell must be obtained. The parameters that must be assigned before energy minimization are the cell parameters and the location and orientation of the molecular model within the crystal; in cases involving conformationally flexible molecules, this must be done separately for all low-energy conformers. Once these are chosen for the test molecule (more rigorously the crystallographic asymmetric unit which usually consists of part or all of a molecule), the remaining (Z-1) symmetry-equivalent entities in the cell are generated using the space group symmetry operations. Typically, numerous hypothetical crystal structures are generated for each space group for which the orientation and conformation of the test molecule have been randomly or systematically selected; in some cases limited intramolecular flexibility about the local conformational energy minima is included during the energy minimization step. The larger the number of hypothetical crystals that are generated, the more fully is configuration space sampled. Most computational methods limit the sampling to the most common space groups, since the generation of large numbers of hypothetical crystals for all 230 crystalline space groups within a single calculation becomes computationally burdensome. This limitation is reasonable, since a survey of organic compounds indicates that 90% of organic crystals are described by only 17 space groups [81].

Regardless of how the initial condition is chosen for the simulation, it is crucial that the system be properly equilibrated to the desired thermodynamic state before time averages are calculated. This can be accomplished for the case of fixed volume simply by integrating the equations of motion for all atoms in the simulation cell and, in some instances, scaling atomic velocities and at periodic intervals during the equilibration in order to more quickly reach the desired temperature. To determine if equilibration has occurred, instantaneous properties such as temperature, pressure, density, potential energy, or kinetic energy should be monitored. It is useful to compute the full stress tensor and multiple measures of temperature to ensure equipartition of energy among the various modes of the system (i.e., phonons and vibrons); this is particularly important if geometric constraints, such as fixed bond lengths or bond angles, are employed in the simulation since conventional initial velocity selection schemes of which we are aware are unable to properly sample the associated manifolds in phase space. When the temporal behavior of the instantaneous properties does not change and the corresponding fluctuations are not large, it can be assumed that equilibration is complete. At this point, thermodynamic and structural properties can be calculated and accumulated for averaging for the remainder of the trajectory integration.

7.6 Development of Interaction Potentials

All MD simulations require a description of the forces acting on the particles in the system. This description is crucial to the quality of the simulation. An inaccurate depiction of the forces experienced by the particles at the conditions of interest will produce spurious results. If, however, the description correctly reflects the forces acting on the particles at the conditions of interest, then the MD results will provide an accurate representation of the classical statistical mechanical behavior of the system. The forces can be derived from semiempirical [82] or empirical [83] functions that describe the potential energy of the system, or they can be calculated directly using quantum mechanical (QM) methods [84] (as will be described in greater detail in the chapter by Kukla). In the latter case, however, the prohibitive cost of a direct QM evaluation of forces during an MD simulation limits the size of the system being simulated and the duration of the trajectory. (Additional care should be taken in the case of QM evaluation of forces, due to numerical issues with obtaining the forces in a time-reversible way [84].) Thus, substantial effort has been devoted to the development of a variety of analytic potential energy functions that accurately represent energetic materials; in the following, we emphasize only the more recent efforts.

It is not difficult to develop reasonably accurate empirical descriptions of the forces for simple materials (e.g., noble gases) or for materials for which the MD simulations will explore a small volume of phase space (e.g., low-temperature simulations). In fact, a few nonreactive empirical interaction potentials (from which forces are derived) have been developed that accurately describe conventional EMs near the ambient state (e.g., Refs. [85–91]). Unfortunately, these interaction potentials are unable to correctly describe the complex chemical and physical behavior of EMs in regimes of high temperature and pressure (e.g., Ref. [92]); thus, more elaborate descriptions are required. Development of such interaction potential is not a trivial task. Conventional energetic materials used for military applications are typically large, polyatomic molecular crystals consisting mainly of carbon, hydrogen, nitrogen and oxygen (with new classes of high-nitrogen materials now being synthesized) [93]. Since many of these molecules are large (ranging in tens of atoms), they have numerous and complicated vibrational degrees of freedom. Additionally, these molecules often have multiple structural conformers, which produce different crystalline polymorphs [94–97]. These materials react rapidly and release energy upon initiation. Moreover, the chemistry associated with either deflagration (a subsonic reaction wave) or detonation (a supersonic wave) is extremely complex, consisting of a strongly coupled web of endothermic and exothermic chemical steps before forming the final products. As indicated above, EMs can undergo a variety of phase transitions with compression and temperature. This suggests that phase transitions could have a role in the detonation or deflagration of an EM. Thus, development of empirical interaction potentials that will correctly capture such complex and diverse behavior in MD simulation requires significant and often arduous effort. It is not surprising that while MD simulations have been performed since the 1970s, it has only been within the past decade that interaction potentials have been developed that allow a more realistic portrayal of the chemical and physical behavior of real EMs.

In the early days of MD simulations of energetic materials, little attention was given to developing interaction potential functions to represent actual chemical systems. Instead, the various studies were designed to explore individual, or a small number of, processes believed to be associated with shock and detonation, with greater emphasis on probing the salient physics as opposed to accurate chemistry. In such studies, extremely simple and highly idealized representations of condensed phase solids were invoked, sometimes with reactions treated in an aphysical fashion [21]. Eventually, models were developed that better represented the chemistry expected in shock initiation, most notably the Reactive Empirical Bond Order (REBO) model developed by Brenner and coworkers [98]. This model, similar to Tersoff-type models that allow for bond formation, bond breaking, or changes in atomic hybridization due to local environment [99], has been subjected to numerous MD studies to investigate reacting shocked solids. Although at the time of its origin REBO was intended to describe detonation chemistry in NO (nitrogen monoxide, a heteroatomic diatomic explosive), REBO does not in fact correspond to any known EM, and its simplicity limits its ability to describe the more complex physics and chemistry characteristic of real energetic materials. Fortunately, efforts have continued since then to produce interaction potentials which more realistically depict both static properties and dynamic events in condensed-phase EMs. These efforts can be partitioned into two categories: nonreactive and reactive models. Both can be used to predict thermophysical properties of the crystals, but only the reactive models can be used to explore chemistry. Within the framework of unreactive models, further subdivision can be made into fully flexible, partially flexible, and fully rigid descriptions of the intramolecular degrees of freedom. Another distinction based on the complexity of nonbonded interactions can be made between many-body potentials (including for instance atomic polarizability) and the more common two-body form with fixed partial atomic charges.

Regardless of which class of model is chosen, the methods for development generally follow the same steps. First, functions are selected that will describe a desired interaction (i.e., Coulombic, dispersion, bond stretching, etc.). Next, parameters are fitted to empirical or quantum mechanical information. In principle, the latter is significantly more desirable since empirical information is often absent, sparse, or inconsistent from one experiment to another. However, care must be taken in the choice of quantum mechanical theory used to generate information for parameterization of the nonbonding terms for EMs. The most common QM method used for study of large systems is density functional theory (DFT) [100,101], due to its modest computational requirements and demonstration of reasonably accurate results for the prediction of intramolecular geometries and some molecular crystal structures. Unfortunately, conventional DFT methods do not properly treat dispersion, which is the main component of the cohesive energy of molecular organic crystals, such as EMs, at low pressures (less than 1 GPa). Therefore, predictions of equilibrium crystal structures at pressures for which dispersion is the dominant interaction will be inaccurate compared to experiment when using intermolecular potentials that are parameterized using DFT results for condensed phase materials at low levels of compression (see, for example, Refs. [102, 103]). However, there have been DFT predictions of crystal structures of compressed EM that are in good agreement with experiment, suggesting that this QM method is appropriate to describe crystals compressed to the point for which nondispersive interactions dominate [104]. (The dispersion error is still present under compression, but the overall dispersion contribution to the total energy, and thus the effect of the error, is greatly reduced compared to Pauli repulsion contribution to the energy.) Therefore, in lieu of reliable QM information for a system, empirical corrections to the model should be made or the user should recognize the limitations of any model that relies on conventional DFT descriptions of cohesive forces of EM at certain conditions (i.e., low pressure).

7.7 Identification of Key MD Simulations to Assess Interaction Potentials

As emphasized in earlier sections, the results of any MD simulation are strongly dependent on the quality of the force field. Too often, however, interaction potentials are used in MD simulations before undergoing extensive testing to assess their limitations and capabilities. Therefore, in this section we will recommend key MD simulations that should be performed to evaluate the interaction potentials. At a minimum, a model interaction potential of a condensed phase EM should be able to predict, in isothermal-isostress (NsT) MD simulations of the ambient conditions phase, the following:

- The correct crystallographic space group symmetry (i.e., this should be conserved during the simulation)
- Crystallographic parameters, including crystal density, to within 3% at the experimental condition
- Independent geometric variables of the unit cell (fractional positions of the centers of mass, relative molecular orientations, etc.)
- Isothermal compression curve over specified intervals of pressure
- Thermal expansion coefficient
- Heats of sublimation and/or fusion

More rigorous tests of the interaction potential would involve using it in NσT-MD simulations of each of the preceding for various polymorphic crystalline forms of the EM. Ideally, the force field will accurately predict the melting point at ambient pressure (the melting curve $T_m = T_m(P)$ is an important quantity that is only rarely known but which can be expected to significantly affect chemical reactions rates and mechanisms of mechanical dissipation) [28]; and thermodynamic phase boundaries between stable crystal polymorphs, the latter of which are particularly complicated since they often involve reconstructive (sometimes called nonmilitary) phase transformations in which simple group–subgroup symmetry relations between the initial and final polymorphs do not exist [105]. This will allow evaluation of local energy minima (and corresponding structures) of the potential energy surface. More telling validation challenges include prediction of tensorial properties such as the elastic tensor, thermal expansion tensor, and thermal transport coefficients for mass, momentum, and energy. It is clearly important to understand where the potential fails, for example by significantly mis-predicting the relative stability of phases or even predicting polymorphs that do not exist experimentally.

In the absence of an experimental determination, generation of candidate initial crystalline structures to use in these simulations can be generated using ab initio crystal structure prediction procedures described above. In one procedure developed specifically for EM research [79], a series of candidate unit cells of different crystal symmetries are generated by orienting a molecule along Euler axes within the cell and generating its symmetry equivalents using the space group symmetry operations. In this way, a large number of unit cells with different space group symmetries and orientations of the molecules within the cells are constructed. In a standard run, tens of thousands of candidate crystal structures are produced, from which a subset of the most dense crystals are subjected to lattice energy minimization. The resulting optimized cells can then be used as initial structures in subsequent NσT-MD simulations to determine stability and energy ranking relative to the experimental structure.

Other MD simulations can be used to explore dynamical details of processes related to or that precede initiation to detonation. These include NPT/NσT-MD simulations over wide intervals of pressure and temperature that span the solid and liquid regions of the phase diagram. MD simulations to predict the pressure-dependent melting points are particularly useful in assessing the predictive capability of a chosen force field model. Finally, the shock Hugoniot of the materials can be evaluated using NVE-, NPT-, or NVT-MD simulations and compared to available experimental data. We note that some of these models are limited in their capacity for prediction, particularly if they are developed for rigid-molecule simulations only, or if they do not include chemical reactivity.

Interesting questions in the case of chemically reactive potentials are the extent to which many-body effects must be included and how to predict, and account for, differences in "zero point energy" between reactants, transition states, and products. Reactants typically include a wide range of vibron and phonon frequencies, whereas typical detonation or deflagration products such as N_2, CO, CO_2, and H_2O have comparatively high vibrational frequencies and consequently significantly different zero-point energies from the reactants. These differences will not be reflected in "bare" classical potentials.

7.8 Interaction Potentials Used in MD Simulations of Energetic Materials

7.8.1 Reactive Models

To our knowledge, only one reactive force field has been developed and used in published MD simulations for condensed phase CHNO EMs; it is known as ReaxFF. We will provide only a brief description of ReaxFF [106], since the majority of its applications in EM research have, to our knowledge, been limited to studying thermal and shock-initiated chemistry [107–109]. While these applications are outside the scope of this chapter, the potential importance of ReaxFF in advancing realistic MD simulations of EMs merits discussion and therefore a short review of published simulations using this force field model for reactive chemistry will be given. This will be followed by more extensive descriptions of nonreactive empirical force fields that have been used within the past few years in MD simulations for predicting properties and behavior of condensed phase EMs over wide intervals of temperature and pressure.

The development of ReaxFF is an ambitious attempt to construct a transferable force field for use in molecular simulations of chemically reacting systems. An overall developmental goal for ReaxFF is to adequately describe any chemical system composed of any element in the periodic table. Another goal of its development is to rely solely on QM information in its parameterization. The philosophical advantage to this approach is that all information used in the parameterization corresponds to a similar, well-defined level of theory uncomplicated by inclusion of experimental information that may have ill-defined accuracies and systematic experimental errors. It is a bond-order-dependent method, similar in spirit to the Tersoff [99] and REBO [98] formalisms mentioned earlier. The force field partitions the total system energy into various valence and nonbonding components, with only the valence terms being bond-order-dependent. The bond orders are calculated for every atom pair within a simulation, and are assumed to be functions only of the interatomic distances. Energetic penalties are imposed for over- and under-coordination on the atoms. The bond-order-independent nonbonding van der Waals and Coulombic interaction terms are calculated for every atom pair and are shielded to eliminate excessive repulsive forces between atom pairs separated by short distances. Environment-dependent atomic charges used in the Coulombic expression are determined using the Electron Equilibration Method (EEM) [110]. The parameterization of ReaxFF is accomplished by fitting to a training set of QM data composed of pertinent structural and reaction path information that characterizes "the atomic interactions under various environments (likely and unlikely [high energy]) each atom can encounter" [108]. To date, ReaxFF has been parameterized for use in molecular simulations of hydrocarbons [106], nitramines [107, 108], peroxide-based explosives [109], silicon/silicon oxides [111, 112], aluminum/aluminum oxides [113], transition metals [114, 115], metal oxide catalysts [116], B-N-H systems [117], alkali metal [118] and alkaline earth systems [119], and ferroelectrics [120].

At this time, ReaxFF is an evolving force field, as new elements are continually being added to the potential energy description. Modifications to the form of ReaxFF are also sometimes necessary to remedy deficiencies in its descriptions of chemical reactions or material properties. Whenever new elements are included, QM information is generated for each element in a variety of different environments and added to the training set for reparameterization. To our knowledge, the majority (if not all) of the condensed phase QM information used in the parameterization of ReaxFF consists of DFT predictions. As discussed above, however, this QM treatment does not properly describe dispersion interactions; thus, parameterizing ReaxFF using only DFT information will clearly diminish the predictive capability of the force field at certain thermodynamic conditions, notably including those of EMs near room temperature and pressure. The subsequent effects of using DFT calculations in the parameterization of ReaxFF is apparent from the results of a study in which ReaxFF was used to generate EOS information for PETN I (Pentaerythrityl tetranitrate, polymorph I) [121]. In this study, EOS information for PETN I for pressures ranging from 0 to 50 GPa at $T = 0$ K was generated using ReaxFF and compared with DFT predictions and experiment. The ReaxFF results were in good agreement with DFT predictions over this pressure interval. Also, ReaxFF was in better agreement with experiment (deviation of cell volume = 5.4%) than all DFT predictions at 0 GPa. This is notable, since information generated using the linear combination of atomic orbitals (LCAO) DFT code SeqQuest [122] was used in the parameterization of ReaxFF, and predictions using this code and the Generalized Gradient Approximation (GGA) [123–126] functional Perdew-Burke-Ernzerhof (PBE) [127] produced a substantially larger deviation from experiment (deviation of cell volume = 10%). It is assumed that future parameterizations of ReaxFF using non-DFT QM information might better represent CHNO materials at low degrees of compression.

ReaxFF was first applied to a conventional CHNO EM in MD simulations of shock-initiated chemistry in the cyclic nitramine RDX [107]; subsequent MD simulations were performed to examine thermal-induced chemistry in the same system [108]. It was later extended to describe another explosive, triacetonetriperioxide (TATP), by inclusion of QM predictions of its unimolecular decomposition [109]. Since then, ReaxFF has been used in MD simulations of other EMs, including β-HMX (octahydro-1,3,5,7-tetranitro-s-tetrazocine), TATB (triaminotrinitrobenzene) [128], PETN [121], nitromethane [129], and RDX-based systems (Estane™/RDX and nano-Al/RDX) [130]. To our knowledge all of the applications of ReaxFF to energetic materials have been used to explore chemical events associated with either cook-off or shock initiation, with the exception of the aforementioned EOS study of PETN I [121].

7.8.2 Nonreactive Models

Most EOS information generated through MD simulation has resulted from the use of nonreactive interaction potentials far less sophisticated than ReaxFF. Usually, these nonreactive interaction potentials focus on accurately depicting the intermolecular interactions within the crystal and, in some cases, energetic differences associated with intramolecular conformational transitions. Most models assume pair-additive terms of simple functions (such as Lennard-Jones or exponential-six) to describe the dispersion interactions, and couple these with a simple Coulombic description of the electrostatic interactions. One exception to this is a very recent formulation and parameterization of a polarizable, many-body force field for pentaerythritol tetranitrate (PETN) [131]. Typically, partial charges centered on atoms in the molecular crystal are determined using quantum mechanical calculations for the isolated molecules or by adjusting these, along with the other intermolecular terms, to reproduce experimental information. Intramolecular interactions can be described by functions that represent various molecular vibrational motions within a molecule such as bond stretching, angle bending, torsions, and wagging or scissor motions. These functions can be parameterized to empirical or ab initio information. Fortunately, intramolecular motion is well described by DFT, and several systems have used both DFT and non-DFT QM treatments in parameterizing these portions of the force field, as detailed below. Also, for single molecules or dimers, accurate correlated electronic structure methods can be used to "benchmark" particular intermolecular interactions.

One of the simplest interaction potentials used to generate EOS information was developed by Sorescu, Rice, and Thompson [85] and parameterized using empirical data for α-RDX at room temperature and pressure. This model is composed of pair-additive terms that describe interactions between atoms of neighboring molecules. The model does not include terms that describe intramolecular interactions and thus does not allow for molecular deformation (or reaction) during an MD simulation; that is, it is a rigid-molecule force field. It was developed for MD simulations in pressure regimes for which molecular deformation would be minimal and has been assessed to determine its ability to describe various explosives in simulations at moderate pressures and temperatures [92]. An extensive series of studies has been performed to evaluate the degree of transferability of this method to other chemical systems [132–136], including rigid-molecule NPT-MD simulations of polymorphs of 2,4,6,8,10,12-Hexanitrohexaazaisowurtzitane (HNIW, or CL-20) [132], HMX [133], and dimethylnitramine (DMNA) [137]. Results from studies on CL-20 and HMX indicate that the force field correctly predicts the order of thermodynamic stability of the different phases of these cyclic nitramines; cell parameters are in agreement with experiment to within a few percent, with little deviation of translational or rotational descriptors from experimentally determined values. For the DMNA simulations [137], crystal densities were within 0.2% of experiment, and the predicted bulk modulus was 5.7% lower than the experiment. However, simulations of the melting of DMNA using this model produced a melting point that was $\sim$15% higher than the experiment. (It is unreasonable to expect a rigid-molecule approximation to a highly flexible system – in this case weakly hindered methyl and nitro rotations are both present – to be accurate for predictions of melting, so this result should not be surprising.) The most rigorous test imposed on this model force field is one in which a total of 174 CHNO crystals were subjected to ab initio crystal prediction [136]. The systems consisted of nitramine, nitroaliphatic, nitroaromatic, and nitrate ester molecules and included a variety of acyclic, monocyclic, and polycyclic/caged species. Eighty-five percent (148 out of 174) of the crystals that were subjected to this methodology produced crystallographic parameters and molecular configurations that matched those of the experimental counterpart. The predicted densities were, on average, higher than the experiment by less than 3%. However, the lattice energy minimizations used in the calculations represent a 0 K result, and do not capture thermal expansion effects due to finite temperature. Similar statistics regarding identification of crystal and structural parameters were produced for each of the chemical CHNO classes studied, namely nitramines, nitroaliphatic, nitroaromatic, and nitrate esters. The uniformity of the statistics for the four categories of CHNO crystals that was identified by the ab initio crystal structure prediction procedure indicates that this interaction potential is transferable among these classes of compounds within the limitations of its ended range of applications.

Another study was performed to determine the room-temperature pressure regime for which the rigid molecule approximation was valid for four EMs (HMX, RDX, CL-20 and PETN) [92]. These molecules are representative of EMs with varying degrees of molecular flexibility. CL-20, a polycyclic caged nitramine, is expected to be the most rigid, followed by the cyclic nitramines RDX and HMX. PETN is an acyclic, "star-like" molecule and is expected to be the most flexible. Molecular Packing (MP) and NPT-MD simulations were performed using the SRT potential function over the interval of pressures for which experimental information was available for each of these systems. Lattice parameters for the RDX, HMX, and CL-20 crystals were in good agreement with experimental values over the entire interval of pressures investigated experimentally. The highest pressures for which experimental data were available for CL-20 and RDX at the time of this study are 2.5 and 3.95 GPa, respectively (an undetermined phase transition occurs in RDX above 3.95 GPa), so it is not surprising that the rigid-molecule approximation is valid within these limits. However, lattice parameters for HMX were in good agreement with the experiment up to the highest pressure (7.47 GPa), indicating that, for simulations at moderate pressures, the rigid molecule approximation is probably valid for similar systems. The rigid-molecule approximation failed for PETN at higher pressures; predicted crystal parameters were in unacceptable agreement with the experiment for pressures higher than a few GPa.

Sorescu, Rice, and Thompson augmented the rigid-molecule form to allow molecular deformations through the addition of intramolecular terms, the parameters of which were determined by fitting to QM (DFT) energies and energy derivatives of an isolated nitromethane molecule [138]. NPT-MD simulations of nitromethane using the augmented SRT potential over the entire temperature existence interval for both solid [138] and liquid [139] phases and over large pressure intervals predicted static and dynamic properties that were in outstanding general agreement with experimental measurements. These included the prediction of an experimentally observed 45° change in methyl group orientation in the high-pressure regime relative to the low-temperature crystal configuration. The model was subsequently used in other MD simulations to predict melting [39, 41, 42, 140] and to explore the structure of the liquid [141]; results for all calculations were in good agreement with experimental values where the latter are available.

Additional flexible force fields that incorporate the SRT intermolecular terms have been used in MD simulations of 1,3,3-Trinitroazetidine (TNAZ) [40] and RDX [142, 143]. For the systems studied by Agrawal et al. [40, 142], intramolecular terms from the generalized AMBER force field [144, 145] were used without modification rather than with the kind of system-specific parameterization that was used for the nitromethane potential described in the preceding paragraph. Predicted densities for both systems were $\sim$10% smaller than experiment, and the crystallographic parameters (unit cell dimensions, space group symmetry, atomic positions, molecular configuration) predicted for TNAZ were in reasonable agreement with experimental values. However, the RDX simulations produced significant deviations from experiment for the atomic positions within the unit cell, and the space group symmetry was not preserved during the simulations. Goto et al. also developed a flexible model of RDX by augmenting the SRT rigid molecular potential with standard functions to describe intramolecular motion. These parameters were developed for protein and nucleic acids [146]. In the Goto et al. study [143], QM calculations using DFT and NPT-MD simulations were used to analyze FT-IR spectra of RDX compressed to pressures up to 50 GPa in diamond anvil cell experiments. As is known, RDX undergoes a phase transition near 4 GPa at room temperature [34], and the crystallographic structure of the new phase (denoted as γ-RDX) had not been resolved at the time of the Goto study. In the absence of definitive crystallographic information, Goto et al. [143] used QM calculations of simulated spectra of various RDX conformers, and hypothesized that γ-RDX corresponds to the conformer in which two of the nitro groups are in a pseudo-equatorial (E) position relative to the ring, and the remaining nitro group is in an axial (A) position (this is denoted AEE; α-RDX, the polymorph at room conditions, has the AAE conformation). The authors then used NPT-MD simulations to assess pressure effects on the vibrational frequencies of their model of γ-RDX, something that could not be captured in the QM calculations of the isolated molecules. The authors performed a sequence of NPT-MD simulations ranging from 0 to 50 GPa for a crystal composed of the proposed γ-RDX conformer of RDX; however, details of the calculations are limited, as are descriptions of the predicted crystal structures. Crucial details of the simulations and results (e.g., the size of the simulation cell, the length of the trajectory, whether space group symmetry was conserved, and the resultant molecular structure) are not provided; the authors only report IR frequencies for the AEE conformer at 10, 30, and 50 GPa. Additionally, they do not give the details of the geometry optimization that appears to have been performed at the end of each NPT-MD simulation; it is unknown whether optimization was performed for the full system, or whether they optimized a single molecule within the field of the crystal to generate the vibrational spectra. A similar and earlier study on the pressure dependence of vibrational frequencies of PETN was performed by one of the authors [147] using a very similar flexible interaction potential; the intermolecular interaction was the same as SRT except the parameters for the N–H intermolecular interaction were refitted. As for the RDX study, computational details were sparse and it is not known how well this interaction potential would reproduce the EOS for PETN.

Other system-specific flexible nonreactive force fields have been developed for nitromethane [148], RDX [149], DMNA [150], energetic dinitro compounds [151], and 3-nitro-1,2,4-triazol-5-one (NTO) [152], all of which extensively relied on QM information in their respective parameterizations of intramolecular terms. The Alper et al. [148] PES for nitromethane was recently used in NPT- and NVT-MD simulations of nitromethane at high temperatures and pressures [153]; the study also included calculations of the shock Hugoniot using this model. The volumes predicted from a series of NPT-MD simulations at room temperature were in reasonable agreement with experiment only at low pressures; significant deviation between calculated volumes and experiment was found for pressures greater than 5 GPa. The authors suggest that the source of the deviation might be due to the use of a Lennard-Jones function that over-describes repulsive forces for the compressed system. This overestimation of repulsive forces would, as demonstrated in this study, also result in significant disagreement of the predicted shock Hugoniot from experimental values.

Temperature-dependent crystallographic parameters for RDX were predicted in NPT-MD simulations at 1 atm and temperatures ranging from 0 to 650 K using an RDX potential developed by Boyd et al. [149]. Beyond this temperature, the crystal melted. While this is substantially higher than the measured melting point of RDX, it is well known that the melting temperature predicted from MD simulations of perfect crystals is an upper bound to the true thermodynamic melting point (see Ref. [42] and references therein). The authors report that the bulk modulus predicted by this model is substantially larger than the experimental value, but note that this property was not used in the parameterization. This, however, suggests that the simulated pressure dependence of the crystal parameters might not be accurately represented using this model.

Hiyoshi et al. [152] used NPT-MD simulations and a CHARMM-type force field [154] to explore pressure effects on the α-polymorph of NTO for pressures up to 50 GPa. Parameters for the intramolecular terms were determined using information generated by DFT calculations for the isolated molecule. Parameters for the nonbonded interactions are those used in CHARMM29, and partial atomic charges used in the Coulombic term were calculated for an isolated NTO molecule at the optimized geometry, except for those atoms involved in hydrogen bonding in the crystal. The latter charges were adjusted iteratively using molecular dynamics until the correct ribbon structure of α-NTO was maintained. The experiments indicate some interesting pressure effects on the crystal in that, overall, Raman bands blue shift with pressure whereas bands corresponding to hydrogen bonds (i.e., carbonyl and amino groups) show a red shift as pressure is increased. The authors used MD simulations to generate radial distribution functions for carbonyl – amino pairs involved in hydrogen-bonding and power spectra of correlation functions of the carbonyl bond length in NTO as functions of pressure; the results of the simulations are consistent with experimental results. As in the Goto et al. study [143] discussed earlier, Hyoshi et al. did not provide enough details of the results of the simulations to establish whether this model reproduces the ambient condition crystal structure; nor did they provide crystallographic information about the system under compression.

Smith and coworkers have developed flexible unreactive force fields for dimethylnitramine (DMNA) [150], HMX [89], bis-2,2-dinitropropyl formal/acetal (BDNPF/A) [151], and PETN [131]. A principal point of distinction of these force fields from others in the same class (fully flexible but unreactive) is that specific emphasis was placed on accurately reproducing the energetics of dihedral displacements, exocyclic rotations, and out-of-plane deformation motions based on comparison to predictions of the highest level electronic structure methods practically feasible at the respective times they were developed. Generally, the size of the target molecules was too large to allow extensive correlated electronic structure calculations; thus, carefully chosen model compounds were considered that collectively captured the most important chemical backbone linkages for the molecule of interest, and these were subjected to intensive study. Simple point charges were adjusted to optimally reproduce the molecular electrostatic potential and molecular dipole moment around isolated gas-phase molecules; a single set of charges was used for a given molecule irrespective of crystal polymorph, and all charges that would be chemically indistinguishable in liquid or gas-phase environments were constrained to have equal values. It was found empirically that polarization effects in the condensed phase could be approximated by scaling partial atomic charges by a factor of 1.25 relative to the calculated gas-phase values. Finally, most nonbonded intramolecular and intermolecular pair interactions were taken from the literature; however, in cases where existing parameters were found to be deficient, correlated electronic structure calculations were used to provide energy profiles for the relevant atom pairs.

A flexible molecule force field whose parameters were derived largely from quantum mechanical calculations was developed by Smith et al. [150] for use in simulations of DMNA in the liquid state. Various physical properties of liquid DMNA calculated from NPT- and NVT-MD are all in very good agreement with experiment. These include solubility parameters, pressure–volume–temperature properties, and correlation time and activation energy for molecular reorientation Although the application of the DMNA force field developed by Smith et al. was first used in MD simulations of the liquid, they subsequently used the model in isothermal-isochoric (NVT)-MD simulations of the DMNA crystal at 295 K [155]; cell shape and volume were sampled during the simulation by interspersing sequences of NsT Monte Carlo moves throughout the trajectories. In this study, the cell edge lengths were found to be within 1% of experimental values, and the predicted crystal density deviated from the experimental value by 4%. Additionally, the shape of the non-orthorhombic unit cell was maintained during the simulation, with the simulated cell angles deviating from 90° by no more than 0.6°. NPT-MD simulations of the melting of the crystal produced a melting point of $\sim$330 K, in outstanding agreement with the experimental values (331 K). Further applications of the model in MD simulations were used to explore atomistic details of mechanisms of melting in both void-containing and perfect crystals of DMNA [37].

The Smith potential developed for DMNA was extended to HMX by Smith and Bharadwaj [89]. In this case, HMX gas-phase conformers and the model compound 1,3-dimethyl-1,3-dinitro methyldiamine (DDMD) were used in conjunction with results for nonbonded parameters taken from the DMNA force field discussed above to determine structures and energetics at the MP2/6–311G**/B3LYP/6-311G** and B3LYP/6-311G** levels of theory for DDMD and HMX, respectively. Among the properties computed during validation of the potential [156] were crystal structures for the three pure crystal polymorphs (β, α, and δ corresponding to monoclinic, orthorhombic, and hexagonal symmetry classes, respectively), linear thermal expansion and compression coefficients, heats of sublimation, and anisotropic sound speeds. A number of additional properties have been reported, including: isothermal second-order elastic coefficients and derived isotropic moduli of all three polymorphs [157]; and temperature dependent self-diffusion coefficients, shear viscosity and thermal conductivity of the liquids at atmospheric pressure [158, 159]. Many of the properties computed for HMX using the Smith potential have been used directly by others for the development and parameterization of mesoscale and continuum scale simulations [10, 11, 160, 161].

Smith and coworkers used similar approaches to those used for DMNA and HMX to parameterize force fields for the energetic compound bis-2,2-dinitropropyl formal/acetal (BDNPF/A, a eutectic liquid at ambient conditions used as a plasticizer) [151] as well as a poly(ester urethane) known as Estane™ [162]. Along with HMX, these two compounds are the main ingredients of the high-performance military explosive known as PBX-9501. In the case of BDNPF/A, most electronic structure calculations were performed at the MP2/aug-cc-pvdz//B3LYP/aug-cc-pvdz level of theory using 2,2-dinitropropane, dimethoxy dimethyl ether, and 2,2-dinitro-3-methoxypropane as model compounds for parameterization of the force field, with an emphasis on accurately determining the backbone dihedral and pendent nitro group energetics; limited calculations were performed for five low-energy conformations of BDNPF at the B3LYP/aug-cc-pvdz level of theory to confirm the accuracy of the potential. Validation simulations of the liquids were performed for all model compounds, BDNPF, and the eutectic. The density and enthalpy of vaporization for 1,1-dinitroethane were both predicted to be less than 3% lower than experiment at 298 K. The density for 2,2-dinitropropane was predicted to be 2.5% lower than experiment while the enthalpy of vaporization is 1% higher than the reported value. The largest error in the validation set was a systematic $\sim$30% underestimation of the self-diffusion coefficient of 2,2-dinitropropane in the interval 328–358 K. The calculated percent difference in the calculated and inferred value for the enthalpy of vaporization of BDNPF at 400 K is 23%, with the predicted value being the larger of the two. The zero frequency shear viscosity of the BDNPF/A eutectic was calculated in the interval 450–700 K; the resulting values were found to extrapolate accurately to measured values in the neighborhood of 323 K. The calculated density of the eutectic liquid at 298 K was found to be within 3% of the experimental value, while the calculated sound speed, 1323 m/s at 328 K, compares favorably to a measured value of 1297.4–1301.9 m s^{-1} with a resulting percent error of 1.7%.

The validation of the Estane™ force field was performed in a fashion similar to that used for BDNPF/A. The same level of electronic theory, MP2/aug-cc-pvdz//B3LYP/aug-cc-pvdz, used for BDNPF/A was employed. Due to the chemical complexity of Estane™, nine different model compounds were used to determine the torsional backbone energetics: methyl, ethyl, and propyl acetate;

methyl propanoate; methyl butanoate; methyl-N-phenyl carbamate; ethyl-N-phenyl carbamate; biphenyl methane; and hexanedioic acid diethyl ester. Specific polar and hydrogen-bonded intermolecular interactions were determined using relevant geometries of approach for dimethylketone dimers $(DMK)_2$, dimethylketone–dimethylamine complexes (DMK–DMA), and dimethylamine–dimethylamine dimers (DMA–DMA). The resulting optimized force field yielded good agreement with all configurations investigated for both the intramolecular and intermolecular interactions. Validation simulations revealed that the predicted density of biphenylmethane is 2% larger than the measured value. The analogous prediction for hexanedioic acid diethyl ester differs from experiment by less than 1%. Finally, the calculated and measured lattice parameters and unit cell volume at 258 K for crystalline dimethyl-4,4′-methylenebis(phenylcarbamate), MDI, are $(a = 5.280\,\text{Å}, b = 9.831\,\text{Å}, c = 30.625\,\text{Å}, V = 1590\,\text{Å}^3, \alpha = \beta = 90.0, \gamma = 91.0)_{\text{calc}}$ and $(a = 5.157\,\text{Å}, b = 9.800\,\text{Å}, c = 31.472\,\text{Å}, V = 1587\,\text{Å}^3, \alpha = \beta = 90.0, \gamma = 93.9)_{\text{expt}}$, respectively.

Very recently, Borodin et al. [131] described the formulation, parameterization, and validation of fully flexible, polarizable, and nonpolarizable, unreactive force fields for alkyl nitrates, including PETN. The quantum-chemistry-based approach used was generally similar to the preceding force field parameterizations due to Smith and coworkers discussed above. The major point of distinction between previous Smith force fields and the new one for PETN is the inclusion of atomic polarizabilities in a many-body formulation of the potential. Specifically, partial atomic charges and atomic polarizabilities were computed at the MP2/aug-cc-pvdz level of theory for several alkyl nitrate model compounds and PETN in the gas phase. Partial charges for the nonpolarizable force field were determined by fitting the dipole moments and electrostatic potential to values for PETN molecules in the crystal phase obtained from molecular dynamics simulations using the polarizable force field. Explicit inclusion of polarizabilities in the many-body potential obviates the need for empirical scaling of the atomic charges prior to use in the condensed phase, while fitting the partial atomic charges in the two-body force field explicitly to the electrostatic potential of PETN molecules in the crystal environment of the polarizable model provides a more rigorous path for obtaining partial charges for the simpler, and computationally cheaper, two-body form. Both the two-body and many-body parameterizations were demonstrated to yield good agreement with available experiment for crystal lattice parameters and liquid density, phase transformation energies and temperatures (vaporization, sublimation, melting point), room temperature elastic tensor, and isothermal compression curves of ethyl nitrate, butyl nitrate, isopropyl nitrate, and PETN. Given the relatively large computational expense associated with use of the polarizable force field, the authors recommended the nonpolarizable form for routine studies of bulk properties, but noted that the polarizable form should be used for studies of interfacial phenomena.

For the most part, the nonreactive potentials discussed heretofore did not use condensed phase QM information in the parameterization of the intermolecular terms; instead, fitting of such terms relied heavily on empirical information. However, there are a few systems for which more computationally costly non-DFT QM treatments have been used in the development of intermolecular interaction potentials for EMs [163, 164]. The first, a study by Gee et al. [163] for TATB, used second-order Møller–Plesset perturbation theory (MP2) [165]. In this study, nine dimer geometries, selected to represent interactions within and between crystallographic axes in the unit cell of TATB at the ambient state, were calculated at the MP2 level using the 6–31 G(d,p) basis set [166]; this information was used to parameterize the intermolecular interaction portion of the function for TATB. NPT-MD simulations of TATB over a wide range of temperatures and pressures were performed using a simple pair-additive intermolecular interaction potential consisting of Coulombic and van der Waals terms, and intramolecular terms describing bond stretches, bond-angle bends, and torsional motions [163]. Calculated temperature and pressure dependencies of the unit cell parameters were in good agreement with measured values over the range of conditions simulated, as were predictions of librational motion of the nitro groups about the C–N bond. While the ab initio calculations used in the fitting were based on dimers and thus did not account for crystal-field effects that might influence the intermolecular forces in the crystal, this work clearly demonstrates that carefully chosen ab initio calculations of dimers and small molecular clusters in configurations that are representative of those of the crystal can be used to develop a reasonably accurate depiction of condensed phase systems.

Sewell and Bedrov [167] have performed a study of the elastic properties of TATB using a force field based on the model of Gee et al. [163]. Pressures and temperatures in the intervals 0–1 GPa and 198–398 K respectively were studied for pure crystals containing 192 molecules as well as defective crystals containing 5% and 10% randomly distributed molecular vacancies. As expected the results indicate very large mechanical anisotropy, with differences approaching a factor of 10 between predicted values of elastic coefficients (C_{11}, C_{22}) corresponding to strains in the strongly intramolecularly and intermolecularly hydrogen-bonded *a*–*b* crystal plane and C_{33} corresponding to deformation parallel to the *c*-axis which is dominated by weak dispersion interactions. From a practical perspective, the most interesting results of this study were the large differences between "isotropic" bulk and shear moduli calculated using Reuss and Voigt bounds corresponding to limits of uniform stress and uniform strain, respectively. These bounds for both the bulk and shear moduli at room temperature and atmospheric pressure differ by more than a factor of two; thus, it is unclear in light of the complicated (probably complex) stress states present in a plastic-bonded explosive what values should be used in efforts to derive physically based constitutive descriptions for TATB-containing composites.

Recently, a six-dimensional potential energy surface for dimers of RDX was fitted using more than 1,000 ab initio points calculated using the highly accurate Symmetry Adapted Perturbation Theory based on the Kohn-Sham description of the monomers (SAPT-DFT) [164]. This PES was subjected to N*s*T-MD simulations of RDX (assuming the rigid molecule approximation) at room conditions and reproduced the three cell edge lengths to within 0.6%; the predicted density deviated from the measured value by 1.5%. Molecular parameters for each of the eight symmetry-equivalent molecules in the Pbca unit cell were averaged over time and all unit cells in the supercell. The maximum deviation of the center-of-mass fractional positions within the unit cell was 0.0021, and the largest deviation of the three Euler angles describing relative orientation of the molecules was 2.5°. Additionally, the crystal space group symmetry (Pbca) was preserved during the simulation. While the latter study indicates a clear success in developing an ab initio-based force field of an EM, the generation of each SAPT-DFT point required ~250 cpu hours on an IBM P4/1.7 GHz processor. However, as computational resources and algorithms advance, we foresee increased dependence of non-DFT QM calculations in force field developments of EMs.

A few recent MD studies of EM have used general purpose, commercially available force fields [168–173] to describe crystalline energetic materials. Pospíšil et al. [168] used the cff_950 force field [174] in NPT-MD simulations of RDX to study the pressure dependence of its decomposition at room temperature. Simulations were performed over a large range of pressures (46–500 GPa), but detailed results of the decomposition were provided only for 220, 225, 230, 350, 400, and 500 GPa. While these authors provide mechanistic details of molecular decomposition under these conditions, they do not report changes in the crystal structures at these conditions. The COMPASS force field [175] has been used in several studies of both pure energetic materials and plastic-bonded explosives (PBX) [169–176]. Gee et al. used the COMPASS force field and NPT-MD simulations to study adhesion of various fluoropolymers with TATB surfaces [176]. The COMPASS force field was used to describe the amorphous bulk fluoropolymer; it appears that the application of mixing rules using the TATB intermolecular parameters developed in Ref. [163] and the COMPASS force field produced the interatomic potential energy parameters for the fluoropolymer-TATB interactions. In order to validate the COMPASS force field for its use in describing fluoropolymers, a series of MD simulations on the bulk amorphous polymers were carried out to produce glass transition temperatures and thermal expansion coefficients for comparison with experiment. The results were in excellent agreement with experiment indicating that, for these types of polymers, the COMPASS force field provides an adequate description. In addition to its use in describing polymeric binders in the PBX [170, 171, 173], the COMPASS force field was used to characterize the pure explosive component [169–173]. Qiu et al. [169], in a combined QM and MD study of crystalline *trans*-1,4,5,8-Tetranitro-1,4,5,8-tetraazadecalin (TNAD), used the COMPASS force field in NPT-MD simulations to explore the temperature dependence of the crystal structural parameters at 1 atm. A corresponding MD study of the pressure dependence of the structural parameters at room temperature was not reported; however, temperature-dependent mechanical properties, including elastic constants, tensile, bulk and shear moduli, Poisson's ratio and Lame coefficients, were given. Crystallographic parameters at ambient pressure over the temperature range of 5–500 K were calculated using NPT-MD simulations. While cell edge lengths at room temperature are reasonable, two of the three cell angles in the unit cell deviate from experiment by more than 10°. The COMPASS force field was also used to describe interatomic interactions for both polymers and crystalline energetic component in MD studies of polymer bonded explosives (PBX) [170, 171, 173]. In each of these, four fluorine-containing polymers were represented by small oligimeric chains with end groups saturated by hydrogen or fluorine atoms. The EM fill was composed of small clusters of either ε-CL-20, TATB or TNAD with the initial configuration consistent with that of the measured crystal structure at room conditions. Two of the studies [170, 171] provided no information as to the accuracy of the force field in predicting structural or mechanical properties of the pure energetic fill or the mixes; rather, they demonstrate dependencies of the mechanical properties on the explosive blend. Also, in the study of TNAD-based PBX [173], NPT-MD predictions of the crystallographic parameters are given for pure TNAD, although the authors state that they only used NVT-MD simulations in that study. It is notable that the values for the crystal parameters at room temperature 1 atm that are reported in this work [173] differ from the earlier-reported values using the COMPASS force field and NPT-MD [169]; the source of the discrepancy is not given. The authors do not report in any of these studies whether the crystalline space group symmetry is maintained for the pure crystals in NPT-MD simulations using the COMPASS force field. However, in Ref. [173], the authors compare translational and orientational parameters of one of the molecules in the TNAD unit cell with an experimental counterpart; the net translation and rotation of the molecular center of mass is $\sim$0.2 Å and $\sim$17°, respectively. We note that in Fig. 2 of Ref. [171], which displays a snapshot of the equilibrated configuration of TATB mixed with poly(vinylidene difluoride) (PVDF), the TATB crystal cluster has undergone significant molecular rearrangement, and appears to have a substantial degree of disorder. It is not known if the disorder is a result of inadequacies in the COMPASS force field or simply introduction of the polymer chains into the simulation cell. However, a simulation of TATB using the COMPASS force field by Gee et al. [163] resulted in a rapid phase transition into an "unknown" TATB crystal polymorph at 175 K. Since it is not clear that the COMPASS force field can adequately describe crystalline energetic materials, we recommend that the evaluation of its performance in NsT-MD simulations of various representative energetic materials at conditions for which experimental crystallographic information is available, to determine if the space group symmetries are even approximately conserved, and whether the crystallographic parameters and atomic arrangements within the unit cells are consistent with experiment. NPT-MD simulations were also used to calculate the elastic properties of pure TATB using the Material Studios software package. The force field used in the simulations and comparison of predicted results with known experimental information (e.g., crystal structure) are given in Ref. [172].

7.9 Obstacles to Success, Identification of Challenges to Overcome

As we have shown in the previous section, several force fields have been rather successful in their ability to predict properties of EM under various conditions; unfortunately, all of them have limited domains of application. Fundamentally, the limiting factor for most of them is lack of inclusion of chemical reaction; all but one

(ReaxFF) [106] is nonreactive. Because a few of them have been parameterized to and used in MD simulations for specific chemical systems, their transferability to other chemical systems is not known and should be carefully assessed prior to use for new systems. Additionally, many of them have been used in only a few applications and thus their capability to predict properties and processes over a wide range of conditions has not been explored. If a defect in a force field is uncovered (e.g., an unphysical local minimum on the potential energy landscape), or if the model needs to be augmented to expand its predictive capacity, the only remedy is to modify the force field, either by reparameterization or through inclusion of new terms.

Ideally, these complications can be avoided through the use of first principles descriptions of the forces in molecular dynamics simulations. We refer to these types of simulations as ab initio or quantum molecular dynamics (QMD) calculations, and they encompass simulations in which atomic forces are calculated and used directly in the integration of the equations of motion [177, 178]. In practice, however, the computational demands of QMD simulations of condensed phase EMS have required the use of QM theories that are semiempirical or that might not well describe EMs at low pressure (i.e., DFT). Only a few DFT-MD studies [179–181] have been used to explore nonreactive states of EM; a few others focused on reactive events [182–184]. Reed et al. [181] used DFT-QMD to explore dynamical effects that might occur at the shock front of solid nitromethane. In this study, molecular collisions between nearest neighbors that might result from shock wave propagation were simulated, as was crystal shearing along a slip plane. Changes in molecular structure, relative positions, and orientations were monitored, as were changes in the HOMO–LUMO gaps as functions of collision velocities. Regarding the latter, the simulations showed that the band gap is affected by the higher velocity collisions. However, the band gap was not lowered enough to result in a significant population of excited electronic states in the crystal, thus suggesting that electronic excitations do not play an important role in the initiation of nitromethane. However, further simulations to more realistically depict dynamic loading of crystalline nitromethane are required before making any conclusions. Tuckerman and Klein [179] used the Car-Parrinello DFT MD [185] approach to predict the molecular structure of nitromethane in both gas and solid phases and to predict the barrier to methyl group rotation at low temperatures. A subsequent study by Megyes et al. [180] also used Car-Parrinello MD to generate radial distribution functions of liquid nitromethane for comparison with experimental diffraction results. The QMD results from both of these studies are in reasonable agreement with experiment; however, these simulations do not allow for any shape or size change of the simulation cell; thus, the performance of this method in predicting crystallographic parameters is not established. However, other DFT studies of nitromethane in which the unit cell was fully optimized have demonstrated that this quantum mechanical treatment is not suitable for prediction of the crystal parameters of nitromethane, although it does reasonably reproduce the structure of the molecules contained within the cell [102, 103]. Therefore, until corrections can be made to conventional DFT to allow a proper description of dispersion interactions, QMD simulations of EMs using DFT should only be performed for compressed crystals [104]. Numerous efforts are underway to remedy this important deficiency in conventional DFT [186–199], and we fully expect that any successful endeavors along these lines will produce an increase in both QMD simulations of EM and in the generation of condensed phase DFT information for use in parameterizing classical force fields.

Acknowledgments BMR is supported by the United State Army. TDS is supported by the United States Department of Energy under Contract No. DE-AC52-06NA25396 with Los Alamos National Security, LLC.

References

1. J. C. Gump and S. M. Peiris, Isothermal equations of state of beta octahydro-1,3,5,7-tetranitro-1,3,5,7-tetrazocine at high temperatures, *J. Appl. Phys.* **97**, 053513 (2005).
2. B. Olinger, B. Roof, and H. H. Cady, The linear and volume compression of β-HMX and RDX, *Proc. Int. Symp. On High Dynamic Pressures* (Paris, CEA, 1978) p. 3.
3. C.-S. Yoo and H. Cynn, Equation of state, phase transition, decomposition of beta-HMX (octahydro-1,3,5,7-tetranitro-1,3,5,7-tetrazocine) at high pressures, *J. Chem. Phys.* **111**, 10229 (1999).
4. M. R. Baer, C. A. Hall, R. L. Gustavsen, D. E. Hooks, and S. A. Sheffield, Isentropic compression experiments for mesoscale studies of energetic composites, *AIP Conf. Proc.* **845**, 1307 (2006).
5. B. Crouzet, D. Partouche-Sebban, and N. Carion, Temperature measurements in shocked nitromethane, *AIP Conf. Proc.* **706**, 1253 (2004).
6. S. G. Bardenhagen, A. D. Brydon, T. O. Williams, and C. Collet, Coupling grain scale and bulk mechanical response for PBXs using numerical simulations of real microstructures, *AIP Conf. Proc.* **845**, 479 (2006).
7. A. D. Brydon, S. G. Bardenhagen, E. A. Miller, and G. T. Seidler, Simulation of the densification of real open-celled foam microstructures, *J. Mech. Phys. Solids* **53**, 2638 (2005).
8. C. A. Bolme, S. D. McGrane, D. S. Moore, and D. J. Funk, Single shot measurements of laser driven shock waves using ultrafast dynamic ellipsometry, *J. Appl. Phys.* **102**, 033513 (2007).
9. For instance: T. R. Gibbs and A. Popolato, *LASL Explosive Property Data* (University of CA, Berkeley, 1980).
10. T. D. Sewell and R. Menikoff, Complete equation of state for β-HMX and implications for initiation, *AIP Conf. Proc.* **706**, 157 (2004).
11. G. A. Ruderman, D. S. Stewart, and J.-I. Yoh, A thermomechanical model for energetic materials with phase transformations, *SIAM J. Appl. Math.* **63**, 510 (2002).
12. R. Menikoff and M. S. Shaw, Review of the Forest Fire Model, *Combust. Theor. Mod.* **12**, 569 (2008).
13. W. G. Proud, M. W. Greenaway, C. R. Siviour, H. Czerski, and J. E. Field, Characterizing the response of energetic materials and polymer-bonded explosives (PBXs) to high-rate loading, *Mat. Res. Soc. Symp. Proc.* **896**, 225 (2006).
14. S. Lecume, C. Boutry, and C. Spyckerelle, Structure of nitramines crystal defects relation with shock sensitivity, *Energetic Materials: Structure and Properties, 35th International Conference of ICT*, Karlsruhe, FRG, p. 2-1 (2004).
15. R. Menikoff, Pore collapse and hot spots in HMX, *AIP Conf. Proc.* **706**, 393 (2004).
16. W. M. Trott, M. R. Baer, J. N. Castaneda, L. C. Chhabildas, and J. R. Asay, Investigation of the mesoscopic scale response of low-density pressings of granular sugar under impact, *J. Appl. Phys.* **101**, 024917 (2007).

17. F. P. Bowden and Y. D. Yoffe, *Initiation and growth of explosion in liquids and solids* (Cambridge University Press, Cambridge, 1952).
18. L. Tran and H. S. Udaykumar, Simulation of void collapse in an energetic material, Part 1: Inert case, *J. Propul. Pow.* **22**, 947 (2006); *ibid.*, Simulation of void collapse in an energetic material, Part 2: Reactive case, **22**, 959 (2006).
19. R. Menikoff, Detonation waves in PBX 9501, *Combust. Theor. Mod.* **10**, 1003 (2006).
20. R. Menikoff, Comparison of constitutive models for plastic-bonded explosives, *Combust. Theor. Mod.* **12**, 73 (2007).
21. D. C. Sorescu, B. M. Rice, and D. L. Thompson, Molecular Dynamics Simulations of Energetic Materials, in P. Politzer and J. S. Murray (Eds.) *Energetic Materials: Part 1. Decomposition, Crystal and Molecular Properties (Theoretical and Computational Chemistry)* (Elsevier Science, Amsterdam, 2003) pp. 125–184.
22. D. A. McQuarrie, *Statistical Mechanics* (Harper & Row, New York, 1976).
23. J.-B. Maillet, M. Mareschal, L. Soulard, R. Ravelo, P. S. Lomdahl, T. C. Germann, and B. L. Holian, Uniaxial Hugoniostat: A method for atomistic simulations of shocked materials, *Phys. Rev. E* **63**, 016121 (2001).
24. R. Ravelo, B. L. Holian, T. C. Germann, and P. S. Lomdahl, Constant-stress Hugoniostat method for following the dynamical evolution of shocked matter, *Phys. Rev. B* **70**, 014103 (2004).
25. J. M. D. Lane and M. Marder, Numerical method for shock front Hugoniot states, *AIP Conf. Proc.* **845**, 331 (2006).
26. E. J. Reed, L. E. Fried, W. D. Henshaw, and C. M. Tarver, Analysis of simulation technique for steady shock waves in materials with analytical equations of state, *Phys. Rev. E* **74**, 056706 (2006).
27. E. J. Reed, L. E. Fried, and J. D. Joannopoulos, A method for tractable dynamical studies of single and double shock compression, *Phys. Rev. Lett.* **90**, 235503 (2003).
28. R. Menikoff and T. D. Sewell, Constituent properties of HMX needed for mesoscale simulations, *Combust. Theor. Mod.* **6**, 103 (2002).
29. A. Strachan and B. L. Holian, Energy exchange between mesoparticles and their internal degrees of freedom, *Phys. Rev. Lett.* **94**, 014301 (2005).
30. Y. Guo, D. L. Thompson, and T. D. Sewell, Analysis of the zero-point energy problem in classical trajectory simulations, *J. Chem. Phys.* **104**, 576 (1996).
31. Z. A. Dreger and Y. M. Gupta, High pressure Raman spectroscopy of single crystals of hexahydro-1,3,5-trinitro-1,3,5-triazine (RDX), *J. Phys. Chem. B* **111**, 3893 (2007).
32. T. R. Park, Z. A. Dreger, and Y. M. Gupta, Raman spectroscopy of pentaerythritol single crystals under high pressures, *J. Phys. Chem. B* **108**, 3174 (2004).
33. J. A. Ciezak, T. A. Jenkins, and Z. X. Liu, Propellants Explosives Pyrotechnics **32**, 472 (2007).
34. P. J. Miller, S. Block, and G. J. Piermarini, Effects of pressure on the thermal-decomposition kinetics, chemical-reactivity and phase-behavior of RDX, *Combust. Flame* **83**, 174 (1991).
35. G. J. Piermarini, S. Block, and P. J. Miller, Effects of pressure on the thermal-decomposition kinetics and chemical-reactivity of nitromethane, *J. Phys. Chem.* **93**, 457 (1989).
36. G. J. Piermarini, S. Block, and P. J. Miller, Effects of pressure and temperature on the thermal-decomposition rate and reaction-mechanism of beta-octahydro-1,3,5,7-tetranitro-1,3,5,7-tetrazocine, *J. Phys. Chem.* **91**, 3872 (1987).
37. L. Zheng, B. M. Rice, and D. L. Thompson, Molecular dynamics simulations of the melting mechanisms of perfect and imperfect crystals of dimethylnitramine, *J. Phys. Chem. B* **111**, 2891 (2007).
38. L. Zheng and D. L. Thompson, Molecular dynamics simulations of melting of perfect crystalline hexahydro-1,3,5-trinitro-1,3,5-s-triazine, *J. Chem. Phys.* **125**, 084505 (2006).
39. A. Siavosh-Haghighi and D. L. Thompson, Molecular dynamics simulations of surface-initiated melting of nitromethane, *J. Chem. Phys.* **125**, 184711 (2006).
40. P. M. Agrawal, B. M. Rice, L. Zheng, G. F. Velardez, and D. L. Thompson, Molecular dynamics simulations of hexahydro-1,3,5-trinitro-1,3,5-s-triazine (RDX) using a combined Sorescu-Rice-Thompson AMBER force field, *J. Phys. Chem. B* **110**, 5721 (2006).

41. L. Zheng, S. N. Luo, and D. L. Thompson, Molecular dynamics simulations of melting and the glass transition of nitromethane, *J. Chem. Phys.* **124**, 154504 (2006).
42. P. M. Agrawal, B. M. Rice, and D. L. Thompson, Molecular dynamics study of the melting of nitromethane, *J. Chem. Phys.* **119**, 9617 (2003).
43. D. Cremer and J. A. Pople, General definition of ring puckering coordinates, *J. Am. Chem. Soc.* **97**, 1354 (1975).
44. C. B. Barber, D. P. Dobkin, H. T. Huhdanpaa, Quickhull algorithm for convex hulls, *ACM Trans. Math. Softw.* **22**, 469 (1996).
45. M. J. Cawkwell, T. D. Sewell, K. J. Ramos, and D. E. Hooks, Shock-induced anomalous plastic hardening in an energetic molecular crystal (*Phys. Rev. B*, submitted).
46. K. Kadau, T. C. Germann, and P. S. Lomdahl, Molecular dynamics comes of age: 320 billion atom simulation on BlueGene/L, *Int. J. Mod. Phys. C* **17**, 1755 (2006).
47. K. Kadau, C. Rosenblatt, J. L. Barber, T. C. Germann, Z. B. Huang, P. Carles, and B. J. Alder, The importance of fluctuations in fluid mixing, *Proc. Nat. Acad. Sci. USA* **104**, 7741 (2007).
48. D. Frenkel and B. Smit, *Understanding Molecular Simulation* (Academic Press, San Diego, 2002).
49. A. Gavezzotti, Are crystal-structures predictable?, *Accounts Chem. Res.* **27**, 309 (1994).
50. P. Verwer and F. J. J. Leusen, Computer simulation to predict possible crystal polymorphs, in *Reviews in Computational Chemistry*, K. B. Lipkowitz and D. B. Boyd (Eds.) (Wiley-VCH, New York, 1998), p. 327.
51. R. J. Gdanitz, Ab initio prediction of molecular crystal structures, *Curr. Opn. Solid State Mater. Sci.* **3**, 414 (1998).
52. A. Gavezzotti, The chemistry of intermolecular bonding: Organic crystals, their structures and transformations. *Synlett* **2**, 201 (2002).
53. T. Beyer, T. Lewis, and S. L. Price, Which organic crystal structures are predictable by lattice energy minimisation?, *Cryst. Eng. Comm.* **44**, 1 (2001).
54. J. P. M. Lommerse, W. D. S. Motherwell, H. L. Ammon, J. D. Dunitz, A. Gavezzotti, D. W. M. Hofmann, F. J. J. Leusen, W. T. M. Mooij, S. L. Price, B. Schweizer, M. U. Schmidt, B. P. van Eijck, P. Verwer, and D. E. Williams, A test of crystal structure prediction of small organic molecules, *Acta Cryst. B* **56**, 697 (2002).
55. W. D. S. Motherwell, H. L. Ammon, J. D. Dunitz, A. Dzyabchenko, P. Erk, A. Gavezzotti, D. W. M. Hofmann, F. J. J. Leusen, J. P. M. Lommerse, W. T. M. Mooij, S. L. Price, H. Scheraga, B. Schweizer, M. U. Schmidt, B. P. van Eijck, P. Verwer, and D. E. Williams, Crystal structure prediction of small organic molecules: a second blind test, *Acta Cryst. B* **58**, 647 (2002).
56. W. T. M. Mooij, B. P. van Eijck, S. L. Price, P. Verwer, and J. Kroon, Crystal structure predictions for acetic acid, *J. Comput. Chem.* **19**, 459 (1998).
57. D. W. M. Hofmann and T. Lengauer, Crystal structure prediction based on statistical potentials, *J. Mol. Model.* **4**, 132 (1998).
58. A. Gavezzotti, Generation of possible crystal-structures from the molecular-structure for low-polarity organic-compounds, *J. Am. Chem. Soc.* **113**, 4622 (1991).
59. H. R. Karfunkel, F. J. Leusen, and R. J. Gdanitz, The ab initio prediction of yet unknown molecular crystal structures by solving the crystal packing problem, *J. Comput.-Aided Mater. Des.* **1**, 177 (1993).
60. D. J. Willock, S. L. Price, M. Leslie, and C. R. A. Catlow, The relaxation of molecular-crystal structures using a distributed multipole electrostatic model, *J. Comput. Chem.* **16**, 628 (1995).
61. D. E. Williams, Ab initio molecular packing analysis, *Acta Cryst. A* **52**, 326 (1996).
62. A. V. Dzyabchenko, T. S. Pivina, and E. A. Arnautova, Prediction of structure and density for organic nitramines, *J. Mol Struct.* **378**, 67 (1996).
63. M. U. Schmidt and U. Englert, Prediction of crystal structures, *J. Chem. Soc. Dalton Trans.* **10**, 2077 (1996).
64. A. M. Chaka, R. Zaniewski, W. Youngs, C. Tessier, and G. Klopman, Predicting the crystal structure of organic molecular materials, *Acta Cryst. B* **52**, 165 (1996).

65. D. W. M. Hofmann and T. Lengauer, A discrete algorithm for crystal structure prediction of organic molecules, *Acta Cryst. A* **53**, 225 (1997).
66. G. M. Day, W. D. S. Motherwell, H. L. Ammon, S. X. M. Boerrigter, R. G. Della Valle, E. Venuti, A. Dzyabchenko, J. D. Dunitz, B. Schweizer, B. P. van Eijck, P. Erk, J. C. Facelli, V. E. Bazterra, M. B. Ferraro, D. W. M. Hofmann, F. J. J. Leusen, C. Liang, C. C. Pantelides, P. G. Karamertzanis, S. L. Price, T. C. Lewis, H. Nowell, A. Torrisi, H. A. Scheraga, Y. A. Arnautova, M. U. Schmidt, and P. Verwer, A third blind test of crystal structure prediction, *Acta Cryst. B* **61**, 511 (2005).
67. P. Erk, Crystal engineering: from molecules and crystals to materials, *NATO Sci. Ser. C* **538**, 143 (1999).
68. B. P. van Eijck and J. Kroon, UPACK program package for crystal structure prediction: Force fields and crystal structure generation for small carbohydrate molecules, *J. Comput. Chem.* **20**, 799 (1999).
69. A. V. Dzyabchenko, V. Agafonov, and V. A. Davydov, A theoretical study of the pressure-induced dimerization of C-60 fullerene, *J. Phys. Chem. A* **103**, 2812 (1999).
70. W. T. M. Mooij, F. B. van Duijneveldt, J. G. C. M. van Duijneveldt-van de Rijdt, and B. P. van Eijck, Transferable ab initio intermolecular potentials. 1. Derivation from methanol dimer and trimer calculations, *J. Phys. Chem. A* **103**, 9872 (1999).
71. W. D. S. Motherwell, Crystal structure prediction and the Cambridge Structural Database, *Nova Acta Leopoldina* **79**, 89 (1999).
72. B. P. van Eijck and J. Kroon, Structure predictions allowing more than one molecule in the asymmetric unit, *Acta Cryst. B* **56**, 535 (2000).
73. T. Beyer and S. L. Price, Dimer or catemer? Low-energy crystal packings for small carboxylic acids, *J. Phys. Chem. B* **104**, 2647 (2000).
74. T. Beyer, G. M. Day, and S. L. Price, The prediction, morphology, and mechanical properties of the polymorphs of paracetamol, *J. Am. Chem. Soc.* **123**, 5086 (2001).
75. J. Pillardy, Y. A. Arnautova, C. Czaplewski, K. D. Gibson, and H. A. Scheraga, Conformation-family Monte Carlo: A new method for crystal structure prediction, *Proc. Nat. Acad. Sci. USA* **98**, 12351 (2001).
76. C. Mellot-Draznieks, S. Girard, G. Ferey, J. C. Schon, Z. Cancarevic, and M. Jansen, Computational design and prediction of interesting not-yet-synthesized structures of inorganic materials by using building unit concepts, *Chem. Eur. J.* **8**, 4103 (2002).
77. E. Pidcock and W. D. S. Motherwell, A new model of crystal packing, *Chem. Commun.* **24**, 3028 (2003).
78. E. Pidcock and W. D. S. Motherwell, A novel description of the crystal packing of molecules, *Cryst. Growth Des.* **4**, 611 (2004).
79. J. R. Holden, Z. Y. Du, and H. L. Ammon, Prediction of possible crystal-structures for C-containing, H-containing, N-containing, O-containing and F-containing organic-compounds, *J. Comput. Chem.* **14**, 422 (1993).
80. D. Q. Gao and D. E. Williams, Molecular packing groups and ab initio crystal-structure prediction, *Acta Cryst. A* **55**, 621 (1999).
81. A. D. Mighell, V. L. Himes, and J. R. Rodgers, Space-group frequencies for organic-compounds, *Acta Cryst. A* **39**, 737 (1983).
82. For example: J. A. Moriarty, L. X. Benedict, J. N. Glosli, R. Q. Hood, D. A. Orlikowski, M. V. Patel, P. Soderlind, F. H. Streitz, M. J. Tang, and L. H. Yang, Robust quantum-based interatomic potentials for multiscale modeling in transition metals, *J. Mat. Res.* **21**, 563 (2006).
83. For example: A. J. Pertsin and A. I. Kitaigorodskii, *The Atom-Atom Potential Method: Applications to Organic Molecular Solids. Springer Series in Chemical Physics 43*. (Springer, Heidelberg, 1987).
84. A. M. N. Niklasson, C. J. Tymczak, and M. Challacombe, Time-reversible ab initio molecular dynamics, *J. Chem. Phys.* **126**, 114103 (2007).
85. D. C. Sorescu, B. M. Rice, and D. L. Thompson, Intermolecular potential for the hexahydro-1,3,5-trinitro-1,3,5-s-triazine crystal (RDX): A crystal packing, Monte Carlo, and molecular dynamics study, *J. Phys. Chem. B* **101**, 798 (1997).

86. D. C. Sorescu and D. L. Thompson, Classical and quantum mechanical studies of crystalline ammonium nitrate, *J. Phys. Chem. A* **105**, 720 (2001).
87. D. C. Sorescu, J. A. Boatz, and D. L. Thompson, Classical and quantum-mechanical studies of crystalline FOX-7 (1,1-diamino-2,2-dinitroethylene), *J. Phys. Chem. A* **105**, 5010 (2001).
88. D. C. Sorescu and D. L. Thompson, Classical and quantum mechanical studies of crystalline ammonium dinitramide, *J. Phys. Chem. B* **103**, 6774 (1999).
89. G. D. Smith and R. K. Bharadwaj, Quantum chemistry based force field for simulations of HMX, *J. Phys. Chem. B* **103**, 3570 (1999).
90. J. Seminario, M. C. Concha, and P. Politzer, A density-functional molecular-dynamics study of the structure of liquid nitromethane, *J. Chem. Phys.* **102**, 8281 (1995).
91. S. W. Bunte and H. Sun, Molecular modeling of energetic materials: The parameterization and validation of nitrate esters in the COMPASS force field, *J. Phys. Chem. B* **104**, 2477 (2000).
92. D. C. Sorescu, B. M. Rice, and D. L. Thompson, Theoretical studies of the hydrostatic compression of RDX, HMX, HNIW, and PETN crystals, *J. Phys. Chem. B* **103**, 6783 (1999).
93. J. P. Agrawal and R. D. Hodgson, *Organic Chemistry of Explosives* (Wiley, Chichester, 2007).
94. H. H. Cady and L. C. Smith, Studies on the polymorphs of HMX, *LANL report LA-MS-2652* (Los Alamos National Laboratory, 1962).
95. H. H. Cady, A. C. Larson, and D. T. Cromer, The crystal structure of α-HMX and a refinement of the structure of β-HMX, *Acta Crystallogr.* **16**, 617 (1963).
96. C. S. Choi and H. P. Boutin, A study of the crystal structure of β-cyclotetramethylene tetranitramine by neutron diffraction, *Acta Cryst. B* **26,** 1235 (1970).
97. R. E. Cobbledick and R. W. H. Small, The crystal structure of the δ-form of 1,3,5,7-tetranitro-1,3,5,7-tetraazacyclooctane (δ-HMX), *Acta Cryst. B* **30**, 1918 (1974).
98. D. W. Brenner, D. H. Robertson, M. L. Elert, and C. T. White, Detonations at nanometer resolution using molecular dynamics, *Phys. Rev. Lett.* **70,** 2174 (1993); *ibid.*, Detonations at nanometer resolution using molecular dynamics,*Phys. Rev. Lett.* **76**, 2202 (1996).
99. J. Tersoff, Empirical interatomic potential for carbon, with applications to amorphous carbon, *Phys. Rev. Lett.* **61**, 2879 (2003).
100. R. L. Martin, *Electronic Structure: Basic Theory and Practical Methods* (Cambridge University Press, New York, 2004).
101. R. G. Parr and W. Yang, *Density-Functional Theory of Atoms and Molecules* (Oxford University Press, New York, 1989).
102. H. Liu, J. J. Zhao, D. Q. Wei, and Z. Z. Gong, Structural and vibrational properties of solid nitromethane under high pressure by density functional theory, *J. Chem. Phys.* **124**, 12450 (2006).
103. E. F. C. Byrd, G. E. Scuseria, and C. F. Chabalowski, An ab initio study of solid nitromethane, HMX, RDX, and CL20: Successes and failures of DFT, *J. Phys. Chem. B* **108**, 13100 (2004).
104. E. F. C. Byrd and B. M. Rice, Ab initio study of compressed 1,3,5,7-tetranitro-1,3,5,7-tetraazacyclooctane (HMX), cyclotrimethylenetrinitramine (RDX), 2,4,6,8,10,12-hexanitrohexaazaisowurzitane (CL-20), 2,4,6-trinitro-1,3,5-benzenetriamine (TATB), and pentaerythritol tetranitrate (PETN), *J. Phys. Chem. C* **111**, 2787 (2007).
105. V. I. Levitas, L. B. Smilowitz, B. F. Henson, and B. W. Asay, Interfacial and volumetric kinetics of the beta ->delta phase transition in the energetic nitramine octahydro-1,3,5,7-tetranitro-1,3,5,7-tetrazocine based on the virtual melting mechanism, *J. Chem. Phys.* **124**, 025101 (2006).
106. A. C. T. van Duin, S. Dasgupta, F. Lorant, and W. A. Goddard III, ReaxFF: A reactive force field for hydrocarbons, *J. Phys. Chem. A* **105**, 9396 (2001).
107. A. Strachan, A. C. T. van Duin, D. Chakraborty, S. Dasgupta, and W. A. Goddard III, Shock waves in high-energy materials: The initial chemical events in nitramine RDX, *Phys. Rev. Lett.* **91**, 098301 (2003).
108. A. Strachan, E. M. Kober, A. C. T. van Duin, J. Oxgaard, and W. A. Goddard III, Thermal decomposition of RDX from reactive molecular dynamics, *J. Chem. Phys.* **122**, 054502 (2005).

109. A. C. T. van Duin, Y. Zeiri, F. Dubnikova, R. Kosloff, and W. A. Goddard III, Atomistic-scale simulations of the initial chemical events in the thermal initiation of triacetonetriperoxide, *J. Am. Chem. Soc.* **127**, 11053 (2005).
110. W. J. Mortier, S. K. Ghosh, and S. Shankar, Electronegativity equalization method for the calculation of atomic charges in molecules, *J. Am. Chem. Soc.* **108**, 4315 (1986).
111. M. J. Buehler, A. C. T. van Duin, and W. A. Goddard III, Multiparadigm modeling of dynamical crack propagation in silicon using a reactive force field, *Phys. Rev. Lett.* **96**, 095505 (2006).
112. K. Chenoweth, S. Cheung, A. C. T. van Duin, W. A. Goddard III, and E. M. Kober, Simulations on the thermal decomposition of a poly(dimethylsiloxane) polymer using the ReaxFF reactive force field, *J. Am. Chem. Soc.* **127**, 7192 (2005).
113. Q. Zhang, Y. Qi, L. G. Hector, T. Cagin, and W. A. Goddard III, Atomic simulations of kinetic friction and its velocity dependence at Al/Al and alpha-Al2O3/alpha-Al2O3 interfaces, *Phys. Rev. B* **72**, 045406 (2005).
114. K. D. Nielson, A. C. T. van Duin, J. Oxgaard, W. Q. Deng, and W. A. Goddard III, Development of the ReaxFF reactive force field for describing transition metal catalyzed reactions, with application to the initial stages of the catalytic formation of carbon nanotubes, *J. Phys. Chem. A* **109**, 493 (2005).
115. J. Ludwig, D. G. Vlachos, A. C. T. van Duin, and W. A. Goddard III, Dynamics of the dissociation of hydrogen on stepped platinum surfaces using the ReaxFF reactive force field, *J. Phys. Chem. B* **110**, 4274 (2006).
116. W. A. Goddard III, A. C. T. van Duin, K. Chenoweth, M. J. Cheng, S. Pudar, J. Oxgaard, B. Merinov, Y. H. Jang, and P. Persson, Development of the ReaxFF reactive force field for mechanistic studies of catalytic selective oxidation processes on BiMoOx, *Topics Catalysis* **38**, 93 (2006).
117. S. S. Han, J. K. Kang, H. M. Lee, A. C. T. van Duin, and W. A. Goddard III, The theoretical study on interaction of hydrogen with single-walled boron nitride nanotubes. I. The reactive force field ReaxFF(HBN) development, *J. Chem. Phys.* **123**, 114703 (2005).
118. S. S. Han, A. C. T. van Duin, W. A. Goddard III, and H. M. Lee, Optimization and application of lithium parameters for the reactive force field, ReaxFF, *J. Phys. Chem. A* **109**, 4575 (2005).
119. S. Cheung, W. Q. Deng, A. C. T. van Duin, and W. A. Goddard III, ReaxFF(MgH) reactive force field for magnesium hydride systems, *J. Phys. Chem. A* **109**, 851 (2005).
120. W. A. Goddard III, O. Zhang, M. Uludogan, A. Strachan, and T. Cagin, The ReaxFF polarizable reactive force fields for molecular dynamics simulation of ferroelectrics, *AIP Conf. Proc.* **626**, 45 (2002).
121. I. I. Oleynik, M. Conroy, S. V. Zybin, L. Zhang, A. C. T. van Duin, W. A. Goddard III, and C. T. White, Energetic materials at high compression: first-principles density functional theory and reactive force field studies, *AIP Conf. Proc.* **845**, 573 (2006).
122. SeqQuest Electronic Structure Code, http://dft.sandia.gov/Quest/
123. D. C. Langreth and J. P. Perdew, Theory of nonuniform electronic systems. 1. Analysis of the gradient approximation and a generalization that works, *Phys. Rev. B* **21**, 5469 (1980).
124. J. P. Perdew and W. Yue W, Accurate and simple density functional for the electronic exchange energy: Generalized gradient approximation, *Phys. Rev. B* **33**, 8800 (1986); *ibid.*, Erratum: Accurate and simple density functional for the electronic exchange energy: Generalized gradient approximation, *Phys. Rev. B* **40**, 3399 (1989).
125. J. P. Perdew, Density-functional approximation for the correlation-energy of the inhomogeneous electron-gas, *Phys. Rev. B* **33**, 8822 (1986); *ibid.*, Correction, *Phys. Rev. B* **34**, 7406 (1986).
126. D. C. Langreth and M. J. Mehl, Beyond the local-density approximation in calculations of ground-state electronic-properties, *Phys. Rev. B* **28**, 1809 (1983); *ibid.*, Erratum: Beyond the local-density approximation in calculations of ground-state electronic properties, *Phys. Rev. B* **29**, 2310 (1984).
127. J. P. Perdew, K. Burke, and M. Ernzerhof, Generalized gradient approximation made simple, *Phys. Rev. Lett.* **77**, 3865 (1996); *ibid.*, Generalized gradient approximation made simple, *Phys. Rev. Lett.* **78**, 1396 (1997).

128. A. C. T. van Duin, S. V. Zybin, K. Chenoweth, L. Zhang, S. P. Han, A. Strachan, and W. A. Goddard III, Reactive force fields based on quantum mechanics for applications to materials at extreme conditions, *AIP Conf. Proc.* **845**, 581 (2006).
129. A. C. T. van Duin, S. V. Zybin, K. Chenoweth, S. P. Han, and W. A. Goddard III, Reactive force fields based on quantum mechanics for applications to materials at extreme conditions. *Lecture Series on Computer and Computational Sciences 4* (Brill Academic Publishers, Amsterdam, 2005) p. 1109.
130. L. Zhang, S. V. Zybin, A. C. T. van Duin, S. Dasgupta, and W. A. Goddard III, Thermal decomposition of energetic materials by ReaxFF reactive molecular dynamics, *AIP Conf. Proc.* **845**, 589 (2006).
131. O. Borodin, G. D. Smith, D. Bedrov, and T. D. Sewell, Polarizable and non-polarizable force fields for alkylnitrates, *J. Phys. Chem. B* **112**, 734 (2008).
132. D. C. Sorescu, B. M. Rice, and D. L. Thompson, Molecular packing and NPT molecular dynamics investigation of the transferability of the RDX intermolecular potential to 2,3,6,8,10,12-hexanitrohexaazaisowurtzitane, *J. Phys. Chem. B* **102**, 948 (1998).
133. D. C. Sorescu, B. M. Rice, and D. L. Thompson, Isothermal-isobaric molecular dynamics simulations of 1,3,5,7-tetranitro-1,3,5,7-tetraazacyclooctane (HMX) crystals, *J. Phys. Chem. B* **102**, 6692 (1998).
134. D. C. Sorescu, B. M. Rice, and D. L. Thompson, A transferable intermolecular potential for nitramine crystals, *J. Phys. Chem. A* **102**, 8386 (1998).
135. D. C. Sorescu, B. M. Rice, and D. L. Thompson, Molecular packing and molecular dynamics study of the transferability of a generalized nitramine intermolecular potential to non-nitramine crystals, *J. Phys. Chem. A* **103**, 989 (1999).
136. B. M. Rice and D. C. Sorescu, Assessing a generalized CHNO intermolecular potential through ab initio crystal structure prediction, *J. Phys. Chem. B* **108**, 17730 (2004).
137. L. Q. Zheng and D. L. Thompson, On the accuracy of force fields for predicting the physical properties of dimethylnitramine, *J. Phys. Chem. B* **110**, 16082 (2006).
138. D. C. Sorescu, B. M. Rice, and D. L. Thompson, Theoretical studies of solid nitromethane, *J. Phys. Chem. B* **104**, 8406 (2000).
139. D. C. Sorescu, B. M. Rice, and D. L. Thompson, Molecular dynamics simulations of liquid nitromethane, *J. Phys. Chem. A* **105**, 9336 (2001).
140. A. Siavosh-Haghighi and D. L. Thompson, Melting point determination from solid-liquid coexistence initiated by surface melting, *J. Phys. Chem. C* **111**, 7980 (2007).
141. T. Megyes, S. Bálint, T. Grósz, T. Radnai, I. Bakó, and L. Almásy, Structure of liquid nitromethane: Comparison of simulation and diffraction studies, *J. Chem. Phys.* **126**, 164507 (2007).
142. P. M. Agrawal, B. M. Rice, L. Zheng, and D. L. Thompson, Molecular dynamics simulations of hexahydro-1,3,5-trinitro-1,3,5-s-triazine (RDX) using a combined Sorescu-Rice-Thompson AMBER force field, *J. Phys. Chem. B* **110**, 26185 (2006).
143. N. Goto, H. Yamawaki, K. Wakabayashi, Y. Nakayama, M. Yoshida, and M. Koshi, High pressure phase of RDX, *Sci. Tech. Energ. Mater.* **66**, 291 (2005).
144. D. A. Case, D. A. Pearlman, J. W. Caldwell, T. E. Cheatham, J. Wang, W. S. Ross, C. L. Simmerling, T. A. Darden, K. M. Merz, R. V. Stanton, A. L. Cheng, J. J. Vincent, M. Crowley, V. Tsui, H. Gohlke, R. J. Radmer, Y. Duan, J. Pitera, I. Massova, G. L. Seibel, U. C. Singh, P. K. Weiner, and P. A. Kollman, *AMBER 7* (University of California, San Francisco, 2002).
145. J. M. Wang, R. M. Wolf, J. W. Caldwell, P. A. Kollman, and D. A. Case, Development and testing of a general amber force field, *J. Comput. Chem.* **25**, 1157 (2004).
146. S. J. Weiner, P. A. Kollman, D. T. Nguyen, and D. A. Case, An all atom force-field for simulations of proteins and nucleic-acids, *J. Comput. Chem.* **7**, 230 (1986).
147. S. Ye, K. Tonokura, and M. Koshi, Theoretical studies of pressure dependence of phonon and vibron frequency shifts of PETN, *Sci. Tech. Energ. Mater.* **64**, 201 (2003).
148. H. E. Alper, F. Abu-Awwad, and P. Politzer, Molecular dynamics simulations of liquid nitromethane, *J. Phys. Chem. B* **103**, 9738 (1999).

149. S. Boyd, M. Gravelle, and P. Politzer, Nonreactive molecular dynamics force field for crystalline hexahydro-1,3,5-trinitro-1,3,5 triazine, *J. Chem. Phys.* **124**, 104508 (2006).
150. G. D. Smith, R. K. Bharadwaj, D. Bedrov, and C. Ayyagari, Quantum-chemistry-based force field for simulations of dimethylnitramine, *J. Phys. Chem. B* **103**, 705 (1999).
151. H. Davande, O. Borodin, G. D. Smith, and T. D. Sewell, Quantum chemistry-based force field for simulations of energetic dinitro compounds, *J. Energ. Mater.* **23**, 205 (2005).
152. R. I. Hiyoshi, Y. Kohno, O. Takahashi, J. Nakamura, Y. Yamaguchi, S. Matsumoto, N. Azuma, and K. Ueda, Effect of pressure on the vibrational structure of insensitive energetic material 5-nitro-2,4-dihydro-1,2,4-triazole-3-one, *J. Phys. Chem. A* **110**, 9816 (2006).
153. H. Liu, J. J. Zhao, G. F. Ji, Z. Z. Gong, and D. Q. Wei, Compressibility of liquid nitromethane in the high-pressure regime, *Physica B: Condens. Mat.* **382**, 334 (2006).
154. B. R. Brooks, R. E. Bruccoleri, B. D. Olafson, D. J. States, S. S. Swaminathan, and M. Karplus, CHARMM: a program for macromolecular energy, minimization, and dynamics calculations, *J. Comput. Chem.* **4**, 187187 (1983).
155. D. Bedrov, O. Borodin, B. Hanson, and G. D. Smith, Comment on "On the accuracy of force fields for predicting the physical properties of dimethylnitramine", *J. Phys. Chem. B* **111**, 1900 (2007).
156. D. Bedrov, C. Ayyagari, G. D. Smith, T. D. Sewell, R. Menikoff, and J. M. Zaug, Molecular dynamics simulations of HMX crystal polymorphs using a flexible molecule force field, *J. Comput. Aid. Mat. Des.* **8**, 77 (2001).
157. T. D. Sewell, R. Menikoff, D. Bedrov, and G. D. Smith, A molecular dynamics simulation study of elastic properties of HMX, *J. Chem. Phys.* **119**, 7417 (2003).
158. D. Bedrov, G. D. Smith, and T. D. Sewell, Thermal conductivity of liquid octahydro-1,3,5,7-tetranitro-1,3,5,7-tetrazocine (HMX) from molecular dynamics simulations, *Chem. Phys. Lett.* **324**, 64 (2000).
159. D. Bedrov, G. D. Smith, and T. D. Sewell, Temperature-dependent shear viscosity coefficient of octahydro-1,3,5,7-tetranitro-1,3,5,7-tetrazocine (HMX): A molecular dynamics simulation study, *J. Chem. Phys.* **112**, 7203 (2000).
160. J. K. Dienes, Q. H. Zuo, and J. D. Kershner, Impact initiation of explosives and propellants via statistical crack mechanics, *J. Mech. Phys. Solids* **54**, 1237 (2006).
161. B. E. Clements, E. M. Mas, J. N. Plohr, A. Ionita, and F. L. Addessio, Dynamic Response of PBX-9501 through the $\beta - \delta$ Phase Transition, *AIP Conf. Proc.* **845**, 204 (2006).
162. G. D. Smith, D. Bedrov, O. Byutner, O. Borodin, C. Ayyagari, and T. D. Sewell, A quantum-chemistry-based potential for a poly(ester urethane), *J. Phys. Chem. A* **107**, 7552 (2003).
163. R. H. Gee, S. Roszak, K. Balasubramanian, and L. E. Fried, Ab initio based force field and molecular dynamics simulations of crystalline TATB, *J. Chem. Phys.* **120**, 7059 (2004).
164. R. Podeszwa, R. Bukowski, B. M. Rice, and K. Szalewicz, Potential energy surface for cyclotrimethylene trinitramine dimer from symmetry-adapted perturbation theory, *Phys. Chem. Chem. Phys.* **9**, 5561 (2007).
165. C. Møller and M. S. Plesset, Note on an Approximation Treatment for Many-Electron Systems, *Phys Rev.* **46**, 618 (1934).
166. W. J. Hehre, L. Radom, P. v. R. Schleyer, and J. A. Pople, *Ab initio Molecular Orbital Theory* (Wiley, New York, 1986).
167. T. D. Sewell and D. Bedrov, Elastic properties of 1,3,5-triamino-2,4,6-trinitrobenzene (TATB), (to be submitted to *J. Chem. Phys.*, September 2008).
168. M. Pospíšil, P. Capková, P. Vavrá, and S. Zeman, Classical molecular dynamics simulations of RDX decomposition under high pressure, *New Trends in Research of Energetic Materials, Proceedings of the 6th Seminar* (Pardubice, Czech Republic, 2003).
169. L. Qiu, H. M. Xiao, W. H. Zhu, J. J. Xiao, and W. Zhu, Ab initio and molecular dynamics studies of crystalline TNAD (trans-1,4,5,8-tetranitro-1,4,5,8-tetraazadecalin), *J. Phys. Chem. B* **110**, 10651 (2006).
170. X. J. Xu, H. M. Xiao, J. J. Xiao, W. Zhu, H. Huang, and J. S. Li, Molecular dynamics simulations for pure epsilon-CL-20 and epsilon-CL-20-based PBXs, *J. Phys. Chem. B* **110**, 7203 (2006).

171. X. F. Ma, J. J. Xiao, H. Huang, X. H. Ju, J. S. Li, and H. M. Xiao, Simulative calculation of mechanical property, binding energy and detonation property of TATB/fluorine-polymer PBX, *Chinese J. Chem.* **24**, 473 (2006).
172. K. Yin, H. Xiao, J. Zhong, and D. Xu, A new method for Calculation of Elastic Properties of Anisotropic material by constant pressure molecular dynamics. *Lecture Series on Computer and Computational Sciences 1* (Brill Academic Publishers, Amsterdam, 2004) p. 586.
173. L. Qiu, W. H. Zhu, J. J. Xiao, W. Zhu, H. M. Xiao, H. Huang, and J. S. Li, Molecular dynamics simulations of trans-1,4,5,8-tetranitro-1,4,5,8-tetraazadecalin-based polymer-bonded explosives, *J. Phys. Chem. B* **111**, 1559 (2007).
174. A. T. Hagler, E. Huler, and S. Lifson, Energy functions for peptides and proteins.1. Derivation of a consistent force-field including hydrogen-bond from amide crystals, *J. Am. Chem. Soc.* **96**, 5319 (1974).
175. H. Sun, COMPASS: An ab initio force-field optimized for condensed-phase applications - Overview with details on alkane and benzene compounds, *J. Phys. Chem. B* **102**, 7338 (1998).
176. R. H. Gee, A. Maiti, S. Bastea, and L. E. Fried, Molecular dynamics investigation of adhesion between TATB surfaces and amorphous fluoropolymers, *Macromolecules* **40**, 3422 (2007).
177. P. B. Balbuena and J. M. Seminario (Eds.), *Molecular Dynamics (Theoretical and Computational Chemistry)* (Elsevier Science, Amsterdam, 1999).
178. D. Marx and J. Hutter, Ab initio molecular dynamics: Theory and Implementation, J. Grotendorst J (Editor) *Modern Methods and Algorithms of Quantum Chemistry* (John von Neumann Institute for Computing, Jülich, 2000) NIC Series **1**, 301.
179. M. E. Tuckerman and M. L. Klein ML, Ab initio molecular dynamics study of solid nitromethane, *Chem. Phys. Lett.* **283**, 147 (1998).
180. T. Megyes, S. Bálint, T. Grósz, T. Radnai, I. Bakó, and L. Almásy, Structure of liquid nitromethane: Comparison of simulation and diffraction studies, *J. Chem. Phys.* **126**, 164507 (2007).
181. E. J. Reed, J. D. Joannopoulos, and L. E. Fried, Electronic excitations in shocked nitromethane, *Phys. Rev. B* **62**, 16500 (2000).
182. M. R. Manaa, L. E. Fried, C. F. Melius, M. Elstner, and T. Frauenheim, Decomposition of HMX at extreme conditions: A molecular dynamics simulation, *J. Phys. Chem. A* **106**, 9024 (2002).
183. M. R. Manaa, E. J. Reed, L. E. Fried, G. Galli, and F. Gygi, Early chemistry in hot and dense nitromethane: Molecular dynamics simulations, *J. Chem. Phys.* **120**, 10146 (2004).
184. S. A. Decker, T. K. Woo, D. Wei, and F. Zhang, Ab initio molecular dynamics simulations of multimolecular collisions of nitromethane and compressed liquid nitromethane, *Proc. 12th Symp. (Intl.) on Detonation* (San Diego, California, 2002) p. 724.
185. R. Car and M. Parrinello, Unified approach for molecular-dynamics and density-functional theory, *Phys. Rev. Lett.* **55**, 2471 (1985).
186. M. Kamiya, T. Tsuneda, and K. Hirao, A density functional study of van der Waals interactions, *J. Chem. Phys.* **117**, 6010 (2002).
187. R. Baer and D. Neuhauser, Density functional theory with correct long-range asymptotic behavior, *Phys. Rev. Lett.* **94**, 043002 (2005).
188. T. Sato, T. Tsuneda, and K. Hirao, van der Waals interactions studied by density functional theory, *Mol. Phys.* **103**, 1151 (2005).
189. H. Iikura, T. Tsuneda, T. Yanai, and K. Hirao, A long-range correction scheme for generalized-gradient-approximation exchange functionals, *J. Chem. Phys.* **115**, 3540 (2001).
190. R. W. Williams and D. Malhotra, van der Waals corrections to density functional theory calculations: Methane, ethane, ethylene, benzene, formaldehyde, ammonia, water, PBE, and CPMD, *Chem. Phys.* **327**, 54 (2006).
191. F. Ortmann, F. Bechstedt, and W. G. Schmidt, Semiempirical van der Waals correction to the density functional description of solids and molecular structures, *Phys. Rev. B* **73**, 205101 (2006).

192. J. G. Angyan, I. C. Gerber, A. Savin, and J. Toulouse, van der Waals forces in density functional theory: Perturbational long-range electron-interaction corrections, *Phys. Rev. A* **72**, 012510 (2005).
193. M. A. Neumann and M. A. Perrin, Energy ranking of molecular crystals using density functional theory calculations and an empirical van der Waals correction, *J. Phys. Chem. B* **109**, 15531 (2005).
194. J. Kleis and E. Schroder, van der Waals interaction of simple, parallel polymers, *J. Chem. Phys.* **122**, 164902 (2005).
195. S. Grimme, Accurate description of van der Waals complexes by density functional theory including empirical corrections, *J. Comp. Chem.* **25**, 1463 (2004).
196. Q. Wu and W. T. Yang, Empirical correction to density functional theory for van der Waals interactions, *J. Chem. Phys.* **116**, 515 (2002).
197. T. Sato, T. Tsuneda, and K. Hirao, A density-functional study on pi-aromatic interaction: Benzene dimer and naphthalene dimer, *J. Chem. Phys.* **123**, 104307 (2005).
198. H. Rydberg, M. Dion, N. Jacobson, E. Schroder, P. Hyldgaard, S. I. Simak, D. C. Langreth, and B. I. Lundqvist, van der Waals density functional for layered structures, *Phys. Rev. Lett.* **91**, 126402 (2003).
199. H. Rydberg, B. I. Lundqvist, D. C. Langreth, and M. Dion, Tractable nonlocal correlation density functionals for flat surfaces and slabs, *Phys. Rev. B* **62**, 6997 (2000).

Chapter 8
Modeling Defect-Induced Phenomena

Maija M. Kuklja and Sergey N. Rashkeev

8.1 Current State of the Field and Its Challenges

Elucidation of dissociation mechanisms, energy localization, and transfer phenomena in the course of explosive decomposition of energetic materials (EMs) are central for understanding, controlling, and enhancing the performance of these materials as fuels, propellants, and explosives. Quality of energetic materials is often judged using two main parameters: sensitivity to detonation and its performance. Low sensitivity is desired to make the material relatively stable to external stimuli, i.e., controllable and able of triggering rapid dissociation only when needed and not accidentally. Performance, on the other hand, is to be high to provide larger heat of the explosive reaction. These parameters do not necessarily correlate with each other and depend on many variables such as molecular and crystalline structures, history of samples, the particle size, crystal hardness and orientation, external stimuli, aging, storage conditions, and others. Mechanisms governing performance are fairly well understood whereas mechanisms of sensitivity are poorly known and need to be much more extensively studied. It is widely accepted though that the thermal decomposition reactions of the materials play a significant role in their sensitivity to mechanical stimuli and their explosive properties [1].

The decomposition of energetic materials can be initiated with a mechanical impact, thermal heating, a shock wave, or a spark. Such events in solids generate molecules in highly excited vibronic and/or electronic states. Clearly, the decomposition of solid explosives under shock, spark, laser, or plasma ignition must include contributions from both ground and excited electronic states. Excitation in the UV can markedly reduce the power requirements for detonation of some secondary explosives. Therefore, establishing the initial steps of high explosive (HE) decomposition is an important goal to pursue. Decomposition triggered by excited electronic states seems appealing because electronic excitation of the system to an unstable potential energy surface can result in rapid dissociation of a molecule and consequent chain reaction.

S.M. Peiris, G.J. Piermarini (eds.), *Static Compression of Energetic Materials*, Shock Wave and High Pressure Phenomena,

Careful studies of all stages of decomposition of energetic materials are needed for solving both practical and fundamental problems. Design of novel materials with prescribed superior properties, establishment of safe conditions for their synthesis and usage, use of shock and detonation techniques in other fields of science and technology require knowledge of mechanisms of thermal decomposition and qualitative data for the description of the processes over a wide range of temperatures, pressures, and different states of matter. The most important practical problem of today is the *safety* of synthesis, processing, storage, use, transportation, and handling of energetic materials including large scale (i.e., large quantities of materials stored at the warehouses, for example). The greatest fundamental challenge is figuring out the relationship between structure, properties, and function in EM. This naturally includes the development of theoretical models of solid-state chemical reactions and understanding of the details of decomposition mechanisms and kinetics as well as conditions that control the energy management in practical materials.

Traditionally, investigations concerning explosive decomposition of energetic materials have been carried out mostly in two main directions. One is the physics of detonation waves describing macro-processes accompanied by explosive decomposition of materials [2–6]. The other is the chemistry of explosives- studying various chemical reactions in terms of molecular energetic barriers or kinetics [1, 7]. While providing important information and most of the understanding of the modern detonation theory, both approaches essentially ignore solid-state effects and behavior of the materials.

Here are a few disturbing examples illustrating the current state of the field of energetic materials. There is an extensive body of studies of the structure and chemical reactions describing all stages and kinetics of explosive decomposition of many energetic materials that are summarized in detailed recent reviews [1, 7–14]; however, there is no consensus of opinion concerning the decomposition mechanism of even one of the simplest compounds, nitromethane. Based on observed CH_3, NO_2, and CH_3O products, by using a molecular beam in conjunction with infrared multiphoton dissociation technique and pyrolysis experiments, it was suggested that below 700°C, the initial stage of the reaction involves cleavage of the C–N bond, which must compete with the rearrangement of CH_3NO_2 to give methyl nitrite and by its subsequent decomposition [15], which was also predicted theoretically [16]. Recent experiments enriched possible scenarios by adding bimolecular [17, 18] and ionic [19] decompositions. Theoretical studies propose that nitro-to-nitrite rearrangement and HONO elimination reactions are the most favorable and are about 15 kcal mol^{-1} lower in energy than C–NO_2 break [20], in some qualitative agreement with recent calculations [21, 22]. This is in sharp contrast, both quantitatively and qualitatively, to the earlier ab initio results [23] in which the rearrangement barrier of 73.5 kcal mol^{-1} was found to be 16.1 kcal mol^{-1} higher than the C–N bond asymptote, concluding that concerted rearrangements on CH_3NO_2 potential surface will not be observed, and to recent calculations [24]. More excitement is augmented by a study of the C–NO_2 rupture via lowest singlet–triplet transitions in a nitromethane molecule and intersystem crossing of ground and excited states [25]. While discrepancies among theoretical efforts are mainly defined by differences in methodologies used, the persistent contradiction between theory and experiment is, in our opinion, due to two main reasons. The first is the complex profile of the potential energy surface of nitromethane [26], which is evidenced by numerous transition states found [22], and represents, most likely, a common feature of many EMs [27, 28]. The second is that even excessively exhaustive computational studies on the decomposition of isolated molecules are unlikely to provide a consistent interpretation of experimental measurements of solid-state processes because decomposition in the condensed phase fundamentally differs from that in gas and/or dilute solutions due to catalysis and intermolecular interactions [1]. This implies that there are plenty of reasons to expect that the inclusion of lattice interactions is imperative to obtaining unambiguous conclusions.

2,4,6-Trinitrotoluene ($C_7H_5N_3O_6$, TNT) achieves its special position among EM because of its compatibility with other materials, low hygroscopicity, low melting point, low cost, relatively low sensitivity to impact and friction, good thermal stability, and high power during explosion [29]. Despite of its long existence (TNT has been available as a pure material since about 1870 and is very widely used) and the fact that specific reactions of initiation have been extensively discussed, the chemical analysis, which reconciles conflicting observations and statements previously made about the chemical mechanisms that control the explosive behavior of TNT was reported only recently [29]. In the meantime, the details of the solid-state decomposition mechanisms and their relationship with the kinetics as well as with catalysis of decomposition products of TNT, not to mention the structure of the samples, are not clear.

Triamino-trinitrobenzine ($C_6H_6N_6O_6$, TATB) manifests a significant stability benchmark because it is remarkably insensitive to high temperatures, shock, and impact [30, 31] yet retains acceptable performance [32]. Like TNT, TATB is a nitroarene, but is not melt castable, nor is it readily soluble in most known solvents, which makes TATB difficult to process [30, 33]. The reasons for such a high stability of TATB are not completely figured out despite the concerted attempts of several groups. Among other efforts to understand the exceptional stability of TATB and perhaps find alternatives, the thermal stabilities of related the methyl-, bromo-, and amino-substituted homologous series were examined [33], which revealed some important trends but still did not provide a consistent answer. The overall balance of the thermal stability of the molecular structures and of the lattice forces leads to homologous series, which show stability trends both parallel and opposed to those in solution [33]. The conclusions of this study suggested that, in condensed phase, intermolecular attractive forces may be more important than the intrinsic molecular structure.

It was noted that an excessive amount of possible correlations in various parameters of related EM do not account for the trend in the shock and impact sensitivity. This was qualified as perfidy in the amino-2,4,6-trinitrobenzene series of explosives [34]. The authors illustrated that the numerous electronic, molecular, crystal, and explosive variables in the series 1,3,5-trinitrobenzene (TNB), 2,4,6-trinitroaniline (MATB), diamino-2,4,6-trinitrobenzene (DATB), and triamino-2,4,6-trinitrobenzene (TATB) give 153 nearly linear correlations and yet do not provide enough information to reveal the sensitivity tendency. It was also suggested from Arrhenius data that C–NO_2 homolysis is the initial decomposition reaction during impact and shock initiation in all cases [1, 34, 35]. The lower activation energy cyclization process of MATB, DATB, and TATB to furazan/furoxan products, as proposed, can be overstepped at shock- and impact-initiation temperatures. However, the fact that C–NO_2 homolysis does not account for the trend in sensitivity is not only because the activation energy is relatively insensitive to the ring substituents [34] but also because intermolecular interactions and materials science issues have to be taken into account.

Extensive literature on the appearance of the mutual influence of atomic groups in various complex organic molecules is available [10, 33]. Combining a number of experimental methods (IR and Raman spectroscopy, X-ray diffraction and neutron diffraction, measurements of dipole moments, etc.) allows for reliable information on the structural features of complex molecules and intramolecular dynamics to be obtained [6, 7, 9, 10]. For instance, the relative extents of loss of NO_2 and NO were determined by electron ionization and tandem mass spectrometry to investigate dissociation processes for molecular ions formed by electron ionization of *para*-substituted nitrobenzene compounds, NO_2, CHO, H, OCH_3 [35]. It was found that electron-acceptor groups favor loss of NO_2, while an electron-donor group favors loss of NO, and that NO_2 to ONO (nitro-to-nitrite) isomerization precedes the loss of NO [35]. At present, a number of stubborn problems related to the donor–acceptor interaction between reaction centers in the molecules of explosives need to be solved. For another illustrative example of the effect of substitutions in homologous series it is worthy to note that it is widely accepted that an α-CH bond ortho to –NO_2 group on an aromatic ring activates the thermal decomposition [1], while the increased number of –NH_2 groups in the aminonitrobezene series contributes to the decrease of the heat of explosion [1]; there are other trends, as well. None of them are able to explain, however, the fact that TNT decomposes about ten times faster in the melt than in the vapor phase [36] since neither the detailed mechanism of hydrogen transfer in the condensed phase, nor the role of catalysis-induced by products, is known. One of the urgent issues to be methodically explored is the interplay of –NO_2 bond scission and –NO_2 to –ONO isomerization, especially in solids.

One of the oldest and most creative, the most explored by many researchers, and at the same time the least elaborated concepts is probably the idea of "hot spots" [37], special sites in a crystalline lattice that are capable of localizing the energy of the shock or impact wave and triggering the chemical reaction. Although it is generally accepted that crystallographic defects play a key role in formation of hot spots, i.e., they are responsible for the sensitivity of energetic materials to detonation initiation, it is not exactly known what kind of defects are important and what chemical mechanisms take place in hot spots [38]. As was postulated by Bowden and Singh, the initiation of explosives could be interpreted in terms of the redistribution of moving dislocations [39]. An abundance of models for hot spot formation to explain the shock and impact initiation, as well as the thermal decomposition of explosives, has been suggested since then, but none of them lead to the development of a comprehensive general theory in a widely agreed upon form partially due to the lack of knowledge about the microstructure of defects in explosive materials [40, 41]. To clarify the situation and to select the proper models, we need to learn more about the electronic and spatial structure of lattice imperfections. This may be provided only by a complete loop of studies ranging from atomistic, quantum-chemical to mesoscale, and continuum levels of the theory in addition to high quality space- and time-resolved experiments to scrupulous analysis of materials. Unfortunately, a careful characterization of samples is rarely performed for the time being, which makes a reliable comparison between theoretical simulations and experimental measurements problematic.

The significant progress produced lately in the field of ultrafast optical techniques and computer modeling clearly indicates some of the possible directions in future research on energetic materials and detonation initiation theory. Recent experimental developments in ultrafast optical techniques in producing waveforms with specified time-dependent amplitude, phase, frequency, and polarization demonstrate considerable progress in understanding the mechanisms and dynamics of photo-induced processes in ionic solids [42] and in reactive organic molecular crystals [43, 44]. The focus of ultrafast spectroscopies is now shifting from observation into applications of light, exploiting specially designed profiles of the above parameters so as to guide the quantum dynamics of a system to a particular target [45]. For example, it has been demonstrated that tailoring of light fields and their interference can be used to control chemical reactions in the gas phase, such as system propagation along a single vibrational mode or reaction along a single chemical pathway [46, 47]. The theory of active control of molecular motion through the use of shaped laser pulses has been developing for several years [48–50]. Femtosecond pump-probe spectroscopy in combination with ab initio calculations of self-trapped excitons and holes were applied to study laser-induced reactions in crystals [51].

A wealth of data on the mechanisms of chemical chain reactions in gases and liquids has been accumulated [52]. A chain reaction occurs due to real migration of active particles (free atoms or radicals) and their collisions and interaction with virgin molecules. The nature of chain reactions in solids is more complex. There is a paucity of experimental data demonstrating the chain-reaction nature of reaction processes in solids. Serious difficulties exist in understanding the nature of the active particles participating in a solid-state chain reaction. Apparently, real migration of reagents such as atoms or radicals over the crystal lattice may occur only as a result of sufficiently slow diffusion processes. This is inconsistent with the experimental data on explosive decomposition of solids where initiation of reaction is thought to occur on a 10–100 ps timescale behind a shock front in some EM materials [53]. Therefore, one can assume that the active particles participating in fast chain reactions leading to an explosion should arise from much faster process, such as electronic excitations of the crystalline lattice: electrons, holes and excitons [54, 55].

Plenty of experimental evidence, indicating that electronic excitations (electrons, holes, excitons) play a key role in the initiation process, exists nowadays [56–64]. A variety of initiation models have been proposed. Williams [65] considered the influence of electronic states and electronic transport on the initiation and propagation of detonation waves in solid explosives. Bowden et al. [66] developed a model for laser-induced selective low-level excitation of a molecular species, which phenomenologically incorporates coherent resonant energy transfer and collision damping. The vibrational energy up-pumping model [53] suggests that the shock wave produces a bath of excited phonons absorbed by the lowest vibrational modes of molecules that make up the crystal. Increased phonon absorption and intramolecular vibrational energy redistribution lead to the excitation of higher frequency modes, eventually leading to an equilibrium transition state, chemical bond breakage, and subsequent chemical reactions [67]. The rate of phonon-to-vibron energy transfer has been related to the sensitivity in common explosives [68]. Dremin et al., [69] noting evidence for similarities between shock decomposition intermediates and those of photochemical processes, proposed electronic excitation as the first molecular response in their multiprocess detonation model. They postulate that detonation is caused by dissociation of molecules within the shock front that can proceed in three ways: through the accumulation mechanism [70, 71], through thermal ionization, and lastly, by electronic excitation [38]. Furthermore, correlation of impact sensitivity with electronic levels of a homologous series of explosive compounds has been reported [72]. Gilman proposed that the compression from the wave front causes local metallization [59, 73], which results in bending of covalent bonds, and thus closes the highest occupied molecular orbital–lowest unoccupied molecular orbital (HOMO–LUMO) gap leading to the increased formation of these delocalized electrons. Sensitivity to initiation of reaction then correlates with the formation of delocalized electrons. It was also suggested that plasmon excitation may trigger fast chemical reactions in a shock front [74]. The idea of the involvement of excited electronic states was buttressed by studies on RDX ($C_3H_6N_6O_6$) and TATB ($C_6H_6N_6O_6$), among other systems, in which initiation sensitivity was found to correlate with details of the electronic structure [75,76]. It is needless to say that despite of loads of scattered data, the theory of detonation initiation including a chapter of the theory devoted to the mechanisms of initiation of chemistry involving electronic excitations is yet to be developed. Those questions undoubtedly represent a challenge, which is appealing to resolve; once addressed, it will have momentous consequences.

In this chapter, we review the results obtained from first principles modeling of various crystallographic defects and local deformations in a series of energetic materials with emphasis on their effect on the electronic structure and energy barriers for chemical decomposition. We believe that these results signify solid fragments towards a comprehensive theory to come in the future. This work, dealing with traditional energetic materials, such as RDX, PETN, and TATB, also presents an analysis of recently synthesized high explosive, FOX-7, and sheds some light on the chemical and mechanical properties of FOX-7 as well as on its failure to become a new prospective insensitive energetic material. Based on ab initio modeling, this review also provides some specific recommendation for the search of new materials with low sensitivity. The theoretical approach used is based on a combination of a periodic model of a crystalline system and a molecular cluster model by means of Density Functional Theory and Hartree-Fock band-structure calculations of perfect and defective materials under a range of conditions.

A brief review of ab initio modeling of structure and properties of energetic materials containing various defects is presented in the next section, which describes an effect of vacancies and voids (Section 8.2.1), dislocation-induced modifications of the electronic structure (Section 8.2.2), orientational molecular defects (Section 8.2.3), and shear strain-assisted decomposition (Section 8.2.4) in energetic crystals. Section 8.3 introduces ideas on electronic excitations and explores their possible role in the initiation of chemistry as evidenced by recent calculations and available experimental data. This section also discusses how the results obtained from quantum-chemical modeling of defects can be linked with experiment. Finally, we present a summary, the main conclusions, and possible avenues for future research in Section 8.4.

8.2 Modeling of Structure and Properties of Energetic Materials Containing Defects

8.2.1 Vacancies, Voids, and Surfaces in RDX

In this section, we illustrate how a solid-state approach applied to a traditional problem of molecular decomposition brings qualitatively and quantitatively new conclusions, which essentially reconciles experimental and theoretical contradictions. We describe here modeling of a single molecular vacancy [77], vacancy dimers [78], and surfaces [79] in RDX crystal. There are two questions we are going to address: the effect of small and large voids on the electronic structure and on the decomposition energy of an RDX molecule (Fig. 8.1).

In the course of the thermal decomposition of solid RDX [80], among all possible reaction pathways, we considered only the N–NO_2 bond rupture channel as the most supported by experiments [81–83]; in fact, the nature of the initial step is still a subject of debate. Corresponding activation energies vary from 24.7 to 52.1 kcal mol^{-1} and reported pre-exponential factors vary from 10^{17} to 10^{20} s^{-1} for overall decomposition [84]. The ab initio study based on gradient-corrected density functional theory found the activation barrier of 34.2 kcal mol^{-1} for the N–NO_2 dissociation mechanism of gas phase RDX [84].

The ab initio Hartree-Fock method combined with two essentially different solid-state models, a molecular cluster and a periodic defect, was applied to this problem. To simulate a molecular vacancy in RDX, one of the eight molecules was removed from the orthorhombic RDX unit cell [77]. Use of periodic boundary conditions (supercell model) results in a model crystal with a concentration of defects as high as 12.5%. By doubling the size of the supercell and changing the positions of vacancies, the corresponding concentration of defects becomes about 6%, one is able to study the effect of defect distribution, in particular the formation of vacancy dimers, on the electronic and optical properties of the solid [78]. In a similar manner we modeled two-dimensional defects, such as fine nano-cracks, placed close to each other [85].

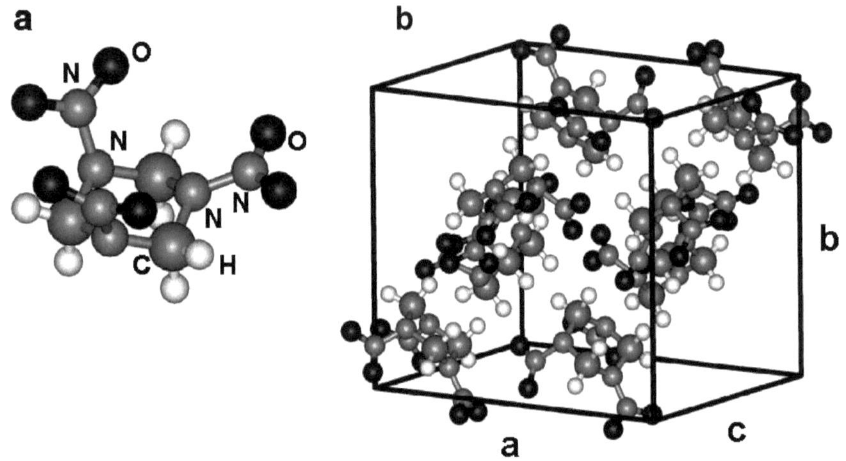

Fig. 8.1 The (**a**) molecular and (**b**) ideal crystalline structures of cyclotrimethylene trinitramine (RDX). The C atoms are depicted by larger gray balls, O are shown in black, N– by smaller gray balls, H– in white

Surfaces were modeled by cutting the crystal in specific directions in accordance with larger crystal faces observed during the crystal growth process. Then, various clusters were constructed to study how the decomposition energy of the molecule placed on a surface differs (if at all) from that placed in a bulk crystal [80].

A range of molecular clusters and unit cells were probed to ensure that the final conclusion does not depend on the computational methods. We established that a molecular vacancy has a low formation energy, and therefore, there should be plenty of vacancies in molecular crystals; vacancies placed close to each other tend to aggregate because of their positive binding energy. This led us to state that the decrease in density leads to mechanical and chemical "softening" of the crystal. It has been recognized that all of the probed imperfections (small or large voids, pores, internal, or free surfaces) only slightly change the electronic structure of the material; neither of them produce local states in the band gap, which is explained by the fact that all of the defects are largely neutral in spite of some electronic density redistribution on the atoms around the void. It has been also found that the energy for RDX N–NO_2 dissociation is strongly dependent on the environment of the molecule. Thus, an isolated RDX molecule with the "crystal geometry" exhibits some difference in energies for the cleavage of the equatorial and the axial bonds against all equal energies for the gas-phase molecule [80]. This is due to bond length changes and symmetry reduction. Next, the equatorial bond of the molecule inside the crystal is sensitive to the crystal field and is characterized by the visibly increased energy compared to the gas-phase dissociation energy. Finally, lowered energy barriers by 8–15 kcal mol^{-1} are obtained for the molecule placed near the surface. Based on energetic considerations, no preference can be made for the equatorial or the axial NO_2 group decomposition. The results obtained are supported by experimental data and previous theoretical investigations.

Two interesting conclusions follow from this study. The molecule in the crystal has three different energetic barriers for cleavage of three N–NO_2 bonds while all three energies are equivalent for the gas-phase molecule. This conclusion is perfectly in accord with the symmetry reduction in the solid phase. Also, a molecule located near the surface can more easily dissociate than a molecule buried in the bulk crystal. This can be understood in terms of missing interactions. The molecule associated with voids or surfaces is more weakly bound to the crystal than the molecule surrounded with all nearest neighbors.

These findings imply that the thermal decomposition will most likely start from breaking bonds in molecules that are placed near voids, surfaces, or dislocation cores due to the reduced energetic barrier caused by the imbalance of intra- and intermolecular interactions. Obviously, the surface-induced effect and dynamics of both the electronic and lattice relaxation due to a shock wave must be taken into account when studying the initiation mechanisms.

8.2.2 The Electronic Structure of RDX Crystals Containing Edge Dislocations

In this section modifications of the electronic structure of RDX induced by an edge dislocation [55, 76, 86–88] are described. The main objective of this research was to study the effect of dislocations on the RDX optical band gap under shock compression.

The calculations are performed by means of the standard Hartree–Fock (HF) method for a periodic system by the CRYSTAL code [89], the 6–21 g split valence set with the modified scaling factor of 1.10 for the outer (most diffuse) Gaussian function was used. In all calculations, internal geometry parameters of the RDX molecule such as bond lengths, angles, and torsion angles were taken from experimental data [90] and fixed, while the lattice constants were varied to minimize the total energy. Electron correlation corrections based on the second-order many-body perturbation theory (MBPT) are included in the calculations to correct the value of the optical band gap of the system, which is overestimated in the HF limit. The calculations have been performed for the crystal structures both in equilibrium and under external hydrostatic compression. The band gap as a function of the crystal compression V/V_0 was carefully analyzed. A response of the ideal and defective RDX crystals to the shock-wave excitation was modeled quantum-mechanically with the rigid molecule approximation. Thus, isotropic compressibility of the crystal in terms of volume-energy dependence was investigated by proportionally decreasing all lattice constants and keeping the structure of the RDX molecules in the crystal unaltered. The pressure was calculated using the low-temperature formula $P = -dU/dV$. The corresponding electronic correlation corrections were obtained for each value of pressure.

Edge dislocations were modeled by assuming that the crystal was composed of stacked parallel layers one-unit-cell thick. The configuration of the edge dislocation was taken from experimental studies. To produce a dislocation with the Burgers vector of [001] one can slide the corresponding crystalline plane with respect to another part of the solid along the z direction. This slip produces a plastic deformation in the material, and the boundary between the slipped and unslipped regions is the *edge dislocation*. Its position is marked by the termination of an extra vertical half-plane of RDX molecules crowded into the upper half of the crystal. In our calculations we modeled the dislocation core by a one-dimensional (polymer) model. The periodic translation was applied along the dislocation line d[010] and no dislocation–dislocation interactions were included (Fig. 8.2).

The correlation correction, which is largely due to polarization effects, reduces the computed pure, perfect RDX gap by 8.42 eV to a value of 5.25 eV. Optical absorption experiments [91] yield 3.4 eV. In particular, it was found that strong intramolecular light absorption at wavelengths below 340 nm (3.4 eV) leads to the formation of NO_2 radicals. It was also revealed that this weak absorption band in the near ultraviolet is not observed in spectra of the solvated or gas-phase RDX compounds [91]. This fact coupled with a value difference of nearly 2 eV between calculation and experiment, leads one to assume that this absorption is associated with the essentially solid-state of matter, with lattice defects. In these calculations,

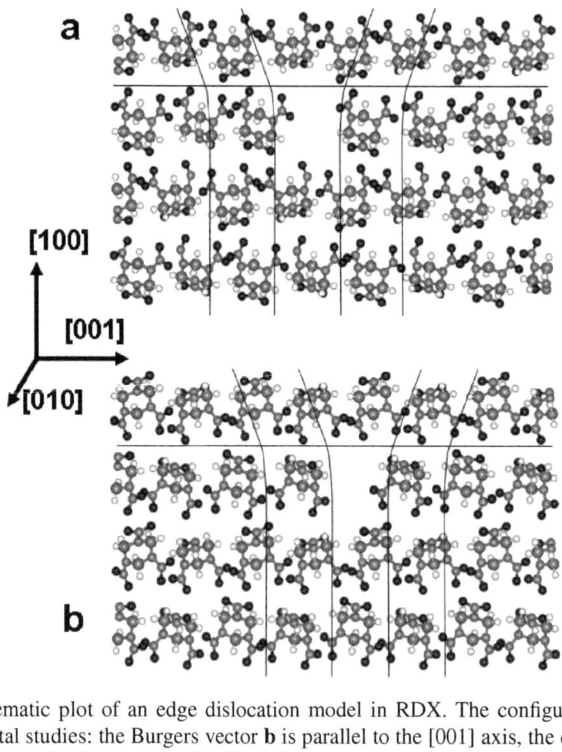

Fig. 8.2 A schematic plot of an edge dislocation model in RDX. The configuration was taken from experimental studies: the Burgers vector **b** is parallel to the [001] axis, the dislocation line d is oriented along the [010] direction. The (**a**) and (**b**) atomic structure around the dislocation line in the two nonequivalent RDX [010] molecular layers

the perfect crystal was modeled while the real solids containing some imperfections were investigated in the experiments. In our study of the dislocation-related deformation of the electronic structure, the band gap of RDX was calculated to be 3.3 eV, much closer to the experimental value (3.4 eV). This creates a new absorption band, which was not observed for either the perfect RDX crystal or the isolated RDX molecule [55, 88]. The results obtained lead to the conclusion that the band gap is most sensitive to the presence of dislocations, which, unlike other defects, produce local energy levels in the forbidden gap [55, 76]. Positions of these dislocation-induced electronic states in the optical band gap are strongly dependent on shear stress and/or molecular motion in the solid [88] (Fig. 8.3).

The dramatic narrowing of the optical band gap, related to the high shear strain, leads to the splitting off of local levels from *both* the top of the valence band and the bottom of the conduction band. These local states, having bonding and antibonding character, respectively, are attributed to the weakest N–NO_2 bond in any RDX molecule located near the dislocation core. From this observation one may suggest that the lowest energy electronic excitation, i.e., promotion of an electron from the highest occupied molecular orbital (HOMO) to the lowest unoccupied molecular orbital (LUMO) may lead to the breaking of the N–NO_2 bond. The nature of the HOMO to LUMO excitation and its relation to the molecular decomposition has also been established in earlier theoretical [92] and experimental [91] studies. The lattice relaxation induced by defects was neglected in this study to reduce computational costs. One can find more details regarding the calculation techniques and approximations used in our studies elsewhere [79, 86–88].

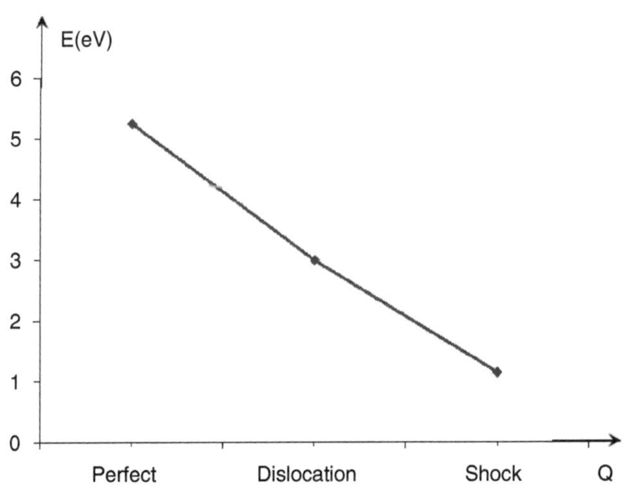

Fig. 8.3 A schematic of the optical band gap in RDX as a function of a generalized geometry configuration Q for perfect equilibrium RDX, for RDX crystal containing an edge dislocation, and for shocked RDX with the edge dislocation

The strain-related change of the optical band gap in RDX may be even more dramatic in the presence of several dislocations that may cluster and form dislocation patterns. The collective behavior of dislocation ensembles is known to play a critical role in many practically important phenomena in metals including work hardening, dislocation patterning in plastically deformed metals, the Portevin-Le Chatellier bands of different kinds in monotonically, or cyclically loaded metals [93–95], Luders fronts [95, 96], persistent slip bands, surface roughening, etc. In particular, the formation of persistent slip bands (they are called persistent because, being removed by surface polishing they reappear in the same location if the cyclic stress is continued) is thought to be a material instability at the mesoscale resulting in the formation of regular arrays of dislocation walls [97]. In spite of the fact that many properties of individual dislocation were successfully studied theoretically [98, 99], as well as in atomistic simulations [100, 101] it is an extremely difficult task to make a connection between an atomistic scale and a mesoscopic level on which dislocation patterning develops even for metals which have much less sophisticated atomic structure than the molecular crystals considered in this publication. Some successful attempts to establish such a connection exist (e.g., in Ref. [102] where both atomistic and mesoscopic simulations of dislocation junctions – Lomer-Cottrell locks – were performed) but are still very rare.

Another approach to dislocation patterning in metals is based on a continuum description, i.e., on the study of nonlinear systems of governing equations for dislocation densities. This "self-organizational" approach considers the formation of dislocation structures as a bifurcation phenomenon leading to solutions which do not belong to the thermodynamical branch [97]. At present, there is no general way to derive these equations starting from individual dislocations and their interactions. The analysis of these governing equations is usually limited to studying their linear stability which describes only primary bifurcations corresponding, e.g., to spatially homogeneous oscillations in time (Hopf instability), or to formation of steady-state spatial waves of dislocation densities (Turing instability).

Due to the rather sophisticated atomistic structures of RDX and other energetic materials, not a single atomistic calculation at the ab initio level for more than one dislocation in these molecular crystals has ever been performed. One could, however, expect a formation of spatial and temporal dislocation patterns in these materials similar to the Hopf and Turing instabilities. In this case, one will observe an accumulation of dislocations in some spatial regions and formation of dislocation piles, which will generate a larger strain in these areas than a single dislocation may produce. This may cause even stronger distortion of some of the atomic bonds located in this region and the appearance of a significant number of additional bonding and antibonding states in the RDX band gap. An extensive body of dislocation analysis in a variety of materials has been performed over the years by Armstrong and Elban [14, 103, 104]; in particular, they demonstrated an interesting contrast in the mechanical properties of molecularly bonded energetic crystals containing dislocations: the crystals are elastically soft, plastically hard, and brittle. They also show that the generation of dislocation clusters at large shock pressures plays an important role similar to the role of anharmonic lattice strains at the dislocation cores – they generate either "in situ" hot spot heating or direct electronic excitation [105].

8.2.3 Molecular Orientational Defects in FOX-7 and TATB

In this section, we explore how the changes of mutual orientation of molecules in molecular crystals and internal stresses attributed to those can affect the electronic structure and decomposition barrier for crystalline diamino-dinitroethylene ($C_2H_4N_4O_4$, FOX-7) (Fig. 8.4a) and triamino-trinitrobenzine ($C_6H_6N_6O_6$, TATB) (Fig. 8.4b). For this purpose, we generate a "orientational-molecule" defects to mimic the structure deformation in the vicinity of a dislocation core, stacking fault, or grain boundary.

The initial hope for FOX-7 to be a promising insensitive energetic material with superior properties [106], and potentially replace RDX ($C_3H_6N_6O_6$) and HMX ($C_4H_8N_8O_8$), the commonly used energetic materials in practical applications has been based on the fact that the chemical radicals in a FOX-7 molecule are the same as in a TATB molecule: $C-NO_2$, $C-NH_2$, and C–C, etc. (see Fig. 8.4c, d). This implies that FOX-7 and TATB would possess similar initiation chemistry and hence the former material can be expected to exhibit the desirable similar stability (i.e., insensitivity to impact and shock) to the latter. The argument is certainly valid provided that all chemistry of the crystal is completely defined by the chemical composition and bond dissociation energies of individual molecules. Some similarity of the layered crystalline structure, the lack of a melting point and related extensive hydrogen bonding network [30, 106] which are characteristic for both crystals, lends some additional support to this line of reasoning.

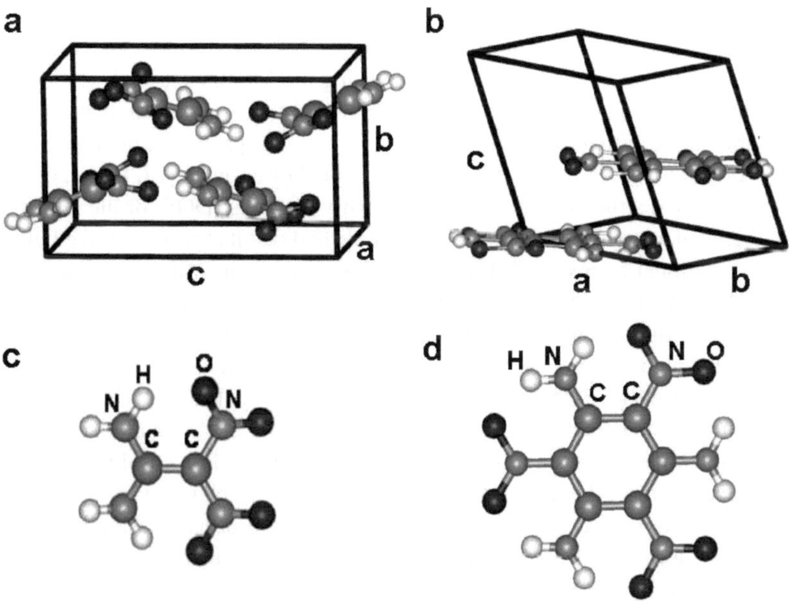

Fig. 8.4 The ideal crystalline (**a**) and (**b**) and molecular (**c**) and (**d**) structures of FOX-7 and TATB. The C atoms are shown by larger gray spheres, N– by smaller gray spheres, O– in black, H– in white

The unimolecular decomposition pathway of the FOX-7 molecule has been recently investigated using density functional theory (DFT). Politzer et al. [107] showed that C–NO_2 dissociation energy is lower than C–NH_2 dissociation energy and is equal to 70 kcal mol^{-1}. Gindulyté et al. [108] proposed that the first step in the decomposition is the C–NO_2 to C–ONO isomerization, which requires 59.7 kcal mol^{-1}, and as they noted agrees with the experimental activation energy (58 kcal mol^{-1} for temperatures between 210° and 250°C) [106]. There is an important point to be considered here; both of these values are related to the dissociation mechanism in the gas phase, while the experimental studies have been performed for the condensed phase. As was mentioned before (see Section 8.1), the dissociation energy of the isolated molecule may or may not be related to that of a molecule in the bulk crystal. Indeed, it has been shown for RDX (see Section 8.2.1) that the activation barrier is different for the solid-state and the isolated (gas phase) molecule and that the decomposition barrier near a defect is lower than in a perfect crystal [80]. Defects formed by change of orientation of molecules in the solid phase may play a significant role in the activation of the decomposition process [109].

The self-consistent field calculations for both molecular and crystalline FOX-7 and TATB were done using VASP codes [110]. Density functional theory in the generalized gradient approximation (GGA) for exchange-correlation, and plane waves with ultrasoft pseudopotentials for C, N, O, and H, were used [111]. The crystalline structure of FOX-7 was determined to have space group $P2_1/n$ with four FOX-7 molecules (56 atoms) per unit cell [112, 113] (Fig. 8.4a). The unit cell of TATB is triclinic with P symmetry and contains two molecules (48 atoms) [114] (Fig. 8.4b). The bonding between layers, which is defined largely by van der Waals interactions, is weak. The energy cutoff for the basis set was set at 24 Ry, and the integrations over the Brillouin zone were done using the Monkhorst-Pack scheme with four **k**-points in the irreducible wedge [115]. The relaxed corresponding structures acceptably agree with the experiment [106, 112, 114–116] and the calculations based on the force field [117], DFT [107, 108, 118], and Hartree-Fock [27, 118–120] methods. The relaxed structure and electronic density of states (DOS) of the defect-free FOX-7 crystal agree well with previously published results based on force field and density functional theory [117, 118] and with our recent investigation based on the Hartree-Fock method [120]. The optimal migration paths and the decomposition barriers were found using the nudged-elastic-band method [121].

We calculated the bond-breaking energies for the N–O, N–H, C–NH_2, and C–NO_2 bonds in the FOX-7 molecule, and we found that the lowest of these energies is the C–NO_2 bond-breaking energy of 72 kcal mol^{-1}, which is in excellent agreement with previous studies [107, 108]. The structure of the ideal FOX-7 crystal looks like a set of parallel corrugated (washboard-shaped) planes (Fig. 8.4a). In each of the layers, molecules are arranged in a checkerboard order, and the NO_2 (NH_2) groups of molecules in a given layer are neighboring with the NH_2 (NO_2) groups of molecules in the next (previous) layer. The band gap is 2.2 eV for the crystal, very close to the HOMO–LUMO gap of 2.3 eV for the molecule. This means that the electronic states of a perfect crystal are mostly defined by intramolecular bonds. The

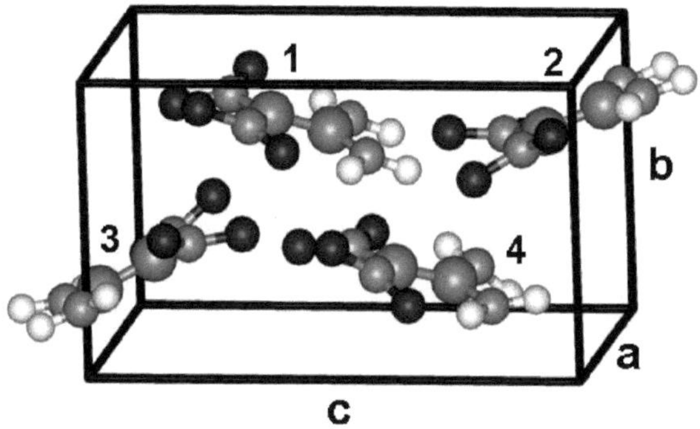

Fig. 8.5 The unit cell for crystalline FOX-7 with the "reversed-orientation-molecule" defect. The molecules 1 and 2 have a normal (NO_2-to-NH_2) intermolecular bonding while the molecules 3 and 4 have an anomalous (NO_2-to-NO_2) intermolecular bonding

intermolecular interactions, however, are responsible for the cohesion in the solid. This is typical for molecular crystals and has been found in a study of RDX (see Sections 8.2.1–8.2.2).

A simple structural defect one might imagine in a FOX-7 molecular crystal is a "reversed-orientation" molecule where one of the molecules is rotated by 180° about an axis perpendicular to the C–C axis. It corresponds to an interchange between the NO_2 and NH_2 groups. In the unit cell that consists of four molecules, there are now two FOX-7 molecules (1 and 2) with normal orientation (the NO_2 groups of the molecule are facing the NH_2 groups of the adjacent molecule) and two molecules (3 and 4) with anomalous orientation (the NO_2 groups of the adjacent molecules are facing each other) (Fig. 8.5). In other words, the system consists of alternating washboard shaped molecular planes each of which consists of layers with normal and anomalous mutual orientation. The total energy of this system is 2.5 eV per cell higher than that of the perfect material. This structure corresponds to a local energy minimum and the barrier to invert the molecule back to its normal orientation by rotation is high.

The electronic structure of the FOX-7 crystal with the reversed-orientation-molecule defect is different from the electronic structure of the perfect crystal (Fig. 8.6). First, the band gap of 1.3 eV is 0.9 eV narrower than the gap of the perfect structure. Second, the additional states that appear in the band gap of the ideal crystal have a dominant contribution from the oxygen atoms associated with the anomalous intermolecular NO_2–NO_2 bonds.

We found that the decomposition barrier for the NO_2 scission from a FOX-7 molecule with "normal" orientation located within the molecular crystal is about 4 eV (92 kcal mol^{-1}), which is much higher than the experimental value of the decomposition threshold of FOX-7 [106]. In other words, some regions of the crystal may become locally overheated due to accumulation of elastic energy before the NO_2 group splits away. The NO_2 scission barrier from a reversed-orientation-molecule defect is 2.6 eV (59 kcal mol^{-1}), which is in excellent agreement with the experimental value [106].

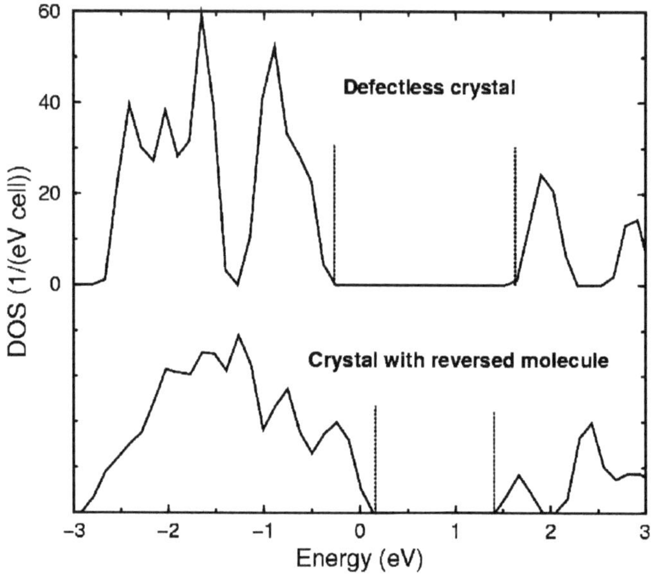

Fig. 8.6 Density of states of ideal FOX-7 unit cell as compared with density of states of FOX-7 unit cell containing reversed orientation molecule

In both the normal and anomalous intermolecular bonding we suggested that the FOX-7 decomposition started with the detachment of one of the NO_2 groups. Reduction of the band gap supports the experimental observations of pre-explosion conductivity and luminescence in metal azides [54,58]. The appearance of new local electronic levels in the gap predicts new absorption peaks in optical spectra and new luminescence lines. This is also consistent with previous experimental [122] and theoretical results for RDX materials [55, 79, 85, 86].

In spite of the similarity of the chemical radicals in FOX-7 and TATB molecules, the difference in their crystalline structure is apparently crucial. The crystalline arrangement of TATB consists of nearly two-dimensional planes parallel to the **ab**-plane (Fig. 8.4b), and the position of the planar TATB molecules in these planes is quite rigid, i.e., in-plane slip motions are energetically costly. Such rigidity is largely related to the high symmetry of the TATB molecule, which results in strong intermolecular and intramolecular hydrogen bonds in the **ab**-plane (the distance between the O and H atoms located at neighboring molecules is 2.23 Å only). Each molecule, however, has a possibility to rotate in-plane around its center, and such a rotation may also generate molecular misorientational defects and produce structural distortion of the two-dimensional planes.

Figure 8.7 shows an example of such a misorientational defect that was constructed by a 60° rotation of one of the TATB molecules around its center in the **ab**-plane. Such a rotation generates three anomalous intermolecular NO_2–NO_2 and three anomalous NH_2–NH_2 bonds (that are typically much weaker than the NO_2–NO_2 bonds). This configuration is metastable, and its energy is only 3 eV higher

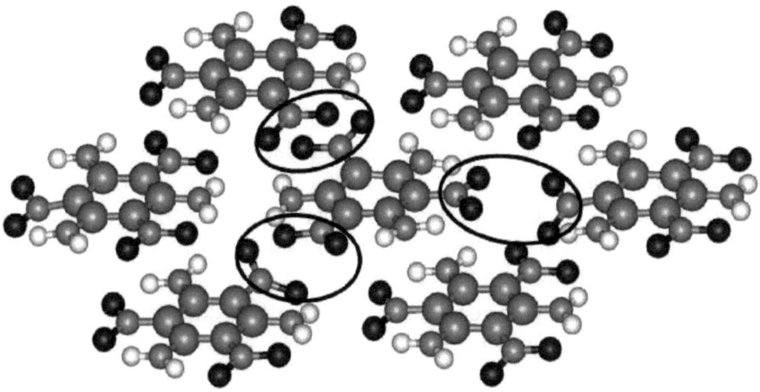

Fig. 8.7 An area of the two-dimensional **ab**-plane in TATB showing the molecular misorientational defect obtained by a 60° rotation of one of the TATB molecules around its center. Each 60° rotation generates three anomalous intermolecular NO_2–NO_2 bonds (encircled by ellipses)

that the energy of an unperturbed 2D- plane, i.e., each NO_2–NO_2 bond costs about 1 eV (compared with 2.5 eV for FOX-7). Therefore, the intermolecular interactions between the NO_2 groups are significantly weaker in TATB than in FOX-7. The rotation and bending of the NO_2 groups (what pushes some of the oxygen atoms away from the plane) can still be observed in TATB but these deformations are not that valuable as in FOX-7, and the NO_2 scission decomposition barrier in TATB does not significantly change in the presence of misorientational defects as it does in FOX-7. Therefore, one can say that TATB is not sensitive to the presence of some defects that may significantly increase the sensitivity in other energetic materials.

8.2.4 Shear-Induced Decomposition in FOX-7 and TATB

In this section, we report the results of ab initio simulations of the shear-induced decomposition in FOX-7 and TATB and reveal the important difference in chemical properties of these two materials. This sheds some light on materials by design issues and provides some suggestions for searching for new materials with prescribed properties.

As we discussed above, the chemical radicals in a FOX-7 molecule are the same as in a TATB molecule: $C-NO_2$, C–NH_2, and C–C, etc. (see Figs. 8.4c, d). Therefore, it was initially expected that FOX-7 and TATB would possess similar initiation chemistry and hence the former material can be expected to exhibit the desirable similar stability (i.e., insensitivity to impact and shock) to the latter. However, in our recent work [123,124] we clearly demonstrated that initiation of chemistry in molecular crystals is a collective phenomenon and cannot be comprehensively understood based on a description of individual molecules and even ideal crystalline structures alone. Hence, solid-state behavior, especially defect-induced incidents, cannot be ignored and should be taken into account.

In order to simulate shear strain displacements we used six-layer molecular slab (or (3+3) slab) supercells (see [125, 126] for details). The slabs were separated by a 10 Å layer of vacuum in the direction perpendicular to the layers. The shear strain was introduced by an instantaneous shift of the three upper layers of the slab relative to the three lower layers in the directions parallel to the layer's plane (i.e., the **ac**-plane for FOX-7 and the **ab**-plane for TATB) on vectors $\Delta = \gamma_a \mathbf{a} + \gamma_c \mathbf{c}$ for FOX-7 ($\Delta = \gamma_a \mathbf{a} + \gamma_b \mathbf{b}$ for TATB). All the γ-parameters were varied in the range between 0 and 1 (the unity corresponds to the in-plane shift on one of the translational vectors). The decomposition barriers were calculated using the nudged elastic band approach [127] for molecules in the two "interfacial" layers (where the strain deformation is maximal) at each vector Δ for both the "instantaneously shifted" or "relaxed" supercells. The relaxation was performed with only one constraint, namely, we fixed the position of one carbon atom attached to the NH_2 group for each molecule in the lower layer in the slab. Such a constraint eliminates the drift of the supercell as a whole in the relaxation process. Also, it is physically justified because the main contribution to the relaxation is related to the NO_2 groups.

We calculated decomposition barriers of the FOX-7 and TATB molecules, perfect crystals, and crystals deformed by the shear to reveal a trend and determine if there is any correlation between the experimentally determined energy of decomposition, which is closely related to sensitivity to impact and shock, and the presence of shear in the solids (see Table 8.1). We focused only on one mechanism among all possible decomposition pathways, that is the C–NO_2 rupture since it is the most supported initial stage of chemistry [1, 128, 129]. Although cyclization reactions of TATB that produce furazans and furoxans were suggested to be lower activation energy processes [34, 130], which may compete with simple C–NO_2 bond scission at ambient conditions, they can be overstepped by the rapid, large temperature rise during shock and impact stimulation [131]. Thus, we do not include cyclization and isomerization reactions caused by hydrogen transfer in our consideration here [132]. It can be seen from Table 8.1 that the energy for molecular dissociation in the gas phase is about 70 kcal mol^{-1} for both FOX-7 and TATB. This falls into the range of 61–70 kcal mol^{-1} reported for a wide variety of compounds [1] and is in general agreement with mass spectroscopy [133, 134], laser-assisted pyrolysis [135], shock tube pyrolysis [136], and theoretical studies [107–109, 119]. This is somewhat higher, however, than the differential calorimetry-based estimate for FOX-7, 58 kcal mol^{-1} [106].

Table 8.1 Energy barrier of the C–NO_2 bond decomposition (kcal mol^{-1})

Material	Isolated molecule	Defect-free crystal	Shear-strained crystal
TATB	71	100–107	100–112
FOX-7	72	92	42–94

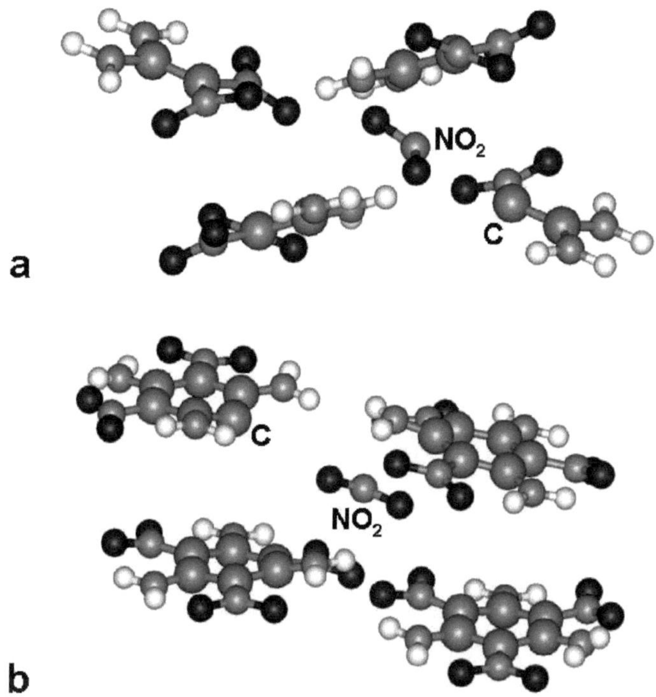

Fig. 8.8 The schematic representation of the NO_2 interstitial position due to the C–NO_2 scission in FOX-7 (**a**) and TATB (**b**)

The bond-breaking barriers in defect-free crystals are significantly higher, 92 kcal mol^{-1} for FOX-7 and 100–107 kcal mol^{-1} for TATB; the crystalline field and extensive hydrogen bonding stabilizes the molecules in the condensed phase. Such a stabilization reflects the fact that after breaking the C–NO_2 bond, the detached NO_2 group should be positioned somewhere in the interstitial between the molecules. In both crystals, the space between the molecular layers is large enough to accommodate the NO_2 group that finds a metastable position in the interstitial but still interacts with other molecules (Fig. 8.8). It is worthy to note that the energy barriers for C–NO_2 bond scission are still very close to each other in both TATB and FOX-7 ideal solids (the range of the barrier values for TATB corresponds to different final states of the system).

A fundamental difference between the crystalline solids TATB and FOX-7 arises when sheer–strain is introduced into the lattices. For the relaxed structures, the C–NO_2 decomposition barrier does not significantly change in comparison with the decomposition barrier in ideal crystals because the relaxation process is mainly defined by the motions (stretching, bending, and rotation) of the most flexible NO_2 groups, and in the relaxed configurations the molecules are not significantly deformed (see Ref. [124] for details). For unrelaxed structures (i.e., immediately after introduction of instantaneous shear strain), the decomposition barrier may be

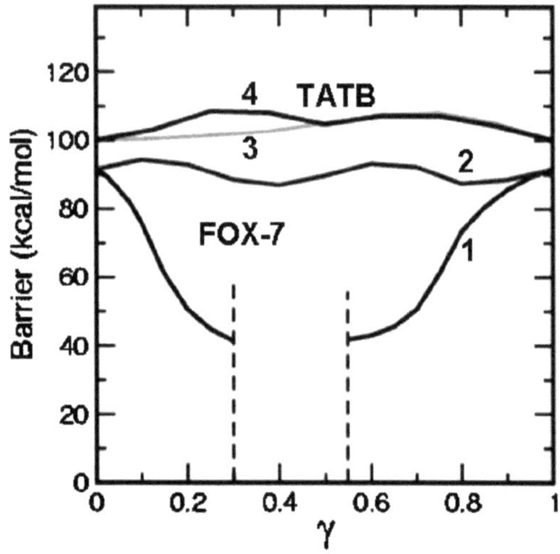

Fig. 8.9 The C–NO_2 scission decomposition barrier as a function of the in-plane shift parameter γ. Curve 1 corresponds to FOX-7 shift along the **c**-axis ($\gamma_a = 0$, γ_c – variable); 2 – FOX-7, **a**-shift (γ_a – variable, $\gamma_c = 0$); 3 – TATB, **a**-shift (γ_a – variable, $\gamma_b = 0$); 4 – TATB, **b**-shift ($\gamma_a = 0$, γ_b – variable). The vertical dashed lines indicate the range where the interfacial FOX-7 molecules overlap for the **c**-shift

significantly affected by the shear strain. In the *c*-direction-shear strained FOX-7, for the parameter γ_c, in the ranges $0 \leq \gamma_c \leq 0.3$ and $0.55 \leq \gamma_c \leq 1$ (for $0.3 < \gamma_c < 0.55$, the NO_2 groups at the neighboring interfacial molecules significantly overlap, and other decomposition channels become possible), the C–NO_2 decomposition barrier at interfacial molecules varies between 42 kcal mol^{-1} (for $\gamma_c = 0.3$ and 0.55) and 92 kcal mol^{-1} (for $\gamma_c = 0$ and 1) as shown in Fig. 8.9. Such a big change in the barrier simply reflects the fact that a significant fraction of the strain energy is accumulated inside the C–NO_2 bonds, which facilitates the detachment of the NO_2 groups. The FOX-7-sub-slab shift along the **a**-translational vector ($\gamma_c = 0$) is energetically more favorable than the shift in the **c**-direction because in such a shift the humps and grooves of the zigzag-shaped layers move parallel to each other and never overlap.

An examination of a crystalline arrangement of TATB in comparison with FOX-7 clearly demonstrates that there is a more or less "rigid" two-dimensional order in the **ab**-plane in which planar TATB molecules are situated with no specific in-plane slips allowed. Such a rigidity is mainly related to the high symmetry of the TATB molecule, which results in strong hydrogen bonds in the **ab**-plane (the distance between the O and H atoms located at neighboring molecules is only 2.23 Å). However, the shear between these layers is not energetically costly because the interlayer overlapping is minimal. This slip motion does not cause any significant effects on the dissociation of TATB molecules (and only slightly affects the electronic structure) (Fig. 8.9).

FOX-7, on the other hand, lacks molecular symmetry and is built out of polar molecules linking with each other by attraction between NO_2 and NH_2 groups located on adjacent molecules. This creates strong bonding in corrugated chains of molecules along the **c**-axis but not in all directions (this is why shear strain along the **a**-direction does not produce so much structural damage). Also, several possible motions of FOX-7 molecules are allowed by symmetry. This results in the appearance of polar surfaces in the structure and allows a richer variety of local structural defects to occur, such as, different conformers [107], or a "flipped" molecule [125]. A major consequence of this defect-induced disorder is attributed to the reduction of the decomposition barrier and in turn, the higher sensitivity of FOX-7 to initiation as compared to TATB. This means that regardless of whether the dissociation starts from the ground state or from one of the excited states, less energy is required to break bonds in a defective crystal than in an ideal crystal. Anyway, ideal crystals of FOX-7 are unlikely to exist [125].

Hence we suggest that there is no sense in continuing to search for better cleaner ways to synthesize FOX-7, hoping that crystals of higher quality with lower concentration of defects will provide lower sensitivity to detonation, because the crystalline lattice of FOX-7 has the ability to easily accommodate certain types of shear slips, reducing the dissociation barrier and causing decomposition. Instead, one may redirect efforts towards looking for another crystalline phase of FOX-7 or other material of similar composition that presumably does not exhibit the reduction of decomposition barriers that accompanies the shear strain and shear-induced defects in the crystalline lattice. In addition, one might search for other materials whose crystalline lattices have close packing of molecules in at least two directions (two planes) and do not allow easy slipping of crystallographic planes or other structural defects that are able to introduce large arrays of disorder in the crystal. In fact, FOX-7 would be as insensitive as TATB if the migration barrier of a zigzag-shaped plane (or a group of planes) in the **ac**-plane would have low anisotropy and low activation energy like the migration of very flat, graphite-like planes in the case of TATB.

8.3 Modeling Initiation of Chemical Reactions

The modeling of chemical reactions in practical energetic materials includes a description of individual imperfections, defect–defect and defect–lattice interactions, a response of the system to the external excitation, and the behavior of materials under high temperature, high pressure and, therefore, represents an extraordinary challenge for both theoretical and experimental studies. Nevertheless, attempts to simulate chemistry in these materials with appropriate models can and should be made. In this section, we examine the possible role of electronic excitations in the initiation of chemical reactions in solids. We try to understand how a fast process, such as an electronic excitation, promoted by lattice defects (especially, edge dislocations or other shear-strain related local deformations) or external stimuli (for example, radiation) and pressure of the shock wave, affects macro-properties of energetic molecular solids and eventually leads to an explosion [76].We will deliberately focus on two different aspects of excited electronic states and their effect on the properties of the explosive material. First, we discuss narrowing of the optical band gap in RDX and an excitonic state, which appears as a consequence of that change in the electronic structure. We think that this subject deserves further consideration since narrowing of the band gap was recently confirmed experimentally in RDX [137] and found theoretically in other materials [138]. Further, we describe how these early ideas were tested on FOX-7 by addressing the question of whether the electronic excitations can modify the dissociation mechanisms and alter the reaction barriers to such an extent that the dominating chemistry changes from the endo- to exothermic type, thus making new pathways for the molecular decomposition available [139]. The scenarios described below are based partly on the available theoretical and experimental results and partly on the extrapolation of these results. Therefore, the models presented should be considered as possible interpretations of the results rather than a complete theory.

Let us consider a very general view on initiation chemistry in high-explosive solids. A transition of the metastable molecules in an explosive crystal (for example, RDX or FOX-7) into the ground state corresponding to the most stable products (CO_2, CO, NO, NO_2, H_2O, HCN, etc. (see, for example, Refs. [1,3,67,140])) results in a large energy release. Under normal conditions, no spontaneous chemical dissociation can happen due to the high energy barrier that separates the metastable state from possible reaction products. An external perturbation of the system, such as a mechanical impact, a spark, a laser beam, shock, or plasma ignition, induces a wave front progressing across the solid. For instance, a shock wave traveling through the RDX crystal creates a local pressure of several GPa (experimental estimates for solid RDX vary from 0.2 [57] to 4 GPa [38]), introducing a large shear stress. The energy carried by the shock front is capable of initiating decomposition and further chemical reactions. It is commonly believed that the first step in the decomposition of most nitro-compounds is a detaching of NO_2 species with the typical dissociation energies from 40 to 70 kcal mol^{-1} [1]. For example, breaking the N–NO_2 bond in the RDX molecule takes about 50 kcal mol^{-1} (~2 eV) [141, 142], and breaking C–NO_2 bond in FOX-7 molecule requires about 58 kcal mol^{-1} [106]. Subsequent decomposition steps release more energy than is required for this process. This additional energy release contributes to the wave front and raises its pressure and temperature, which, in turn, lead to a chemical chain reaction and an explosion.

In this very general picture it is not all clear how the energy of the shock front can be transformed into chemical energy as the energy of compression due to impact is not enough to break bonds [76]. It was shown earlier that a localization of all the energy of the impact wave (i.e., the energy stored in a compressed RDX crystal) on just *one* N–NO_2 stretching mode has a very low probability; hence the bond-breaking in the RDX molecule while being on the ground-state potential energy surface is also a highly improbable event [76]. Therefore, we need to look for a process that selectively places the available energy (that is pumped up into the material as a result of an external perturbation) in only a few desired modes. Alternatively, there should exist processes that can locally change the energetics and/or reduce decomposition barriers. There is another important question. How many bonds per unit time must be broken initially in order to start the chain reaction? This is, in fact, the same question as how hot are the hot spots [143], reformulated in terms of activation energy instead of thermodynamic temperature. Indeed, temperature is related to a reaction rate. The reaction rate may also be described in terms of the number of parameters per unit time achieving a certain activation energy, most likely through collisions. The number of collisions per unit time and the energy they impart will, of course, be related to the temperature.

8.3.1 Electronic Excitations in RDX

Let us now consider a way by which the reaction goes through the *excited* state potential energy surface. The main modifications of the band structure of RDX are illustrated in Fig. 8.3. As was discussed in Section 8.2, dislocations and, probably, other complex defects (see, for example, Sections 8.2.3–8.2.4) reduce the perfect calculated band gap of 5.25 eV to roughly 3 eV because of the local states in the band gap, which is in good agreement with the experimental observations of the optical absorption spectrum. The system is in the ground state until the impact wave arrives. Volumetric contraction, induced by a shock wave progressing through the solid, leads to an additional narrowing of the optical gap [76]. In other words, a shock wave drives the system to a point from which the molecule has a good chance of being excited. The probability for this excitation under shock condition is, clearly, higher than at the equilibrium due to the locally reduced gap and a high sensitivity of this local deformation of the electronic structure to external stimuli. The locally increased temperature also contributes to the increased probability for excitation. The number of electronic transitions in the shock front passing through N molecules can be estimated by the simplified formula as $w = N \cdot \exp(-E_g/kT)$. Assume a temperature of 600 K (kT = 0.046 eV, in fact, the temperature in hot spots can be much higher) and the size of the RDX sample is 1 mm^3. Our ab initio calculations of the compressibility show that a pressure of $\sim$15–20 GPa (or even less) locally reduces the gap, E_g, to about 1 eV for RDX containing an edge dislocation. Using the RDX molar volume, 1.233×10^{-4}m^3 (which can be found from the known density, 1,800 kg m^{-2}, and molecular weight, 222.12), the number of RDX molecules in a sample of 1 mm^3 size, $\sim$4.9$\times$ 10^{18}, and typical concentration of dislocations per unit area, 5$\times$10^{12} cm^{-2} [144], one can find that about 8.5$\times$10^{16} molecules are located in the immediate vicinity of the dislocation cores in a 1 mm^3 sample. Then, a simple estimate by the formula mentioned above shows that about 3.1$\times$ 10^7 such molecules will be excited by the impact front. The estimated number of excited molecules is not large compared to the number of molecules in the solid. However, it can be very significant relative to the number of locally broken bonds necessary for chain reaction initiation.

The nature of electronic excitations in crystalline solid nitromethane under conditions of shock loading and static compression has been investigated [145]. The authors argue that the narrowing of the band gap observed is not enough to produce a significant population of excited states in the crystal. We would like to note that we used here the energy gap corresponding to the *optical* (or vertical) electronic transition. In fact, the *thermal* excitation of electrons requires even less energy due to nuclear relaxation in the excited state. The set of calculations discussed here do not use full nuclear relaxation. Since molecular vibrations (10–100 fs) are fast compared to the timescale of reaction initiation (10–100 ps [44]), nuclear relaxation is potentially important. Nuclear relaxation can either inhibit (ground state relaxation) or promote (excited state non-vertical transition) electronic excitations. Furthermore, a dynamic response of the lattice to excitation by a shock/impact wave in terms of probability and population of the electronic excited states is expected to be much more pronounced. Therefore, the more reliable estimate of the probability of electronic excitations should include all the contributions mentioned above, which is not a trivial problem. Also, a comparison to mass spectroscopy experimental data would be highly desirable. For now, we have to conclude that external stimuli in combination with lattice defects drive the system to the excited state with a nonzero probability. Our recent quantum mechanical molecular dynamics study of the optical band gap of FOX-7 as a function of shear strain and temperature illustrates the complexity of the crystalline lattice response to a mechanical and thermal perturbation [124, 126] (Fig. 8.10).

Further, let us see what happens with the system once it is excited and transferred into the *excited* state potential energy surface. There are three possibilities. First, the system can relax back to the ground state with the energy release in the form of a photon. This corresponds to a radiative transition accompanied by optical luminescence. In this process the system becomes unavailable for initiation from an excited electronic state because it is once again on the ground-state potential surface. It can only be subject of a subsequent excitation. Second, the system can undergo a fast nonradiative transition to the vibrationally hot state [25, 38, 44] thus speeding up the vibrational excitation process. The subsequent decomposition will then start from the vibrationally excited state [25]. Third, there is an actual dissociation of the molecule from the excited state. As mentioned above, when the system (in our example, an RDX crystal) is excited, an electron is removed from HOMO to LUMO,

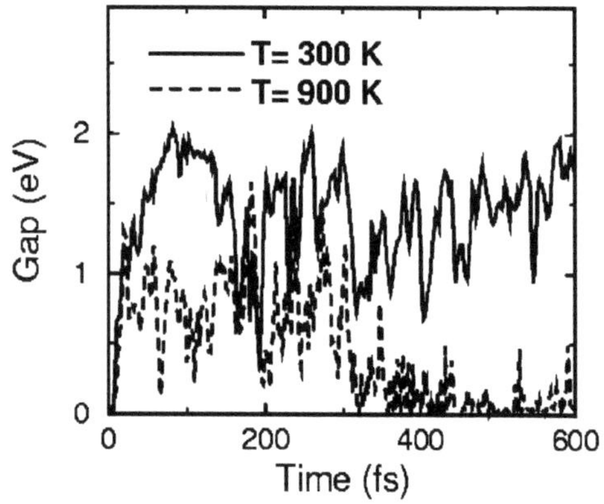

Fig. 8.10 Relaxation of the optical band gap of FOX-7 exposed to a shear-strain deformation at two different temperatures

and the energy is localized on a single N–NO_2 stretching mode. If the electronic excitation is long-lived (triplet excited states are known to live much longer than the period of nuclear vibrations), then this excess energy can be released in a form of nuclear motion [146] and, with a high probability, result in the breaking of the N–NO_2 bond followed by the release of products having high kinetic energy. The energy released will be expended in heating the sample and promoting further excitations and subsequent dissociations. The process repeats itself, increasing the local pressure and temperature around the site. Thus, even a moderate reduction of the band gap can trigger initiation.

Our proposed model of initiation triggering by electronic excitations associated with the dislocation presence in the crystal was here illustrated with RDX. Consistent results concerning the electronic structure were also obtained for PETN [$C(CH_2ONO_2)_4$] crystal. Local electronic states induced by edge dislocations are very sensitive to any molecular displacements around the dislocation core. New optical absorption bands attributed to these local states are predicted to be a strong function of a mechanical deformation of the crystal [79, 88]. This is in good agreement with shock compression measurements of optical spectra of PETN [147]. PETN represents an intermediate case between primary and secondary high explosives with respect to sensitivity to detonation. A number of experimental and theoretical data described in the literature lead us to believe that the suggested model of initiation may be general for many solid molecular explosives. A detailed discussion in favor of this proposal is available [76].

There is experimental evidence of the proposed excitonic mechanism, which deserves to be mentioned here.

It has been reported that eliminating or severely reducing the number of moving dislocations during rapid deformation will reduce the number and intensity of shock-induced hot spots within the crystal. Carefully prepared crystals of the explosives RDX, TNT, HMX, and PETN, with limited defect content and few dislocations, are nearly impossible to ignite at shock pressures often in excess of 40 GPa [148, 149]. Similar crystals of the same materials but with high defect/dislocation content will ignite at a shock pressure of 2 GPa or less [150].

The initiation of RDX by laser excitation under modest pressure conditions (up to 5.0 GPa) suggests that sample initiation occurred in a discrete region of the crystals and not uniformly across the sample [122]. From this, it was concluded that absorption must have occurred at point or line defects. In that study, theory favored absorption at a dislocation or a vacancy associated with a dislocation core. As was discussed above, the band gap of the RDX crystal containing this complex defect is very sensitive to pressure and significantly reduced ($\sim$2 eV with the pressure in the GPa range). Thus, it is directly demonstrated that the energy associated with this defect is consistent with the absorption of the green excitation wavelength observed experimentally, whereas normally RDX is transparent to green light [122]. This study in fact demonstrates a real possibility for the laser-induced excitation of selective modes of the sample.

A dislocation-induced effect in early stages of slow decomposition in the electric field has been experimentally studied for AgN_3 whiskers [54, 55]. It was observed that the gas (a product of chemical reaction) evolved mostly from dislocation etch pits. Next, mobility of dislocations, caused by their magnetic moment, was used for a significant reduction of the dislocation density in the crystal, which is, electrical cleaning of the sample. Further, electric current was applied to the cleaned samples for 20 h, and neither etch pits nor gas emission were observed during all that time. Then, the initial properties of the sample were restored and gas emission regions near etch pits were observed again. The obtained results suggest that the regions of increased chemical reactivity, responsible for the initial stages of decomposition, are associated with dislocations. Furthermore, a series of experiments were performed to study kinetics of the pre-explosion conductivity of AgN_3 crystals using YAG:Nd^{3+} laser pulse initiation (1,064 nm, 10 mJ, 30 ps), the methodology of which is described elsewhere [54, 58]. Two groups of samples were probed: one set of crystals with dislocation density $\sim 2 \cdot 10^3 cm^{-2}$, and another set of "pure" crystals with dislocation density less than $10^2 cm^{-2}$. A comparison of the kinetics of pre-explosion conductivity for these two groups of crystals show that chemical decomposition in crystals with a low density of dislocations develops noticeably slower than in the highly defective samples.

8.3.2 Charge Trapping and Electronic Excitation as a Precursor of Initiation in FOX-7

Elucidation of initiation mechanisms in FOX-7 is of significant interest for understanding sensitivity trends in energetic materials because of two reasons. First, FOX-7 is an attractive model system because of its relatively small molecule (Fig. 8.4c) and simple unit cell (Fig. 8.4a). Second, FOX-7 is chemically related to TATB (Fig. 8.4c, d) and therefore, a careful comparative analysis of both materials could potentially disclose a guide to desirable control of the explosive's sensitivity. In this section, we describe recent results on the modeling of chemical reactions for the unimolecular decomposition of a single FOX-7 molecule in gas phase, in order to understand the differences between the charged or excited states of the molecule and its equilibrium state. It was revealed that charging and excitation may not only reduce the activation barriers for decomposition reactions, but also change the dominating chemistry from endo- to exothermic type [139].

The differential scanning calorimetry experiments of FOX-7 in condensed state suggest that the thermal decomposition energy is approximately 58 kcal mol^{-1} [106]. Politzer [107] and Gindulytė [108] simulated possible unimolecular decomposition pathways of FOX-7 from its ground state and proposed different scenarios. Only a limited amount of experimental data on FOX-7 [106] and conflicting conclusions derived from theoretical calculations [107, 108] hamper establishing the mechanisms of sensitivity to detonation initiation and of the related initial step(s) of the decomposition of this material.

First, the relaxed structures of the ground state FOX-7 molecule, ionized $[FOX\text{-}7]^+$ and negatively charged $[FOX\text{-}7]^-$ radicals, and the lowest triplet excited state have been calculated. Positively charged ion has two stable configurations with similar energies and different geometries, which we denote h_a and h_p. The electronic structure of the FOX-7 molecule and its fragments as well as the barriers for molecular decomposition were calculated using Density Functional Theory (DFT) and the hybrid B3LYP functional as implemented in the Gaussian03 code [151] and 6–31+G(d,p) Gaussian-type basis sets. The details of the methodology and the corresponding structures of FOX-7 are described in great detail elsewhere [139].

Then, we simulated relevant decomposition processes, such as rupture of the C–NH_2, C–NO_2, and C=C bonds as well as CONO- and HONO-isomerizations (see Fig. 8.11). We found that C–NH_2 and C=C bond ruptures as well as HONO isomerization are not favorable decomposition pathways of FOX-7. However, the presence of an additional electron may change the energetics and make those reactions feasible [139]. Thus, dissociation of a $[FOX7]^-$ radical via C=C break requires only 63.8 kcal mol^{-1}, much lower than 120.2 kcal mol^{-1} in the ground state. Also, in $[FOX\text{-}7]^-$, the trapped electron is distributed over nitro-groups, which provides extra attraction between protons and oxygen atoms. This makes the exothermic HONO arrangement (Fig. 8.12a) possible with the small energy gain of 3.25 kcal mol^{-1} and a relatively low barrier of 29.76 kcal mol^{-1}.

Next, we found that the C–NO_2 bond fission reaction is an endothermic reaction that requires 66.98 kcal mol^{-1}, which is in good agreement with 70 kcal mol^{-1} from previous calculations [107] and experimental results for nitro-compounds [1]. In the case of positively charged $[FOX\text{-}7]^+$, the cleavage of a C–NO_2 bond becomes exothermic with the energy gain of 27.67 kcal mol^{-1} for the h_a configuration and 8.08 kcal mol^{-1} for the h_p configuration while the barriers for these reactions are 54.71 and 59.10 kcal mol^{-1}, respectively. The decomposition of negatively charged

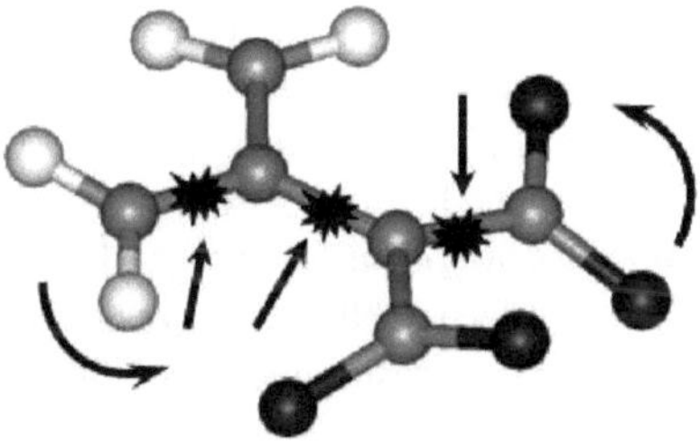

Fig. 8.11 A schematic plot of possible unimolecular decomposition mechanisms in FOX-7: breaking of C–NH_2, C–NO_2, and C=C bonds, or –NO_2 to -–ONO rearrangements

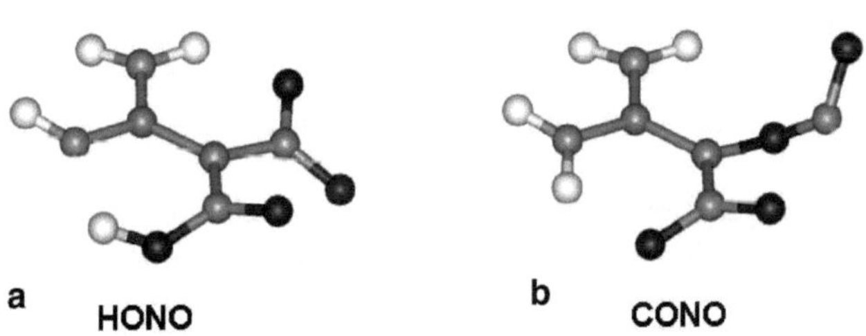

Fig. 8.12 The molecular structures of (**a**) HONO and (**b**) CONO isomers in FOX-7

$[FOX\text{-}7]^-$ is endothermic with a dissociation energy of 14.48 kcal mol^{-1} and a barrier of only 20.62 kcal mol^{-1}. Such a significant reduction of the barrier is due to a larger repulsion between nitro-groups accommodating an extra-electron. In the triplet state of the FOX-7 molecule, the C–NO_2 bond cleavage is also endothermic with a calculated dissociation energy of 12.44 kcal mol^{-1} and a barrier of 32.97 kcal mol^{-1}. We confirm that the homolytic C–NO_2 cleavage is a plausible initial decomposition reaction in FOX-7. In addition, we reveal that the barrier for this reaction becomes significantly lower in the electronically excited and charged states of the molecule.

The presence of NO registered in final products of the decomposition of nitro compounds suggests that the isomerization of C–NO_2 to a C–ONO and/or other intra-molecular rearrangements should take place prior to the decomposition [1, 29] (Fig. 8.12b). In our simulations, formation of a stable CONO isomer in the neutral FOX-7 molecule is an exothermic reaction with a small energy gain of 4.1 kcal mol^{-1}. The calculated barrier for this transformation is 63.5 kcal mol^{-1} and the transition state is characterized by a geometry in which one nitro-group is rotated with respect to the molecular plane. After the CONO-isomer is formed, the CO–NO bond can be cleaved releasing the NO species (see Fig. 8.12b). The calculated barrier for this reaction is 27.4 kcal mol^{-1} and the overall energy gain is 6.6 kcal mol^{-1}, which is consistent with the previously reported results [108].

Importantly, we found that the NO species can detach spontaneously once the molecule is charged or excited. Indeed, an additional electron can be trapped by the CONO fragment. The vertical electron affinity of the unrelaxed negatively charged CONO isomer of the FOX-7 molecule is 22.6 kcal mol^{-1}. This configuration then relaxes with the spontaneous NO detachment and overall energy gain of 23.8 kcal mol^{-1} (with respect to the energy of the negatively charged CONO isomer of the FOX-7). Alternatively, an ionization of the molecule (calculated vertical Ionization Potential is 191.6 kcal mol^{-1}, or 8.31 eV) also results in the spontaneous NO detachment with an energy gain of 27.7 kcal mol^{-1} (with respect to the energy of the positively charged CONO isomer). It follows from our calculations that isomerization of C–NO_2 to C–ONO does not occur for charged FOX-7 molecules.

A singlet-triplet excitation of the CONO isomer results in a bonding-antibonding transition at the CO–NO bond. As a result, the CO–NO bond weakens and the NO group can then detach without a barrier.

Further, we note that since the barriers of the C–NO_2 to CONO isomerization and the C–NO_2 rupture are comparable, both of these processes are likely to take place at the early stage of the decomposition, which is consistent with the observed existence of both NO and NO_2 products in the mass spectra of FOX-7 [106]. We predict that once the CONO isomer is formed, trapping of a negative or positive charge as well as photo-excitation will result in a spontaneous detachment of the NO molecule.

Finally, our calculations demonstrate that the excitation and charging of the molecule can have a dramatic effect on the decomposition process by facilitating some mechanisms of dissociation and precluding the others [139]. For example, as discussed above, the formation of CONO isomers can take place in the neutral molecule, but the negatively charged FOX-7 molecule does not support nitro-nitrite rearrangement and therefore, breaks this channel of decomposition. On the other hand, trapping an electron by a CONO-rearranged FOX-7 molecule leads to the spontaneous detachment of NO species with an energy gain.

It is obvious that the decomposition of FOX-7 in the solid state is much more complex [124–126] and is controlled by cooperative behavior, which involves the excitation processes and structural inhomogeneities in a crystalline lattice. Explicit modeling of the solid state chemical reactions for decomposition of molecules placed on structural defects is currently underway.

8.4 Summary and Discussion of Perspectives

A series of structural and electronic defects have been considered here by means of first-principle calculations. We noted two important outcomes that the defects and simulated deformations have in energetic materials: the band gap narrowing and the reduction of the dissociation barrier. Our observation is that the band gap exhibits quite a complex behavior right after external stimuli. When the system is let to relax, it tends to retain its equilibrium band gap and dissociation barriers, yet the relaxation time is large enough for chemistry to occur and the relaxation dynamics, most likely, reflect the complexity of the initiation process. This should be scrupulously studied further on several classes of EM with significantly different types of critical chemical bonds to draw conclusions.

Among all considered structural perturbations in the crystalline lattice, dislocations, local stresses, and shear-strain-induced deformations generate localized electronic states in the band gap. At the same time, hydrostatic compression alone (even to a very high degree), molecular vacancies, vacancy complexes, voids, and surfaces barely modify the electronic structure of the materials. However, they make the materials chemically and mechanically softer, unlike dislocations, which make the material mechanically harder and chemically softer. An important conclusion from this observation is that the internal interfaces that are generally associated with the presence of dislocations, grain boundaries, or shear strains are good candidates for the main ingredients of hot spots since they assist in the decomposition of molecules by affecting mechanical, chemical, and electronic properties of materials. This combined power can force the system to various cross-sections of the potential energy surface and open up unusual pathways for the decomposition. From our study, the molecules placed on structurally perturbed spots in the lattice, will readily decompose right after the external excitation.

Another deserving point we would like to raise here is that the explosive decomposition in condensed EM cannot be comprehensively understood based on gas-phase experiments or modeling of isolated molecules because cooperative behavior and interactions between molecules are apparently crucial and certainly control solid-state processes. This argument has been mentioned in literature by many researchers (see for example, [33, 34, 38, 55]), yet not enough information is currently available. Hence, solid-state studies should be pursued at all possible levels of theory and experiment. In order to illustrate only one aspect of this multicomponent approach, the initiation of the chemical reaction by means of excited electronic states in RDX has been described here from a solid-state physics viewpoint (see Section 8.3.1). In particular, an attempt to look for the relationship between the defect-induced modification of the electronic structure, electronic excitations, and chemical reactions under shock conditions in RDX has been made. Ab initio band structure calculations were performed to model the effect of the strong compression induced by a shock wave on crystals with and without edge dislocations (see Section 8.2.2). The subsequent thermal decomposition via N–NO_2 pathways of solid RDX has been simulated. Both an excitonic mechanism for the earliest stages of initiation suggested earlier with RDX and an explicit modeling of all possible initiation pathways in FOX-7 from equilibrium and excited states are discussed here in conjunction with recent developments in fast spectroscopies.

We would like to draw attention to some fundamental aspects of the problem, which extend beyond simply the elucidation of the mechanism(s) of EM decomposition. We believe that this chapter convincingly demonstrates a very interesting possible realization of chemical reactions in solid state, wherein the overall decomposition can be triggered (or even controlled) by manipulating the particular *local* atomic arrangement in the vicinity of lattice imperfections and hence the electronic structure of the system rather than by its macro-parameters such as temperature or pressure. The localization of electronic excitations at definite sites (presumably hot spots) of the crystalline lattice leads to the birth of active radicals at the site. Thus, a slow process of migration of heavy real particles (usually, in a diffusion process) is replaced with the much faster migration of electronic excitations (quasi-particles).

As of yet, no complete microscopic theory for the early stages of initiation currently exists. Many questions, for example, the problem of how the energy of the shock wave is transferred into the individual molecules, and how the released energy produces the multiplication of electronic excitations to support a chemical reaction, remain unanswered. Also, the question of how many bonds must be broken within a certain period of time to trigger a chain reaction, and how sufficient populations of electronic excited states need to be created, still need to be answered.

The results of this review establish some significant fragments of a general picture of the explosive decomposition of solids. First, electronic excitations of the crystal lattice may trigger and control the decomposition process for high explosives. The key role here belongs to the processes of regeneration and multiplication of excitations, which still need to be clarified. If this multiplication is a result of thermally activated processes, the classical thermal explosion takes place [9,55,143]. If this multiplication is not related to thermal activation, we are dealing with the hot spot chain explosion mechanism [52,55].

Second, the earliest stages of decomposition, such as initiation, are associated with the lattice defects and the modifications of the electronic structure induced by the shock wave progressing through the solid. Dislocations and shear strains here play the particular role of inducing localized states and consequent rapid narrowing of the band gap locally, in the vicinity of molecules placed on dislocations or interfaces.

The decomposition energy reduction favors the dissociation of those molecules. The reduction of the band gap facilitates the electronic excitation generation in these regions (for example, thermal or optical initiation). This is consistent with the hot spot hypothesis of initiation.

Next, once the system is electronically excited, it can dissociate through a "slow" forbidden transition from the triplet state to the singlet state. Or it can undergo a fast nonradiative transition to the vibrationally hot state and then rapidly dissociate. Also, it can radiatively relax to the ground state to become a subject for the next excitation. As was shown on the example of FOX-7, charging or excitation of the molecule or its isomers (see Section 8.3.2 and [139] for details) can lead to a spontaneous decomposition with energy release. At later stages, therefore, the development of the decomposition is determined by the regeneration of all kinds of excitations across the bulk of the crystal because other mechanisms are also involved here. As extra energy becomes available as a result of the first decomposed species, other molecules (with higher decomposition barriers) that are situated farther from imperfections are ready to dissociate. This way the chemical reaction can propagate through the material.

Finally, one would like to stress that the recent advances in the field of ultrafast spectroscopies can provide a very powerful method for investigations of fast chemical reactions in situ. This is extremely important for many different types of fast processes, such as explosive decomposition. For example, a possible selective excitation of vibrational modes or a reaction along the single pathway seem really appealing not only for elucidation of initiation mechanisms, but also for the highly desired control of the process. Femtosecond and nonlinear optics opens up new perspectives in applying the methods of solid-state physics to the technology and physics of explosives. The use of ultrafast methods to tailor selected properties of the materials by the controlled creation of lattice defects and deformations appears to be a very promising tool for a successful solution of safety, aging, and handling issues pertaining to explosives.

Acknowledgments This work is supported in part by the ARO MURI under Grant #W911NF-05-1-0266 and by a grant of computer time from the DoD High Performance Computing Modernization Program at the Maui High Performance Computer Center (MHPCC), Naval Oceanographic Office (NAVO) and the US Army Engineer Research and Development Center (ERDC). M.K. is grateful to NSF for support under the IRD Program. Any appearance of findings, conclusions, or recommendations, expressed in this material are those of the authors and do not necessarily reflect views of the National Science Foundation.

References

1. T. Brill, and K. James, Kinetics and mechanisms of thermal decomposition in nitroaromatic explosives, *Chem. Rev.* **93**(8), 2667 (1993)
2. C.M. Tarver, R.D. Breithaupt, J.W. Kury, *J. Appl. Phys.* **81**, 7193 (1997)
3. C.M. Tarver, J.W. Kury, R.D. Breithaupt, *J. Appl. Phys.* **82**, 3771 (1997)
4. C.M. Tarver, P.A. Urtiew, W.C. Tao, *J. Appl. Phys.* **78**, 3089 (1995)

5. J.W. Kury, R.D. Breithaupt, C.M. Tarver, *Shock Waves* **9**, 227 (1999)
6. G.I. Kanel, S.V. Razorenov, A.V. Utkin, V.E. Fortov. *Shock-Wave Phenomena in Condensed Matter*, (in Russian) (Yanusz -K, Moscow, 1996) p. 407
7. *Decomposition, Combustion and Detonation Chemistry of Energetic Materials,* Ed. by T.B. Brill, T.P. Russel, W.C. Tao, and R.B. Warde, *Mat. Res. Soc. Symp. Proc.* **418,** 257–264, MRS, Pittsburgh, Pennsilvania (1996)
8. L.E. Fried, M.R. Manaa, P.F. Pagoria, R.L. Simpson, Design and synthesis of energetic materials, *Ann. Rev. Mater. Res.* **31**, 291–321 (2001)
9. V.I. Tarzhanov, Preexplosion Phenomena in Prompt Initiation of Secondary Explosives (Review), *Comb. Explo. Shock Waves* **39**(6), 611–618 (2003)
10. L.P. Smirnov, Chemical physics of decomposition of energetic materials. Problems and prospects, *Russ. Chem. Rev.* **73**(11), 1121–1141 (2004)
11. S. Zeman, New aspects of initiation reactivities of energetic materials demonstrated on nitramines, *J. Hazard. Mater.* **132**(2–3), 155–164 (2006)
12. P. Politzer, S. Boyd, Molecular Dynamics Simulations of Energetic Solids, *Struct. Chem.* **13**(2), 105–113 (2002)
13. D. Thompson, T. Brill, R. Shaw (Eds.), *Overviews of Recent Research on Energetic Materials*, World Scientific, New Jersey (2004)
14. R.W. Armstrong, W.L. Elban, Review materials science and technology aspects of energetic (explosive) materials, *Mater. Sci. Technol.* **22**(4), 381–395 (2006)
15. A.M. Wodtke, E.J. Hintsa, Y.T. Lee, *J. Phys. Chem.* **90**, 3549 (1986); ibid *J. Chem. Phys.* **84**, 1044 (1986)
16. M.J.S. Dewar, J.P. Ritchie, J. Alster, *J. Org. Chem.* **50**, 1031 (1985)
17. N.C. Blais, R. Engelke, S.A. Sheffield, *J. Phys. Chem. A* **101**, 8285 (1997)
18. J.M. Winey, Y.M. Gupta, *J. Phys. Chem. A* **101**, 10733 (1997)
19. Y.A. Gruzdkov, Y.M. Gupta, *J. Phys. Chem. A* **102**, 2322 (1998)
20. A. Gindulyté, L. Massa, L. Huang, J. Karle, *J. Phys. Chem. A* **103**, 11040 (1999)
21. Minh Tho Nguyen, Hung Thanh Le, Balazs Hajgato, Tamas Veszpremi, M. C. Lin, *J. Phys. Chem. A* **107**, 4286–4291 (2003)
22. W.F. Hu, T.J. He, D.M. Chen, F.C. Liu, *J. Phys. Chem. A* **106**, 7294 (2002)
23. M.L. McKee, *J. Am. Chem. Soc.* **108**, 5784 (1986).
24. R.P. Saxon, M. Yoshimine, *Can. J. Chem.* **70**, 572 (1992)
25. M.R. Manaa, L.E. Fried, *J. Phys. Chem A* **102**, 9884 (1998); *ibid.* **103**, 9349 (1999)
26. F.J. Zerilli, J. Hooper, M.M. Kuklja, *J. Chem. Phys.* **126**, 114701 (2007)
27. F.J. Zerilli, M.M. Kuklja, Ab initio Equation of State of an Organic Molecular Crystal: 1,1-diamino-2,2-dinitroethylene, *J. Phys. Chem. A* **111**(9), 1721 – 1725 (2007)
28. F.J. Zerilli, M.M. Kuklja, Ab initio 0K isotherm for organic molecular crystals, *AIP Conf. Proc.* **706**(1), 123–126 (2004)
29. T.B. Brill, K.J. James, *J. Phys. Chem.* **97**, 8759–8763 (1993)
30. S.F. Rice, R.L. Simpson, "The Unusual Stability of TATB: A Review of the Scientific Literature," Lawrence Livermore National Laboratory, Livermore, CA, Report UCRL-LR-103683 (July, 1990)
31. B.M. Dobratz, The Insensitive High Explosive Triaminotrinitrobenzene (TATB): Development and Characterizations -1888 to 1994; Los Alamos National Laboratory: Los Alamos, NM (1995)
32. T.R. Gibbs, A. Popolato, USL *Explosive Property Data* University of California Press, Berkeley (1980)
33. J. Oxley, J. Smith, H. Ye, R.L. McKenney, P.R. Bolduc, *J. Phys. Chem.* **99**, 9593 (1995)
34. T.B. Brill, K.J. James, *J. Phys. Chem.* **97**, 8152–8758 (1993)
35. T.B. Brill, K.J. James, R. Chawla, G. Nicol, A. Shukla, J.H. Futrell, *J. Phys. Org. Chem.* **12**(11), 819–826 (1999)
36. Maksimov Yu. Ya. *Russ. J. Phys. Chem.* **45**, 441 (1971)
37. F.P. Bowden, Y.D. Yoffe, in *Initiation and Growth of Explosion in Liquids and Solids*, Cambrige, University Press, London (1952), pp. 64–65

38. A.N. Dremin, *Chem. Phys. Rep.* **14**(12), 1851–1870 (1995)
39. F.P. Bowden, K. Singh, *Proc. Roy. Soc. (London) A* **227**, 22 (1954); see as well Ref. 1 and references therein
40. A.N. Dremin, S.D. Savrov, V.S. Trofimov, K.K. Shvedov, in *Detonacionnye volny v kondensirovannyh sredah (Detonation waves in condensed matters),* Nauka, Moscow (1970)
41. A.N. Dremin, *Chem. Phys. Rep.* **14**(12), 1851–1870, (1995)
42. H. Kawashima, M.M. Wefers, K.A. Nelson, *Ann. Rev. Phys. Chem.* **46**, 627 (1995)
43. W. Wang, D.D. Chung, J.T. Fourkas, L. Dhar, K.A. Nelson, *J. Phys IV*, **5**, 289 (1995)
44. D. Dlott, *Ann. Rev. Phys. Chem.* **50**, 251–278 (1999)
45. A.L. Shluger, J.L. Gavartin, M.A. Szymanski, A.M. Stoneham, *Nucl. Instr. Meth. Phys. Res. B* **166–167**, 1 (2000)
46. M. Shapiro, P. Brumer, *J. Chem. Soc. Faraday Trans.* **93**, 1263 (1997)
47. R.J. Gordon, S.A. Rice, *Ann. Rev. Phys. Chem.* **48**, 601 (1997)
48. R. Kosloff, S.A. Rice, P. Gaspard, S. Tersigni, D.J. Tannor, *Chem. Phys.* **139**, 201 (1989)
49. C.J. Badeen, J. Che, K.R. Wilson, V.V. Yakovlev, V.A. Apkarian, C.C. Martens, R. Zadoyan, B. Kohler, M. Messina, *J. Chem. Phys.* **106**, 8486 (1997)
50. K. Sundermann, R. de Vivie-Riedle, *J. Chem. Phys.* **110**, 1896 (1999)
51. A.L. Shluger, K. Tanimura, *Phys. Rev. B.* (Condensed Matter) **61**(8), 5392, (2000)
52. N.N. Semenov, *Chain Reactions, (in Russian)*, Moskva, Nauka (1986), p. 534
53. A. Tokmakoff, M.D. Fayer, D.D. Dlott, *J. Phys. Chem.* **97**, 1901 (1993)
54. B.P. Aduev, E.D. Aluker, G.M. Belokurov, Yu. A. Zakharov, A.G. Krechetov, *JETP* **89**, 906 (1999)
55. M.M. Kuklja, B.P. Aduev, E.D. Aluker, V.I. Krasheninin, A.G. Krechetov, A. Yu. Mitrofanov, *J. Appl. Phys.* **89**, 4156 (2001)
56. J. Sharma, B.C. Beard, in *Structure and Properties of Energetic Materials MRS Symposium Proceedings*, D.H. Liedenberg, R.W. Armstrong, J.J. Gilman (Eds.), vol. 296, MRS, Pittsburgh, Pensylvania (1993)
57. T.R. Botcher, H.D. Landouceur, T.R Russel, *AIP Conf. Proc.* **429**, 989 (1998)
58. B.P. Aduev, E.D. Aluker, G.M. Belokurov, A.G. Krechetov, *Chem. Phys. Rep.*, **16**, 1479 (1997); *ibid.* **17**, 469 (1998)
59. J.J. Gilman, *Chem. Propul. Inf. Agency* **589**, 379 (1992)
60. W.L. Elban, R.G. Rosemeir, K.C. You, R.W. Armstrong, *Chem. Propul. Inf. Agency* **404**, 19 (1984)
61. G.E. Duval, in *Shock Waves in Condensed Matter*, Y.M. Gupta (Ed.), Plenum Press, New York (1985), p. 1
62. J.K. Dienes, *Chem. Propul. Inf. Agency* **404**, 19 (1984)
63. F.J. Owens, J. Sharma, *J. Appl. Phys.* **51**, 1494 (1979)
64. H.-S. Im, E.R. Bernstein, *J. Chem. Phys.* **113**, 7911 (2000)
65. F. Williams, *Adv. Chem. Phys.* **21**, 289 (1971)
66. C.M. Bowden, J.D. Stettler, N.M. Witriol, *J. Phys. B: Mol. Phys.* **10**, 1789 (1997)
67. C.M. Tarver, *J. Phys. Chem. A* **101**, 4845 (1997)
68. L.E. Fried, A.J. Ruggerio, *J. Phys. Chem.* **98**, 9786 (1994)
69. A.N. Dremin, V.Y. Klimenko, O.N. Davidove, T.A. Zoludeva, Multiprocess detonation model; The Ninth Symposium (International) on Detonation, Portland, Oregon, ONR, Arlington, VA (1989), vol. I, p. 724.
70. A.N. Dremin, V.Yu. Klimenko, K.M. Mikhailyuk, V.S. Trofimov, 7th Symposium (Int.) on Detonation, J.M. Short (Ed.), Naval Surface Weapon Center, White Oak, MD, Dahlgren, VA, NSWC MP 82–334 (1981), p. 789
71. V.Yu. Klimenko, A.N. Dremin, in *Detonation, Chemical Physics of Combustion and Explosion Processes (in Russian),* Institute of Chemical Physics, Chernogolovka, Russia (1980), p. 69
72. J. Sharma, B.C. Beard, M.J. Chaykovsky, *J. Phys. Chem.* **95**, 1209 (1991)
73. J.J. Gilman, *Science* **274**, 65 (1996)
74. J.J. Gilman, *Phil. Mag. B* **79**, 643 (1999)

75. A.B. Kunz, *Phys. Rev. B.* **53**, 9733 (1996)
76. M.M. Kuklja, E.V. Stefanovich, A.B. Kunz, *J. Chem. Phys.* **112**, 3417 (2000)
77. M.M. Kuklja, A.B. Kunz, *J. Phys. Chem. Solids* **61**, 35 (2000)
78. M.M. Kuklja, A.B. Kunz, *J. Phys. Chem. B* **103**, 8427 (1999)
79. M.M. Kuklja, A.B. Kunz, An effect of hydrostatic compression on defects in energetic materials: *ab initio* modeling, in *Multiscale Modelling of Materials*, V.V. Bulatov, T.D. Rubia, R. Pjillips, E. Kaxiras, N. Ghoniem (Eds.), *Mater. Res. Soc. Symp. Proc.* **538**, 347–352 (1999)
80. M.M. Kuklja, *J. Phys. Chem. B* **105**, 10159 (2001)
81. C.A. Wight, T.R. Botcher, *Am. Chem. Soc.* **114**, 830 (1992)
82. T.R. Botcher, C.A. Wight, *J. Phys. Chem.* **97**, 9149 (1993); T.R. Botcher, C.A. Wight, *J. Phys. Chem.* **98**, 5541 (1994)
83. M.D. Pace, W.B. Moniz, *J. Magnet. Reson.* **47**, 510 (1982)
84. C.J. Wu, L.E. Fried, *J. Phys. Chem. A* **101**, 863 (1997)
85. M.M. Kuklja, A.B. Kunz, *AIP Conf. Proc.* **505**, 401 (2000)
86. M.M. Kuklja, A.B. Kunz, *J. Appl. Phys.* **87**, 2215 (2000)
87. M.M. Kuklja, A.B. Kunz, *J. Appl. Phys.* **86**, 4428 (1999)
88. M.M. Kuklja, A.B. Kunz, *J. Appl. Phys.* **89**, 4962 (2001)
89. R. Dovesi, V.R. Saunders, C. Roetti, M. Causà, N.M. Harrison, R. Orlando, E. Aprà, *CRYSTAL95 User's Manual*, University of Torino, Torino (1996)
90. C.S. Choi, E. Prince, *Acta Crystallogr.* B **28**, 2857 (1972)
91. P.L. Marinkas, *J. Luminescence* **15**, 57 (1977)
92. M.K. Orloff, P.A. Mullen, F.C. Rauch, *J. Phys. Chem.* **74**, 2189 (1970)
93. P.G. McCormick, The Portevin-LeChatelier effect in an Al-Mg-Si alloy, *Acta Metall.* **19**, 463 (1971)
94. P.G. McCormick. The Portevin-Le Chatelier effect in an Al-Mg-Si alloy loaded in torsion. *Acta Metall.* **30**, 2079 (1982)
95. S.N. Rashkeev, M.V. Glazov, F. Barlat, *Comput. Mater. Sci.* **24**, 295 (2002)
96. S. Dj. Mesarovic, *J. Mech. Phys. Solids* **43**, 671 (1995)
97. D. Walgraef, E.A. Aifantis, *Int. J. Eng. Sci.* **24**, 1351 (1985); *ibid: Int. J. Eng. Sci.* **24**, 1789 (1986); *ibid: J. Appl. Phys.* **58**, 688 (1985)
98. J.P. Hirth, J. Lothe, *Theory of Dislocations*, Wiley, New York (1982)
99. F.R.N. Nabarro, *Theory of Crystal Dislocations*, Dover, New York (1987)
100. V. Vitek, A. Gonis, P.E.A. Turchi, J. Kudrnovsky (Eds.), *Stability of Materials*, Plenum Press, New York (1996), p. 53
101. A. Seeger, P. Veyssiere, L.P. Kubin, J. Castaing (Eds.), *Dislocations 1984*, CNRS, Paris (1984), p. 141
102. V.V. Bulatov, F.F. Abraham, L.P. Kubin, B. Devincre, S. Yip, *Nature* **391**, 669 (1998)
103. R.W. Armstrong, C.S. Coffey, W.L. Elban, Adiabatic heating at a dislocation pile-up avalanche *Acta Metallurgica* **30**, 2111–2116 (1982)
104. J. Sharma, R.W. Armstrong, W.L. Elban, C.S. Coffey, H.W. Sandusky, *Appl. Phys. Lett.* **78**, 457 (2001)
105. R.W. Armstrong, W.L. Elban, *Dislocations in Solids*, F.R.N. Nabarro, J.P. Hirth (Eds.), Elsevier, Oxford (2004), Vol. 12, p. 403
106. H. Östmark, A. Langlet, H. Bergman, U. Wellmar, U. Bemm, FOX-7 – A New Explosive with Low Sensitivity and High Performance; *11th Symp Detonation Proceedings*, Office of Naval Research, ONR 33300-5 (1998), pp. 807–812.
107. P. Politzer, M.C. Concha, M.E. Grice, J.S. Murray, P. Lane, D. Habibollazadeh, *J. Mol. Struct.* (THEOCHEM) **452**, 72 (1998)
108. A. Gindulyté, L. Massa, L. Huang, J. Karle, *J. Phys. Chem. A* **103**, 11045 (1999)
109. E.J. Reed, J.D. Joannopoulos, L.E. Fried, *Phys. Rev. B.* **62**, 16500 (2000)
110. G. Kresse, J. Hafner, *Phys. Rev. B* **48**(13), 115 (1993); G. Kresse, J. Furthmüller, *Phys. Rev. B* **54**(11), 169 (1996); G. Kresse, J. Furthmüller, *Comput. Mater. Sci.* **6**, 15 (1996)

111. M.C. Payne, M.P. Teter, D.C. Allan, T.A. Arias, J.D. Joannopoulos, *Rev. Mod. Phys.* **64**, 1045 (1992)
112. U. Bemm, H. Östamark, *Acta Cryst.* **C54**, 1997 (1998)
113. R. Gilardi, private communication (2001)
114. H. Cady, A. Larson, *Acta Cryst.* **18**, 485 (1965)
115. D.J. Chadi, M.L. Cohen, *Phys. Rev. B* **8**, 5747 (1973)
116. M.M. Kuklja, F.J. Zerilli, S.M. Peiris, *J. Chem. Phys.* **118**, 11073 (2003)
117. D.C. Sorescu, J.A. Boatz, D.L. Thompson, *J. Phys. Chem. A* **105**, 5010 (2001)
118. C.J. Wu, L.H. Yang, L.E. Fried, *Phys. Rev. B* **67**, 235101 (2003)
119. K.K. Baldrige, J.S. Siegel, *J. Am. Chem. Soc.* **115**, 10782 (1993)
120. M.M. Kuklja, F.J. Zerilli, *J. Phys. Chem. A* **110**, 5173–5179 (2006)
121. H. Jonsson, G. Mills, K.W. Jacobsen, *Classical and Quantum Dynamics in Condensed Phase Systems*, B.J. Berne, G. Cicotti, D.F. Coker (Eds.), World Scientific, River Edge, NJ (1998)
122. A.B. Kunz, M.M. Kuklja, T.R. Botcher, T.P. Russel, *Thermochimica Acta. Special edition: Energetic Materials*, **384**, 279–284 (2002)
123. M.M. Kuklja, S.N. Rashkeev, F.J. Zerilli, *Appl. Phys. Lett.* **89**, 071904 (2006)
124. M.M. Kuklja, S.N. Rashkeev, *Appl. Phys. Lett.* **90**, 151913 (2007)
125. S.N. Rashkeev, M.M. Kuklja, F.J. Zerilli, *Appl. Phys. Lett.* **82**, 1371 (2003)
126. M.M. Kuklja, S.N. Rashkeev, *Phys. Rev. B* **75,** 104111 (2007)
127. G. Mills, H. Jonsson, G.K. Schenter, *Surf. Sci.* **324**, 305 (1995)
128. J. Sharma, W.L. Garrett, F.J. Owens, V.L.Vogel, *J. Phys. Chem.* **86**, 1657 (1982)
129. J. Sharma, J.W. Forbes, C.S. Coffey, T.P. Liddiard, *J. Phys. Chem.* **91**, 5139 (1987)
130. C. Wu, L. Fried, *J. Phys. Chem. A* **104**, 6447–6452 (2000)
131. T.B. Brill, K.J. James, *J. Phys. Chem.* **97**, 8752–8758 (1993)
132. Our recent results indeed show that hydrogen transfer may play an important role in TATB and FOX-7 decomposition pathways, especially under low temperature conditions; they will be reported in the further communications.
133. S. Meyerson, R.W. Vander Haar, E.K. Fields, *J. Org. Chem* **37**, 4114 (1972)
134. V.G. Matveev, V.V. Dubikhin, G.B. Nazin, *Izv. Akad. Nauk SSSR, Ser. Khim* **675** (1978)
135. A.C. Gonzalez, C.W. Larson, D.S. McMillen, D.M. Golden, *J. Phys. Chem.* **89**, 4809 (1985)
136. W. Tsang, D. Robaugh, W.G. Mallard, *J. Phys. Chem.* **90**, 5968 (1986)
137. N. Goto, H. Yamawaki, K. Tonokura, K. Wakabayashi, M. Yoshida, M. Koshi, Chemical Reactions and Other Behaviors of High Energetic Materials under Static Ultrahigh Pressures, *Mater. Sci. Forum* **465/466**, 189–194 (2004)
138. Materials of APS SCCM 2007 meeting in Hawaii, AIP Conference Proceedings, M. L. Elert, M. D. Furnish, R. Chau, N. Holmes, J. Nguyen (Eds.), 2007.
139. A.V. Kimmel, P.V. Sushko, A.L. Shluger, M.M. Kuklja, *J. Chem. Phys.* **126**, 234711 (2007)
140. R.J. Doyle Jr., J.E. Campana, *J. Phys. Chem.* **89**, 4251 (1985)
141. M.E. Grice, D. Habibollahzadeh, P. Politzer, *J. Chem. Phys.* **100**, 4706 (1994) and references therein
142. C.J. Wu, L.E. Fried, *J. Phys. Chem. A* **101**, 8675 (1997) and references therein
143. C.M. Tarver, *AIP Conf. Proc.* **620**, 42 (2002)
144. J. Sharma, C.S. Coffey, A.L. Ramaswamy, R.W. Armstrong, Atomic Force microscopy of hot spot reaction sites in impacted RDX and laser heated AP, *Decomposition, Combustions and Detonation Chemistry of Energetic Materials*, T.B. Brill, T.P. Russel, W.C. Tao, R.B. Warde (Eds.), *Mat. Res. Soc. Symp. Proc.* **418**,257–264, MRS, Pittsburgh, Pennsilvania (1996)
145. E.J. Reed, J.D. Joannopoulos, L.E. Fried, *Phys. Rev. B.* **62**, 16500 (2000)
146. C.M. Tarver, L.E. Fried, A.J. Ruggiero, D.F. Calef, 10th International Detonation Symposium, 1993, Boston, MA, Office of Naval Research, Arlington, VA, ONR 33395-12 (1995), pp. 3–10
147. Y.A. Gruzdkov, Y.M. Gupta, J.J. Dick, Time-Resolved Absorption Spectroscopy in Shocked PETN Single Crystals, in *Shock Compression of Condensed Matter-1999*, M.D. Furnish, L.C. Chhabildas, R.S. Hixson (Eds.), AIP conference proceedings, Vol. 505, pp. 929–932 (1999)

148. C.S. Coffey, *Structure and Properties of Energetic Materials*, D.H. Liedenberg, R.W. Armstrong, J.J. Gilman (Eds.), Mat. Res. Soc. Proc. **296**, Pittsburgh, Pennsylvania (1993), pp. 63–73
149. J.J. Dick, *Appl. Phys. Lett.* **44**, 859 (1984)
150. T.P. Liddiard, J.W. Forbes, D. Price, *Proc. of the 9th Symposium on Detonation*, ONR, Arlington, VA (1989), p. 1235
151. J. Frisch, et al. GAUSSIAN 03 Pittsburgh, PA: Gaussian, (2003)

Index

ab initio modeling, 296, 297

absorption corrections for x-radiation, 19
activation barrier, 297, 304, 316
activation volume of RDX, 55, 59
ammonium nitrate (AN), 99
ammonium perchlorate (AP), 110, 111, 117, 123
  diffraction spectra, 110
  static compression, 111
amorphous nitrogen, 92
anisotropic sensitivity, 242
anthracene, 228, 231–234, 247
Astin, 4
azide, 22–25, 75, 76

band structure calculations, 296, 320
basis set, 304, 317
Bassett, W. A., 104
benzene, 19–21
Birch-Murnaghan EOS formalism, 104, 105
Block, S., 15, 17, 25
Bridgman opposed anvils, 7, 8, 34
Bridgman, P. W., 5, 7, 23
bromine, 19, 21
bulk EM representation, 261
bulk modulus, 102, 104–106, 109–111, 121
Bunting, E. N., 2, 3, 5, 6, 9, 14

Cady, H., 104, 108
calibration of ruby R-line pressure shift, 25
  precise, 25
  preliminary, 25
carbon disulfide, 21
carbon tetrachloride, 20, 21
Cassegrain optics for FTIR, 52
charge trapping, 316
chemical decomposition, 221, 223
chemical reaction, 292, 294–296, 311, 312, 316, 319–321
CL-20, 103, 111, 112, 117, 118, 120, 123, 216
  decomposition by laser initiation, 216
  products of decomposition, 216
  static compression, 112
COMPASS force field, 278, 279
compressibilities, 23–25
Compressibility of Explosive Liquids, 64
  NM, 67
CONO formation, 317, 318
crystal structure, 235, 243
crystallography of RDX phases, 46, 47
CsF transition, 18
cubic gauche structure, 76, 77, 81, 82, 89, 94

Debye temperature, 260
decomposition kinetics, 206, 209
decomposition of AP, 210
decomposition rate constants, 55, 58, 59
decomposition reaction, 291, 294, 316
defect, 297, 304–309, 311, 315, 320
deformation, 300, 301, 303, 308, 313–315
density functional theory, 296, 297, 304, 317
detonation initiation, 294–296, 316
deviatoric stress, 228, 230, 231
di-vacancy (vacancy dimmer), 297
diamond, 76, 79–86, 89–94
diamond anvil cell (DAC), 78, 79, 82, 92, 220, 229, 230, 247
  Bassett, 41
  dynamic diamond cell, 68
  Holzapfel, 41
  Mao-Bell, 39
  Merrill-Bassett, 42
  NIST, 1–3, 12, 33, 38, 39, 41, 45, 52, 69

dilatometry, 176
  dynamic compression, 175
  Hugoniot, 176, 177
  shock compression, 177
dimer, 233, 234
dislocation, 222, 229, 233, 234, 241, 257, 294, 297, 299–303, 311, 313, 315, 316, 319, 320
dissociation, 291, 292, 294, 296–298, 303, 304, 308, 310–312, 314, 315, 317–319, 321
dynamic compression, 221, 223

elastic properties of TATB, 277
electronic properties, 297, 312, 319
electronic structure, 228, 231–233, 246, 296–299, 301, 305, 310, 312, 313, 315, 317, 319, 320
energetic barrier, 292, 299
energetic material (EM), 128, 132, 134, 219–221, 223, 225, 226, 234, 235, 241, 247, 255–262, 264–269, 278–281, 291, 292, 294–297, 302, 303, 307, 311, 316, 319
energy transfer, 221
equation of state (EOS), 98, 100–111, 113, 115, 120, 121, 123, 124, 255, 256, 258, 261, 269, 270, 272
Estane 5703, 128, 131, 132, 156, 160, 163, 171, 172, 180, 187, 189
Estane^{TM} force field, 275
excimer, 233
excitations, 291, 295–297, 299, 301, 302, 311–316, 318–321
excited state, 292, 311, 313–316, 320
exciton, 223, 295
excitonic mechanism, 315, 320
experimental kinetic measurement methods, 203
  combustion front propagation (CFPR), 209
  single-shot laser initiation, 211
  time-resolved absorbance, 210
  time-to-explosion, 208
explosives, 75, 76, 98, 100, 103–105, 108–110, 112, 114, 120, 127–130, 132, 133, 153, 161, 163, 178, 180, 181, 184, 190, 191, 193
  high, 99, 100, 103, 112
  low, 99

first principles, 296, 319
fixed-point pressure scale, 39, 67
fluorescence, 225, 229, 231–233, 246
fluorescing materials, 25
force fields, 258, 262, 266–270, 272–281
Forman, R., 25
formulations, 129, 131, 132, 134, 152, 153, 160, 176
  binders, 129, 131, 132, 134, 153, 160
  equation of state, 134, 152, 176
FOX-7 (diamino-dinitroethylene, $C_2H_4N_4O_4$), 296, 303–312, 314, 316–321
freezing pressures of liquids, 23, 25, 31
FTIR, 2, 46, 48, 52, 53, 56, 69

gas gun, 181, 223
  spectroscopy, 181
Gasket Technique, 14, 15
General Services Administration, 9
glass transition pressure, 31

Helmholz free energy, 101, 102
high energy density material, 75, 77
high explosives, 220, 291, 296, 312, 315, 320
high pressure, 76, 77, 85, 87, 88, 90–93, 220, 229, 231, 232, 234–236, 241, 243, 244, 247
High Pressure Diamond Optics, 14
high pressure phases, 98, 112, 114, 116
hindered shear, 242, 245
HMX, 99, 103, 105–107, 115, 121, 123, 209, 212–214
  absorbance change, 212
  advanced interface model, 213
  by time-resolved-absorbance, 210
  chain-type model, 213
  decomposition mechanism, 214
  EOS parameters, 123
  high pressure polymorphs, 105
  pressure/volume data, 104
  rate determining step, 214
HNIW, 103, 111, 117
  compression, 111
  high pressure polymorphs, 111, 117
  phase diagram, 118
  XRD patterns, 121
HONO formation, 317
hot spots, 203, 205, 207, 208, 257
Hot-Molecular Zone (HMZ), 203
Hugoniot, 104, 105, 110, 268
hydrogen transfer, 294, 308
hydrostatic limits, 28
hydrostatic pressures, 15, 22, 28, 31, 69
hydrothermal diamond cell (HDAC), 104

ice VI, 15, 16, 19, 68
inter-molecular interactions, 293, 294, 299, 305, 307

interactive potentials, 259, 264–267, 269–272, 277
intra-molecular interactions, 299, 318
invention, 1–3, 69
ionic reactions, 242, 245
IR spectra of RDX phases, 48, 55
IR spectroscopy, 69, 110, 113
ir spectroscopy, 7, 17
isomerization, 294, 304, 308, 317, 318
isothermal EOS parameters, 108
isotherms, 106, 122, 124

Kel-F 800, 128, 129, 132, 133, 153, 154, 156, 158, 159, 163, 166, 171, 174, 184–187, 189
key MD simulations, 266
kinetic models, 204, 206, 212, 213
  advancing interface, 206, 212, 213
  chain-type nucleation, 207, 212, 213
  Lee-Tarver model, 207
  sigmoid, 204, 206, 212

La transition, 18
laser heating, 76, 78, 82, 83, 85–87, 89, 94
laser spectroscopy, 225
Lee-Tarver model, 210
Lippincott, E. R., 2, 3, 5, 6, 14

McMurdie, H. F., 4, 5, 9, 17
MD simulations, 255, 256, 259, 261, 262, 264–274, 277–280
mechanical anisotropy, 241, 242
Menikoff, R., 257
Merrill-Bassett cell, 209
mesoscale, 256, 258, 275
metastable polymeric nitrogen, 94
methanol:ethanol:water mixture, 30, 31, 44
methods of MD, 258
  Hamiltonian, 259
  Lagrangian, 259
  Newtonian, 259
molecular conformers, 242, 243
molecular crystals, 219–221, 228, 229, 231, 234, 236, 247
molecular dissociation, 308
molecular dynamics (MD), 254, 258, 273, 276, 280, 314
molecular orientation, 303
molecular structure, 238, 241, 243, 244

Nd:YAG laser, 214, 216
nitro-nitrite isomerization, 294, 304
nitrogen, 75–94
nitroglycerine, 99
Nobel Prize, 5
nonhydrostatic pressures, 22, 29, 53
nonhydrostaticity, 227–231
nonmolecular nitrogen, 76–78, 89, 92
nonreactive force fields, 273
nonreactive models, 269

Olinger, B., 104–109, 116
optical absorption, 300, 313, 315
optical fluorescence system, 32
optical polarizing microscopy, 2, 24, 43
optical properties, 297
optical spectroscopy, 225
orientational defects, 303

periodic model, 296
PETN, 109, 110, 116, 123, 239, 241–247
  static compression, 110, 117
PETN (pentaerythritol-tetranitrate, $C_5H_8N_4O_{12}$), 296, 315
phase diagrams, 45, 100, 115, 117, 118, 120, 123
  NM, 50
  RDX, 45
phase transition, 226, 235–238, 240
photoluminescence, 245, 246
Picatinny Arsenal, 22
Piermarini, G. J., 4, 213
plane waves, 304
polymer materials, 133, 142
polymeric nitrogen, 76–79, 82, 83, 86, 87, 91, 92, 94
polymers, 127–129, 132, 133, 135, 142, 143, 146–150, 152–155, 157, 158, 161–163, 166–168, 170–175, 177, 178, 180, 182, 184, 186, 188, 190, 191, 193–195
polymorphism, 234–236, 247
polynitrogen, 75
polytetrafluoroethylene, 131
potassium nitrate transition, 21
potential energy surface, 258, 260, 267, 277
powder x-ray diffraction, 48
pressure measurement, 16, 24, 28, 32, 36, 43, 45, 65, 69
propellant, 77
pseudopotentials, 304

Raman, 163, 167
  high Pressure, 163, 167
Raman spectroscopy, 114, 115, 225, 226, 236, 240, 243, 247
RDX, 99, 103, 105–107, 114, 123, 235–240, 247
  high pressure polymorphs, 105

RDX (cyclotrimethylene-trinitramine, $C_3H_6N_6O_6$), 296–306, 312–315, 320
RDX decomposition, 51, 54, 55, 57, 58, 214
reaction energy, 291, 311, 313, 317, 318, 320
reaction initiation of solid HEs, 204
  coalescence, 206
  induction period, 205
    deceleration, 205
    growth process, 205
    ingestion, 206
  nucleation, 205
reaction rates, 203, 206, 207, 212, 213, 215, 217
reactive model (ReaxFF), 265, 268, 269, 280
reversed molecule, 305, 306
ruby, 229, 230, 232
Ruby fluorescence, 2, 24, 28, 30, 33, 36, 45, 46, 69

shear strain, 297, 301, 308–311, 314, 319, 320
shear stress, 228, 229
shock compression, 219–221, 225, 228, 231–233, 235, 238, 239, 241, 244, 245, 247
shock Hugoniot, 255, 256, 259, 267, 273
shock initiation, 221, 238, 241
shock wave, 219–221, 223–230, 235, 238, 241, 245, 247, 291, 296, 299, 311–313, 320
shock-induced decomposition, 223, 245
single crystal x-ray diffraction, 14–17, 20, 21, 23, 32, 42
single crystals, 2, 17, 21, 31, 45, 47, 61, 65
singly bonded nitrogen, 75
slip planes, 245
Smithsonian Institution, 7
sodium perchlorate, 99
specific volume of NM, 67
static compression, 220, 221, 227–229, 231, 233, 235, 236, 238, 239, 243, 245
static high pressure, 203, 204, 209, 211, 214, 216, 217, 220, 229, 231, 235, 236, 241, 243, 244, 247
static high pressure MD simulations, 260
Stokes Falling Ball method, 30, 69
strain energy, 102
stress, 219–221, 224, 226–233, 238, 240, 241, 243, 245–247, 301–303, 312, 319, 321
structural defects, 229
superhard, 76, 93, 94
superpressed state of H_2O and D_2O, 68, 69
surface, 291–293, 297–299, 302, 311–314, 319
Sylgard 184, 163, 167, 171, 172, 175, 189
symmetry-adapted perturbation theory (SAPT), 277, 278
synchrotron, 87

TATB, 99, 108, 109, 115, 121, 123, 215, 216
  absorbance, 215, 216
  Compression, 109
  diffraction spectra, 116
  high pressure polymorphs, 121
  laser-initiated decomposition, 215, 216
  photolysis, 216
TATB (triamino-trinitrobenzine, $C_6H_6N_6O_6$), 293, 294, 296, 303, 304, 306–311, 316
thermal decomposition rates, 52, 55, 61, 63
THV 500, 156, 158, 159, 162, 163, 166, 182, 185, 186
time resolved spectroscopy, 225, 226, 240
TNT, 293, 294, 315
trinitrotoluene (TNT), 99, 124
type I diamond, 10
Type II diamond, 8, 9

ultrafast spectroscopy, 295, 321
ungasketed technique, 31, 34
uniaxial strain, 225, 228, 229, 238, 241, 243
University of Maryland, 5, 6

vacancy, 297, 298, 315, 319
Van Valkenburg, A., 2–5, 14
vibrational structure, 231
vibron, 77, 81, 84–86, 90
viscosity of liquids, 30
volume thermal expansion, 107

Wachtman, J. B., 25
Weir, C. E., 2–4

x-ray diffraction, 78–83, 85, 87–90
x-ray precession camera, 16, 18, 20, 21, 23

zinc sulfide transition, 44

MIX
Papier aus verantwortungsvollen Quellen
Paper from responsible sources
FSC® C105338

If you have any concerns about our products,
you can contact us on
ProductSafety@springernature.com

In case Publisher is established outside the EU,
the EU authorized representative is:
Springer Nature Customer Service Center GmbH
Europaplatz 3, 69115 Heidelberg, Germany

Printed by Libri Plureos GmbH
in Hamburg, Germany